48.95

D1281064

St. Olaf College

JAN 5 1981

Science Library

Variational Inequalities and Complementarity Problems

Variational Inequalities and Complementarity Problems

Theory and Applications

Edited by

R. W. Cottle
Department of Operations Research, Stanford University

F. Giannessi
Department of Mathematics, University of Pisa

J-L. Lions
Collège de France

A Wiley–Interscience Publication

JOHN WILEY & SONS
Chichester · New York · Brisbane · Toronto

QA
316
.V37

Copyright © 1980 by John Wiley & Sons Ltd.

All rights reserved.

No part of this book may be reproduced by any means, nor transmitted, nor translated into a machine language without the written permission of the publisher.

British Library Cataloguing in Publication Data:

Variational inequalities and complementarity problems.
1. Calculus of variations—Congresses
2. Inequalities (Mathematics)—Congresses
3. Maxima and minima—Congresses
I. Cottle, R W II. Giannessi, F III. Lions, Jacques Louis
515'.64 QA316 79–40108

ISBN 0 471 27610 3

Printed and bound in Great Britain at The Pitman Press, Bath

This book is dedicated to the memory of Guido Stampacchia

05146

Contributing Authors

C. BAIOCCHI	*Dept of Mathematics, Univ. of Pavia, Strada Nuova 65, 27100 Pavia, Italy.*
H. BERESTYCKI	*Analyse Numérique, Univ. of Paris VI, Tour 55/65 5E, 4 place Jussieu, 75230 Paris Cedex 05, France.*
H. BREZIS	*Analyse Numérique, Univ. Pierre et Marie Curie, Tour 55/65, 4 place Jussieu, 75230 Paris Cedex 05, France.*
I. CAPUZZO DOLCETTA	*Dept of Mathematics, Univ. of L'Aquila, L'Aquila, Italy.*
R. CONTI	*Dept of Mathematics, Univ. of Florence, Viale Morgagni 67/A, Florence, Italy.*
R.W. COTTLE	*Dept of Operations Research, Stanford University, Stanford, California 94305, U.S.A.*
M. CUGIANI	*Dept of Mathematics, Univ. of Milan, Via C. Saldini 50, Milan, Italy.*
G.B. DANTZIG	*Dept of Operations Research, Stanford University, Stanford, California 94305, U.S.A.*
A. FRIEDMAN	*Dept of Mathematics, Northwestern University, Evanston, Illinois 60201, U.S.A.*
F. GIANNESSI	*Dept of Mathematics, Univ. of Pisa, Piazza dei Cavalieri 2, 56100 Pisa, Italy.*
F.J. GOULD	*Graduate School of Business, Univ. of Chicago, 5836 Greenwood Avenue, Chicago, Illinois 60637, U.S.A.*
P.L. JACKSON	*Dept of Operations Research, Stanford University, Stanford, California 94305, U.S.A.*
D. KINDERLEHRER	*School of Mathematics, Univ. of Minnesota, 127 Vincent Hall, Minneapolis, Minnesota, U.S.A.*

C.E. LEMKE — *Dept of Mathematical Sciences, Rensselaer Polytechnic Institute, Troy, New York 12181, U.S.A.*

J.L. LIONS — *Collège de France, place Marcelin Berthelot, 75015 Paris, France.*

J.P. LIONS — *E.N.S., 45 rue d'Ulm, 75230 Paris Cedex 05, France.*

L. MCLINDEN — *Dept of Mathematics, Univ. of Illinois at Urbana, Champaign, Urbana, Illinois 61801, U.S.A.*

U. MOSCO — *Dept of Mathematics, Univ. of Rome, Città Universitaria, 00185 Roma, Italy.*

M. RAOUS — *C.N.R.S., Laboratoire de Mécanique et d'Acustique, 31 chemin Joseph Aiguier, 13274 Marseille Cedex 2, France.*

R.T. ROCKAFELLAR — *Dept of Mathematics, Univ. of Washington, Seattle, Washington 98195, U.S.A.*

F. SCARPINI — *Dept of Mathematics, Univ. of Rome, Città Universitaria, 00185 Roma, Italy.*

C.P. SCHMIDT — *Graduate School of Business, Univ. of Chicago, 5836 Greenwood Avenue, Chicago, Illinois 60637, U.S.A.*

J. SIJBRAND — *Mathematical Institute, Univ. of Utrecht, Budapestlaan 6, 3508 TA Utrecht, The Netherlands.*

R. TEMAM — *Analyse Fonctionnelle et Numérique, Univ. of Paris Sud, Bâtiment 425, 91405 Orsay, Paris, France.*

R.J.B. WETS — *Dept of Mathematics, Univ. of Kentucky, Lexington, Kentucky 40506, U.S.A.*

Preface

This volume contains the Proceedings of an International School of Mathematics devoted to Variational Inequalities and Complementarity Problems in Mathematical Physics and Economics. Held from 19 to 30 June 1978, the site for the meeting was the 'Ettore Majorana' Centre for Scientific Culture in Erice, Sicily. The course was conceived and directed by R.W. Cottle, F. Giannessi, and the late G. Stampacchia as a mechanism for promoting interaction between two related, but seldom communicating, branches of mathematical science.

The school was attended by approximately 70 men and women from 15 countries. In the tradition of the Ettore Majorana Centre for Scientific Culture, the participants were classified either as 'invited speakers' or as 'students', though in fact the latter were not students in the ordinary sense; they were all Ph.D. holders with established publication records. Several of them contributed lectures to the programme and to these proceedings.

The school was deeply saddened by the untimely death of Professor Stampacchia on 27 April 1978. His passing represents an enormous loss to the scientific community in general and, in particular, to the present enterprise he did so much to bring forth. This volume, dedicated to his memory, begins with a paper by J.L. Lions entitled 'The Work of G. Stampacchia in Variational Inequalities'. Appropriately presented out of the otherwise alphabetical ordering of the papers, it clearly shows the greatness of Professor Stampacchia's contributions to the field.

By now it has become a cliché to observe that the recent decades have recorded a great development in the theory and techniques of optimization and that the 'computer revolution' has played a major role in this by making it possible to implement such techniques for everyday use and by stimulating new effort aimed at the solution of more complex problems. The study of variational inequalities and complementarity problems is part of this development inasmuch as optimization problems can often be 'reduced' to the solution of variational inequalities and complementarity problems. But it is important to remark that these subjects pertain to more than just optimization problems, and therein lies much of their attractiveness.

A variational inequality is usually posed with respect to a Hilbert space V with dual space V^*, a non-empty closed convex set K in V, a bilinear form $a(\cdot\,,\cdot)$ on $V \times V$ and an element $f \in V^*$. One wants to find

(1) $$u \in K \quad \text{such that} \quad a(u, v-u) \geq (f, v-u) \quad \text{for all } v \in K$$

In a setting of this generality, the complementarity problem is customarily stated with respect to a closed convex cone K in V and a mapping $F : V \to V^*$. One then seeks

(2) $$u \in K \quad \text{such that} \quad F(u) \in K^* \text{ (polar of } K) \quad \text{and} \quad (F(u), u) = 0$$

With K so restricted, the variational inequality

(3) $$u \in K \quad (F(u), v-u) \geq 0 \quad \text{for all } v \in K$$

and the complementarity problem

(4) $$u \in K \quad F(u) \in K^* \quad (F(u), u) = 0$$

have the same solutions, if any. However, it is to be noted that when the bilinear form $a(\cdot\,,\cdot)$ is *not* symmetric, the variational inequality (1) does not correspond to a minimization problem. It is of interest to study this case, for it permits one to treat free-boundary problems which can be formulated in the framework of variational inequalities but not in the framework of minimization problems. Likewise, in the context of economic equilibrium problems, it is of interest to study complementarity problems with fairly general mappings F which are not derived from Lagrangian functions associated with underlying minimization problems.

Although variational inequalities and complementarity problems have much in common, historically there has been little direct contact between researchers in these two fields. Despite some notable exceptions, it can be said that the people who work in variational inequalities tend to be educated in the tradition of more classical applied mathematics, even if they use very modern tools such as computers. Models of physical problems, differential equations and topological vector spaces are common elements of their work. The complementarity people—again with some exceptions—lie closer to other branches of mathematical science, such as operations research and combinatorics. Their efforts are closely related to mathematical programming which is often (though not always) motivated by management or economic problems and is mainly finite-dimensional. In these proceedings, one will find papers exemplifying these differences, as well as others which help to bridge the gap between the two fields.

Many of the papers here—particularly those concerned with variational inequalities—deal with free-boundary problems. These arise in the study of flow through porous media, hydrodynamic lubrication, and elasto-plastic analysis. The papers on complementarity problems deal with algorithms, existence theory, and the relationships between complementarity problems, variational inequalities, mathematical programming, and monotone operator theory. Still others cover calculus of variations, stochastic optimization and control, convex analysis, and systems of non-linear equations. The editors hope this volume will contribute to further exchanges between researchers in the theory and applications of variational inequalities and complementarity problems.

We express our sincere thanks to all those who took part in the school. Their invaluable discussions have made this book possible. Special mention should once more be made of the Ettore Majorana Centre in Erice which offered its facilities and stimulating environment for the meeting. We are all indebted to the Italian National Research Council (C.N.R.), Public Education Ministry, and IBM-Italy for their financial support. We thank Alberto Vallerini Company (Pisa) for the final typing of the manuscripts. Finally, we want to express our special thanks to John Wiley & Sons for their unfailing cooperation.

R.W. Cottle (Stanford)
F. Giannessi (Pisa)
J.L. Lions (Paris)

Contents

1 **The Work of G. Stampacchia in Variational Inequalities** (*J.L. Lions*) 1
1. Introduction 1
2. Variational inequalities 1
3. Application to potential theory 3
4. A number of variants and extensions 5
5. The obstacle problem 8
6. An abstract regularity theorem. Application to the obstacle problem 11
7. An inequality for the obstacle problem 13
8. Elasto-plastic problem 15
9. Hodograph method and VI 17
10. Fourth-order VI 19
11. Infiltration in porous media 21
12. Conclusion 22

2 **Variational Inequalities and Free-Boundary Problems** (*C. Baiocchi*) 25
1. The simplest elliptic free-boundary problem. Interconnections with variational inequalities 25
2. Some other examples of FB problems 28
3. Variational approach for problem 2 31

3 **Existence of a Ground State in Non-Linear Equations of the Klein-Gordon Type** (*H. Berestycki and P.L. Lions*) 35
1. Introduction 35
2. Proof of theorem 1.1 37
3. Discussion on the hypotheses and indications on other methods 45

4 **Some Variational Problems of the Thomas-Fermi Type** (*H. Brézis*) 53
1. Introduction 53

2. The variational formulation and the Euler-Lagrange formulation. Conditions for equivalence 56
3. A direct approach for solving the Euler-Lagrange equation 61
4. A min-max principle for the Lagrange multiplier 69

5 Implicit Complementarity Problems and Quasi-Variational Inequalities (*I. Capuzzo Dolcetta and U. Mosco*) 75
1. Introduction 75
2. The strong ICP 75
3. QVIs and the weak ICP 76
4. The Young-Fenchel equation and QVIs in duality 78
5. Existence of solutions 79
6. Remarks on stability, numerical approximation, and algorithms 83

6 A Minimum Time Problem (*R. Conti*) 89
1. Introduction 89
2. Trasfer Sets 89
3. The function T 91
4. Regularity of T 92
5. The synthesis problem 93

7 Some Recent Developments in Linear Complementarity Theory (*R.W. Cottle*) 97
1. Introduction 97
2. Complementary submatrices and cones 97
3. Existence: a characterization of the class $\mathbf{P}_0 \cap \mathbf{Q}$ 99
4. Stability: topological properties of $\mathbf{Q}$ 100
5. Degeneracy: least-index rules for its resolution 100
6. Computational complexity: worst-case behaviour of complementarity algorithms 102

8 Some Stochastic Strategies in Optimization Problems (*M. Cugiani*) 105
1. Introduction 105
2. Deterministic strategies 106
3. Probabilistic methods 113
4. The clustering approach 114
5. The ψ function approach 115
6. Global optimization via stochastic models 120

9 Pricing Underemployed Capacity in a Linear Economic Model (*G.B. Dantzig and P.L. Jackson*) 127
1. Introduction 127
2. The method 127
3. Summary 134

10 **The Dam Problem with Variable Permeability** (*A. Friedman*) 135
1. The Stationary problem 135
2. Non-steady flow 138

11 **Stochastic Control with Partial Observations** (*A. Friedman*) 143
1. Introduction 143
2. The problem 143
3. Imbedding the problem and deriving the quasi-variational inequalities 145

12 **Theorems of Alternative, Quadratic Programs, and Complementarity Problems** (*F. Giannessi*) 151
1. Introduction 151
2. Theorems of alternative for non-linear systems 151
3. Applications to extremum problems and to variational inequalities 159
4. Applications to vector extremum problems 165
5. Concave auxiliary function for quadratic programs 167
6. Decomposition of convex quadratic programs 169
7. Decomposition of non-convex quadratic programs 175
8. Decomposition of a class of complementarity problems 177
9. Numerical Examples 181

13 **An Existence Result for the Global Newton Method** (*F.J. Gould and C.P. Schmidt*) 187

14 **The Smoothness of Free Boundaries Governed by Elliptic Equations and Systems** (*D. Kinderlehrer*) 195
1. Introduction 195
2. Hodograph mappings 198
3. The obstacle problem: a single second-order equation 201
4. The confined plasma 202
5. The double membrane 203
6. The smoothness of the liquid edge 207
7. Higher order equations: the deflection of a thin plate 207
8. An extension of a theorem of Hans Lewy 210

15 **A Survey of Complementarity Theory** (*C.E. Lemke*) 213
1. Introduction 213
2. Some equivalences 214
3. The linear complementarity problem 217
4. Pivoting 218
5. Some classes of matrices. Existence and uniqueness in LCP 220
6. Classes of matrices and CPS 223
7. Some relations between LCP and LP 223
8. Models and applications of LCP 224

9. CAM developments 227
10. Simplicial approximation methods (SAM) 227
11. Solutions to systems of equations 230
12. Continuation methods 235

16. **An Estimate of the Lipschitz Norm of Solutions of Variational Inequalities and Applications** (*P.L. Lions*) 241
1. Introduction 241
2. The main result 242
3. An application to the elastic-plastic problem 246

17 **The Complementarity Problem for Maximal Monotone Multi-functions** (*L. McLinden*) 251
1. Introduction 251
2. Problem statement and examples 252
3. Existence and properties of trajectories of approximate solutions 259
4. Stability, continuity, and generic properties 268

18 **On Some Non-Linear Quasi-Variational Inequalities and Implicit Complementarity Problems in Stochastic Control Theory** (*U. Mosco*) 271
1. Introduction 271
2. Strong regularity of weak solutions 272
3. Proof of proposition 1 277
4. Iterative algorithms 281

19 **On Two Variational Inequalities Arising from a Periodic Visco-elastic Unilateral Problem** (*M. Raous*) 285
1. Introduction 285
2. The mechanical problem 285
3. The constitutive law 288
4. Functional scheme 289
5. The equations 290
6. Discretization and numerical treatment 293
7. Example 296
8. Other aspects and references 301

20 **Lagrange Multipliers and Variational Inequalities** (*R.T. Rockafellar*) 303
1. Introduction 303
2. The general problem 304
3. Discretization 306
4. Complementarity 309
5. Linearity versus non-linearity 312
6. Generalization to multifunctions 315
7. Penalty-duality methods 316
8. Proximal method of multipliers 319

21 **A Numerical Approach to the Solution of Some Variational Inequalities** (*F. Scarpini*) 323
1. Introduction 324
2. Algorithms 326
3. Unilateral Dirichlet problems related to some operators . . 334

22 **Error Estimates for the Finite Element Approximation of Some Variational Inequalities** (*F. Scarpini*) 339
1. Introduction 339
2. A first example of stationary variational inequality 339
3. A second example of stationary variational inequality . . . 342
4. A parabolic evolution inequality of type I 343
5. A parabolic evolution inequality of type II 345

23 **Bifurcation Analysis of the Free Boundary of a Confined Plasma** (*J. Sijbrand*) 349
1. Introduction 349
2. Method of solution 350
3. Example 354

24 **Mathematical Problems in Plasticity Theory** (*R. Temam*) 357
1. Introduction 357
2. The model problem for limit analysis 357
3. The model problem for deformation theory 367

25 **Convergence of Convex Functions, Variational Inequalities, and Convex Optimization Problems** (*R.J.B. Wets*) 375
1. Introduction 375
2. Convergence of convex functions 375
3. Convergence of sequences of closed convex sets 377
4. Characterizing e-convergence 380
5. Relations between e-convergence and p-convergence 388
6. A further characterization of e-convergence 394
7. Convergence of the infima 398
8. Convergence of optimal solutions 400

Subject Index 405

Chapter 1

The Work of G. Stampacchia in Variational Inequalities

J.L. Lions

1. INTRODUCTION

An introductory survey on variational inequalities should have been made here by G. Stampacchia.

All those of you who knew him, who had the pleasure to share with him long and stimulating discussions, who knew his warm personality, will share my emotion and my sorrow.

In what follows, I will try to present some of his main ideas and his main contributions *in the field of variational inequalities,* the main topic of the meeting mentioned in the preface and where he was looking forward to participating and lecturing.

Therefore, I will not speak of his previous contributions; a general report with a complete bibliography will be presented by E. Magenes in the *Bollettino dell'Unione Matematica Italiana*.

In the field of partial differential equations and functional analysis, in 1958 he published a survey with E. Magenes (*Annali Scuola Normale Superiore Pisa*, **12** (1958), 247-357), which had a very deep influence on the teaching of partial differential equations (PDE), and he made very important contributions to the study of second-order elliptic operators, in particular those without any smoothness hypothesis on the coefficients. It was while he was working on deep questions of regularity of solutions when coefficents are only assumed to be bounded and measurable, and on problems of potential theory, that he was led, at the beginning of the 60's, to variational inequalities.

2. VARIATIONAL INEQUALITIES

The following result is now classical [1]: let V be a Hilbert space on $\mathbb{R}$; let $a(u, v)$ be a continuous bilinear form on V, which is *not necessarily sym-*

metric, and which is V-elliptic, i.e. which satisfies

$$(1) \qquad a(v,v) \geqslant \alpha \|v\|^2, \quad \alpha > 0, \quad \forall v \in V$$

($\|\ \|$ denotes the norm in V.) Let K be a closed convex subset of V, $K \neq \phi$, and let $v \to (f, v)$ be a continuous linear form on V; then, there exists a unique element $u \in K$ such that

$$(2) \qquad a(u, v-u) \geqslant (f, v-u) \qquad \forall v \in K$$

This (2) is what is called a *variational inequality* (in short VI).
Let us remark that:
(i) if $K = V$, (2) is equivalent to

$$(3) \qquad a(u,v) = (f,v), \qquad \forall v \in V$$

and the above result gives the Lax-Milgram lemma;
(ii) if a is symmetric (i.e. $a(u, v) = a(v, u)\ \forall u, v \in V$) then (2) is equivalent to

$$(4) \qquad \frac{1}{2} a(u, u) - (f, u) = \min_{v \in K} \left[\frac{1}{2} a(v, v) - (f, v)\right]$$

The idea of the original proof of Stampacchia is as follows:
(I) the result is immediate, according to (ii) above, if a is symmetric;
(II) if (2) is proven for $a(u, v)$, it will also be proven for $a(u, v) + p(u, v)$ where $p(u, v)$ is a not too large perturbation of $a(u, v)$;
(III) with this in mind, one introduces

$$(5) \qquad \begin{aligned} \alpha(u, v) &= \tfrac{1}{2}[a(u, v) + a(v, u)] \\ \beta(u, v) &= \tfrac{1}{2}[a(u, v) - a(v, u)] \end{aligned}$$

and, for $0 \leqslant \theta \leqslant 1$,

$$(6) \qquad a_\theta(u, v) = \alpha(u, v) + \theta \beta(u, v)$$

By virtue of (I), the result is true for $\theta = 0$; using (II) one checks that the result is true for $a_\theta(u, v)$, $0 \leqslant \theta \leqslant \theta_0$, where θ_0 is a constant depending only on a, and one proceeds in this way.

The main application that Stampacchia had in mind at the beginning of this theory was to *potential theory*; he was at that time giving a series of lectures in Leray's seminar [2, 3]. Let us give one example extracted from one of his works (see [2]).

3. APPLICATION TO POTENTIAL THEORY

Let Ω be a bounded open set of $\mathbb{R}^n$; we consider the classical Sobolev spaces

$$H^2(\Omega) = \left\{ v \,\middle|\, v, \frac{\partial v}{\partial x_1}, \ldots, \frac{\partial v}{\partial x_n} \in L^2(\Omega) \right\}$$

$$H_0^1(\Omega) = \{v \,|\, v \in H^1(\Omega), v = 0 \text{ on } \Gamma\}$$

Let $a_{ij}(x)$ be a family of functions such that

(7) $$a_{ij} \in L^\infty(\Omega),$$

(8) $$\sum_{i,j=1}^{n} a_{ij}(x)\, \xi_i\, \xi_j \geq \alpha \sum_{i=1}^{n} \xi_i^2, \quad \alpha > 0, \quad \xi_i \in \mathbb{R}$$

and let us define, $\forall\, u, v \in H^1(\Omega)$:

(9) $$a(u, v) = \sum_{i,j} \int_\Omega a_{ij}(x) \frac{\partial u}{\partial x_j} \frac{\partial v}{\partial x_i}\, dx$$

We do *not* assume symmetry ($a_{ij} \neq a_{ij}$ in general) and we do *not* assume regularity on the a_{ij} .

Let us define (in a vague manner for the moment) the set K as follows: let E be a closed subset of Ω and let us set

(10) $$K = \{v \,|\, v \in H_0^1(\Omega), v \geq 1 \text{ on } E\}$$

The precise meaning of '$v \geq 1$ on E' is as follows: we say that $v \geq 1$ on E in the sense of $H_0^1(\Omega)$ if there exists a sequence of smooth functions u_m in $H_0^1(\Omega)$ such that:

(i) $u_m \to v$ in $H_0^1(\Omega)$
(ii) $u_m \geq 1$ on E

If K is not empty, then there exists a unique element $u \in K$ satisfying

(11) $$a(a, v - u) \geq 0 \quad \forall\, v \in K$$

3.1. Interpretation of (11)

Stampacchia shows that

(12) $$a(u, v) = \int_\Omega v\, d\mu, \quad \forall\, v \in H_0^1(\Omega) \cap C^0(\Omega)$$

where

(13) $d\mu$ is a positive measure, with support in the boundary ∂E of E

The fact that one has (12) with a positive *measure* is very simple: let φ be any smooth function with, say, compact support in Ω, and such that $\varphi \geqslant 0$; then if u is the solution of (11), it is clear that $v = u + \varphi$ belongs to K, so that by using this choice of v in (11) we obtain

(14) $$a(u, \varphi) \geqslant 0 \qquad \text{for every } \varphi \geqslant 0$$

The result follows using a theorem of Schwartz (every distribution which is greater than or equal to 0 is a (positive) *measure*).

The main point consists of showing that μ has its support in ∂E.

One shows first that

(15) $$u = 1 \qquad \text{on } E$$

(in the sense of $H_0^1(\Omega)$). In order to do that, Stampacchia used a simple technique, but a very powerful one, which is now one of the classical tools of the theory of partial differential equations. Let us define

(16) $$w = \inf\{u, 1\}$$

One checks that $w \in K$ and that

(17) $$a(w, w - u) = 0$$

Indeed, either $u \geqslant 1$ and then $w = 1$, or $u \leqslant 1$ and then $w - u = 0$, so that

$$a_{ij}(x) \frac{\partial w}{\partial x_j}(x) \frac{\partial (w-u)}{\partial x_i}(x) = 0$$

in either case. We can take $v = w$ in (11), and from (11) and (17) we deduce that

$$a(w - u, w - u) \leqslant 0$$

Since[1]

$$a(w - u, w - u) \geqslant \alpha \|w - u\|^2$$

it follows that

$$w = u$$

(1) If we set $\|v\|^2 = \sum_{i=1}^{n} \int_\Omega \left(\frac{\partial v}{\partial x_i}\right)^2 dx$

hence (15) follows.

Let us take now φ as a smooth function in $H_0^1(\Omega)$, with support in $\complement E$; then $v = u \pm \varphi \in K$, and taking $v = u \pm \varphi$ in (11) gives

$$(18) \qquad a(u, \varphi) = 0, \quad \forall\ \varphi \text{ with compact support in } \complement E$$

Then (13) follows from (15) and (18).
The measure μ is called the *capacitary measure* of E with respect to $a(u, v)$ and to Ω, and $\mu(1)$ in the corresponding *capacity* of E.

In reference [2](see also reference [3]), Stampacchia proceeds to study the properties of this capacity. He introduces, among other things, the notion of regular points with respect to A and shows that this notion is in fact *independent* of A (in the class of elliptic operators), so that it is equivalent with the Wiener condition(relative to $A = -\Delta$). (A is the second-order elliptic operator associated with a.)

The techniques and the ideas of Stampacchia gave rise to several interesting contributions in potential theory(2).

4. A NUMBER OF VARIANTS AND EXTENSIONS

Let us return now to (8). A number of theoretical questions immediately present themselves.

A first natural question is connected with (10); if one considers, instead of $\frac{1}{2} a(v, v) - (f, v)$, a *general* convex function $J(v)$ defined in a *Banach* space, one is lead to a V I of the form (if J is differentiable).

$$(J'(u), v - u) \geqslant 0, \quad \forall\ v \in K$$
$$u \in K$$

It is then natural to replace the operator J' by a *monotonic operator.*

This led to the paper of Hartman and Stampacchia [4] where they study VI in reflexive Banach spaces, for non-linear partial differential operators of the types of those introduced (in increasing order of generality) by Minty(3), Browder(4), and Leray and Lions(5). The good abstract notion for *abstract* operators A leading to 'well set' elliptic V I

$$(19) \qquad (A(u), v - u) \geqslant 0, \quad \forall\ v \in K$$
$$u \in K$$

(2) R. M. Hervé. *Ann. Inst. Fourier,* **14** (1964), 493-508; M. Hervé, R. M. Hervé. *Ann. Inst. Fourier* **22** (1972), 131-145; A. Ancona. J. Mat. Pures et Appl., **54** (1975), 75-124.

(3) G. Minty. *Duke math. J.,* **29** (1962), 341-6.

(4) F. Browder. *Bull. Am. Math. Soc.*, **71** (1965), 780-5.

(5) J. Leray and J. L. Lions. *Bull. Soc. Math. Fr.*, **93** (1965), 97-107.

was introduced by Brezis[6] with the notion of *pseudo-monotonic* operators.

As always in the work of Stampacchia, there is a *motivation* for the 'abstract' part of the work[4]; we shall return to that.

Another question, motivated by the so-called 'unilateral boundary conditions' arising in elasticity (or 'Signorini's problem'; see Fichera[7]) is whether the coerciveness hypothesis (7) can be relaxed. This has been studied by Stampacchia and Lions [5, 6]. Let us mention here *one* result: suppose that $a(u, v)$ is given as in section 2 but that it satisfies, instead of (7), the much weaker condition:

$$a(v, v) \geq 0, \qquad \forall\, v \in V \tag{20}$$

We *assume* that (8) allows *at least one* solution, and we denote by X the set of *all* solutions; one checks immediately that X is a *closed* convex *set*; let $b(u, v)$ be a continuous bilinear form on V such that

$$b(v, v) \geq \beta \,\|v\|^2, \qquad \beta > 0, \qquad \forall\, v \in V \tag{21}$$

Let $v \to (g, v)$ be a continuous linear form on V; for every $\varepsilon > 0$, there exists (according to the result (1) of section 2) a unique $u_\varepsilon \in K$ such that

$$a(u_\varepsilon, v - u_\varepsilon) + \varepsilon b(u_\varepsilon, v - u_\varepsilon) \geq (f + \varepsilon g, v - u_\varepsilon), \qquad \forall\, v \in K \tag{22}$$

Then, as $\varepsilon \to 0$, $u_\varepsilon \to u_0$ in V, where u_0 is the solution of

$$\begin{gathered} b(u_0, v - u_0) \geq (g, v - u_0), \qquad \forall\, v \in X \\ u_0 \in X \end{gathered} \tag{23}$$

This result is used in reference [6], among other things, to solve the unilateral problem.

Still another natural question is the *evolution analogue* of (2): find a function $t \to u(t)$, where t is the *time*, such that[8]

$$u(t) \in K \tag{24}$$

$$\left(\frac{\partial u(t)}{\partial t}, v - u(t)\right) + a(u(t), v - u(t)) \geq (f(t), v - u(t)) \qquad \forall\, v \in K \tag{25}$$

$$u(t)\,\big|_{t=0} = u(0) = u^0 \qquad \text{is given (in } K\text{)} \tag{26}$$

(6) H. Brezis. *Ann. Inst. Fourier*, **18** (1968), 115-75.
(7) G. Fichera. *Mem. Accad. Naz. Lincei*, **8** (1964), 91-140.
(8) We do not define in detail the function spaces where u can be taken.

When $K = V$, (25) reduces to

$$\left(\frac{\partial u(t)}{\partial t}, v\right) + a(u(t), v) = (f(t), v), \quad \forall\ v \in V \tag{27}$$

It is the variational form of 'abstract' parabolic equations.
This problem has been introduced in [6] ; it was considerably extended and deepened in the work of Brezis[9]; many examples arising from mechanics have been studied[10]; this problem is also connected with *non-linear semi-groups*[11].

One difficulty which arises in connection with (25) is in the *definition* of what we mean by a *solution* of a VI and an important remark is now in order: let A be a non-linear operator from a reflexive Banach space V into its dual V', and let us assume that A is *monotonic*, i.e.

$$((A(u) - A(v), u - v) \geqslant 0, \quad \forall\ u, v \in V \tag{28}$$

Then if u is a solution of the VI

$$\begin{aligned} &(A(u), v - u) \geqslant (f, v - u), \quad \forall\ v \in K \\ &u \in K \end{aligned} \tag{29}$$

one has

$$\begin{aligned} &(A(v), v - u) \geqslant (f, v - u), \quad \forall\ v \in K \\ &u \in K \end{aligned} \tag{30}$$

This is obvious, since

$$(A(v), v - u) = (A(u), v - u) + (A(v) - A(u), v - u) \geqslant (A(u), v - u)$$

(using (28)); but the *reciprocal* property is true, provided A is hemi-continuous (i.e. $\lambda \to (A(u + \lambda v), w)$ is continuous $\forall\ u, v, w \in V$). Indeed, if $\hat{v}$ is given in K, and if we choose in (30)

$$v = (1 - \theta) u + \theta \hat{v}, \quad \theta \in]\ 0,1]$$

we obtain, after dividing by θ:

$$(A((1 - \theta) u + \theta \hat{v}), \hat{v} - u) \geqslant (f, \hat{v} - u) \tag{31}$$

(9) H. Brezis. *NATO Summer School*, Venice, June 1968; H. Brezis. *J. Math. Pures et Appl.*, **51** (1972), 1-168.

(10) G. Duvaut and J. L. Lions. '*Les inéquations en Mécanique et en Physique*.' Dunod, Paris (1972).

(11) H. Brezis. '*Opérateurs maximaux monotones et semi groupes de contractions dans les espaces de Hilbert*.' North-Holland, Amsterdam (1973).

By virtue of the hemi-continuity, we can let $\theta \to 0$ in (31) and we obtain (29) (with $\hat{v}$ instead of v).

This remark allows one to define *weak* solutions, or generalized solutions, of VI; it is used in the paper with Lewy [7] (we shall return to that) and it can also be used for (25) (to 'replace' $\partial u/\partial t$ by $\partial v/\partial t$).

5. THE OBSTACLE PROBLEM

In section 4 we indicated very briefly *some* of the problems in variational inequalities which were under study in the years 1966-68; it was at about this time, may be a little earlier, that Stampacchia started working on a problem which is simple, beautiful and deep – and which led to important discoveries some of them being reported in this book.

This is the so-called 'obstacle' problem. Let us consider $a(u, v)$ to be given by (9) and let us define

$$K = \{v \mid v \in H_0^1(\Omega), v \geqslant \psi, \psi \text{ given in } \Omega\} \tag{32}$$

Of course, one has to specify in (32) the class where ψ is given, so that, in particular, K is not empty; the function ψ represents the *obstacle.* The corresponding VI (2) has a unique solution and the problem is as follows.
(1) How to interpret the VI?
(2) What are the regularity properties of the solution u?

In solving (1) and (2) a free-boundary problem will appear and the next question will be the following.
(3) What are the regularity properties of the free boundary?

Let us explain the basic idea of the work with Brezis [8] in a simple particular case. According to (30) the VI can be written(12).

$$(Av, v-u) \geqslant (f, v-u), \qquad \forall\, v \in K \tag{33}$$

where K is given by (32). Let us assume that

$$\begin{aligned} &\psi \in H^1(\Omega) \qquad \psi \leqslant 0 \qquad \text{on } \Gamma \\ &\text{and} \\ &A\psi \leqslant 0 \end{aligned} \tag{34}$$

Everything is based on a particular choice of v in (33).
For $\varepsilon > 0$ we *define* u_ε as the solution in $H_0^1(\Omega)$ of

(12) $Av = -\sum \dfrac{\partial}{\partial x_i}\left(a_{ij}(x)\dfrac{\partial v}{\partial x_j}\right) \in H^{-1}(\Omega)$ (dual space of $H_0^1(\Omega)$)

$$(35)\qquad \begin{aligned} \varepsilon A u_\varepsilon + u_\varepsilon &= u \quad \text{in } \Omega \\ u_\varepsilon &= 0 \quad \text{on } \Gamma \end{aligned}$$

Let us allow for the moment – *this is the crucial point* – that

$$(36)\qquad u_\varepsilon \in K \qquad (\text{i.e. } u_\varepsilon \geqslant \psi \text{ in } \Omega)$$

Then one can choose $v = u_\varepsilon$ in (33) and after dividing by ε it gives

$$(Au_\varepsilon, Au_\varepsilon) \leqslant (f, Au_\varepsilon)$$

Hence, it follows that

$$(37)\qquad \| Au_\varepsilon \|_{L^2(\Omega)} \leqslant \| f \|_{L^2(\Omega)}$$

It is a simple matter to check that $u_\varepsilon \to u$ in $H_0^1(\Omega)$ as $\varepsilon \to 0$, so from (37) one obtains that

$$(38)\qquad Au \in L^2(\Omega)$$

Therefore, if we set $Au = \hat{f}(\hat{f} \neq f$ in general!), one can think of u as being given by the solution of the Dirichlet's boundary value problem

$$Au = \hat{f} \qquad (\hat{f} \in L^2(\Omega)), \qquad u = 0 \qquad \text{on } \Gamma$$

It follows that if the coefficients of A are smooth enough and if the boundary Γ of Ω is smooth enough, then

$$(39)\qquad u \in H^2(\Omega)$$

that is

$$\frac{\partial^2 u}{\partial x_i\, \partial x_j} \in L^2(\Omega), \qquad \forall\, i, j$$

Let us verify now that (36) holds. We write (35) as

$$(40)\qquad \varepsilon A(u_\varepsilon - \psi) + (u_\varepsilon - \psi) + \varepsilon A\psi = u - \psi$$

and we take scalar products with $(u_\varepsilon - \psi)^-$ (where, in general, $v^- = \sup(-v, 0)$). We obtain, since $(u_\varepsilon - \psi)^- \in H_0^1(\Omega)$ and since $a(v, v^-) = -a(v^-, v^-)$, $(v, v^-) = -(v^-, v^-)$:

$$(41)\qquad \begin{aligned} &-\varepsilon a((u_\varepsilon - \psi)^-, (u_\varepsilon - \psi)^-) - \| (u_\varepsilon - \psi)^- \|^2_{L^2(\Omega)} + \varepsilon (A\psi, (u_\varepsilon - \psi)^-) = \\ &= (u - \psi, (u_\varepsilon - \psi)^-) \end{aligned}$$

Since $A\psi \leqslant 0$, then $(A\psi, (u_\varepsilon - \psi)^-) \leqslant 0$, and (41) gives:

$$(u-\psi, (u_\varepsilon-\psi)^-) + \|(u_\varepsilon-\psi)^-\|^2_{L^2(\Omega)} + \varepsilon a((u_\varepsilon-\psi)^-, (u_\varepsilon-\psi)^-) \leqslant 0$$

But $(u-\psi, (u_\varepsilon-\psi)^-) \geqslant 0$ and therefore $(u_\varepsilon-\psi)^- = 0$, i.e. $u_\varepsilon \geqslant \psi$.

The above analysis can be extended, as we show below. Before doing so, let us apply (38) to the interpretation of the VI. One shows easily that u is characterized by

$$\begin{aligned} &Au - f \geqslant 0 \\ &u - \psi \geqslant 0 \\ &(Au-f)(u-\psi) = 0 \quad \text{in } \Omega \end{aligned} \tag{42}$$

and of course

$$u = 0 \quad \text{on } \Gamma$$

Consequently, there are two sets in Ω:

$$\begin{aligned} &\text{the } \textit{coincidence} \text{ set, where} \quad u = \psi \\ &\text{the } \textit{equilibrium} \text{ set, where} \quad Au = f \end{aligned} \tag{43}$$

At least in the two-dimensional case, one can think of this problem as giving the displacement of a membrane subjected to forces f and required to stay above the obstacle ψ.

The membrane touches the obstacle on the coincidence set. The two regions are separated by a 'surface' S, which is a free surface; S is not given, and on S one has two 'boundary' conditions. If $\psi \in H^2(\Omega)$, one has

$$\begin{aligned} &u = \psi \\ &\text{and} \\ &\frac{\partial u}{\partial x_i} = \frac{\partial \psi}{\partial x_i} \quad \forall i \text{ on } S \end{aligned} \tag{44}$$

A natural and important question is now: under suitable hypotheses on f and on ψ, is it true that

$$Au \in L^p(\Omega) \tag{45}$$

This is, of course, important for the regularity of u in spaces like

$$W^{2,p}(\Omega) = \left\{ v \;\middle|\; v, \frac{\partial v}{\partial x_i}, \frac{\partial^2 v}{\partial x_i \partial x_j} \in L^p(\Omega) \right\}$$

for p large. For the study of this problem, a more 'abstract' presentation is in order, always following the work of Brezis and Stampacchia.

6. AN ABSTRACT REGULARITY THEOREM. APPLICATION TO THE OBSTACLE PROBLEM

We consider the VI

$$(46)\qquad \begin{aligned} &(Av, v-u) \geq (f, v-u), \qquad \forall\, v \in K \\ &u \in K \end{aligned}$$

where $K \subset V$, and V is a Hilbert space. The situation extends to cases where A is non-linear and where V is a reflexive Banach space. Let us consider a space X such that

$$(47)\qquad V \subset X \subset V'$$

with continuous imbedding, each space being dense in the following one.

Example 1
$V = H_0^1(\Omega)$, $V' = H^{-1}(\Omega)$ and $X = L^p(\Omega)$, for p large enough. The problem considered by Stampacchia and Brezis is: when can we conclude that

$$Au \in X\ ?$$

One introduces a *duality mapping* J from $X \to X'(V \subset X' \subset V')$, i.e. a (non-linear) mapping from $X \to X'$ such that

$$(J(u), u) = \|J(u)\|_{X'} \|u\|_X$$

$\|J(u)\|_{X'}$ is a strictly increasing function of $\|u\|_X$ and goes to $+\infty$ as $\|u\|_X \to \infty$.

Example 2
If $X = L^p(\Omega)$, $J(u) = |u|^{p-2}\, u$.

If X is a Hilbert space that we identify with its dual ($X = L^2(\Omega)$ in the example), then J = identity. The crucial hypothesis is now:

$$(48)\qquad \begin{aligned} &\text{One can find a duality mapping } J \text{ from } X \to X' \text{ such that} \\ &\forall\, \varepsilon > 0 \text{ and } \forall\, u \in K, \text{ there exists } u_\varepsilon \text{ such that} \\ &u_\varepsilon \in K, \quad Au_\varepsilon \in X \quad \text{and} \quad u_\varepsilon + \varepsilon J A u_\varepsilon = u \end{aligned}$$

One can then take $v = u_\varepsilon$ in (46) and using the properties of J one obtains that

$$\|Au\|_X \leq \text{constant}$$

Hence, it follows that

$$Au \in X \tag{49}$$

Application to the problem.
We take $J(u) = J_p(u) = |u|^{p-2}u$ and we consider the equation (48), i.e. since

$$J^{-1} = J_{p'} \quad \text{and} \quad \frac{1}{p} + \frac{1}{p'} = 1$$

then

$$Au_\varepsilon + J_{p'}\left(\frac{u_\varepsilon - u}{\varepsilon}\right) = 0 \tag{50}$$

We want to show that $u_\varepsilon \geqslant \psi$. We use the same technique as in section 5, i.e. we multiply by $(u_\varepsilon - \psi)^-$. We obtain

$$-a((u_\varepsilon - \psi)^-, (u_\varepsilon - \psi)^-) + (A\psi, (u_\varepsilon - \psi)^-) + \left(J_{p'}\left(\frac{u_\varepsilon - u}{\varepsilon}\right), (u_\varepsilon - \psi)^-\right) = 0 \tag{51}$$

But

$$\left(J_{p'}\left(\frac{u_\varepsilon - u}{\varepsilon}\right), (u_\varepsilon - \psi)^-\right) = -\frac{1}{\varepsilon^{p'-1}} \int_\Omega |u_\varepsilon - u|^{p'-2}[(u_\varepsilon - \psi)^-]^2\, dx$$

$$-\frac{1}{\varepsilon^{p'-1}} \int_\Omega |u_\varepsilon - u|^{p'-2}(u - \psi)(u_\varepsilon - \psi)^-\, dx$$

so that (51) gives (since $A\psi \leqslant 0$):

$$\int_\Omega |u_\varepsilon - u|^{p'-2}[(u_\varepsilon - \psi)^-]^2\, dx \leqslant 0 \tag{52}$$

Therefore, either $u_\varepsilon(x) - u(x) = 0$, and hence $u_\varepsilon(x) \geqslant u(x) \geqslant \psi(x)$, or $(u_\varepsilon(x) - \psi(x))^- = 0$, i.e. $u_\varepsilon(x) \geqslant \psi(x)$.

It will follow that, under reasonable assumptions, the solution u of the obstacle problem satisfies

$$u \in W^{2,p}(\Omega) \tag{53}$$

Simple one-dimensional examples show that one cannot obtain L^p estimates for higher-order derivatives. One can study the regularity in *Schauder* spaces; we refer the reader to the book of Kinderlehrer and Stampacchia(13) and to other chapters of this book. See also the report at the International Congress of Mathematicians, Vancouver, 1974, made by Kinderlehrer.

(13) An introduction to variational inequalities and their applications. Academic Press. To appear.

Remark Due to the physical interpretation of the obstacle problem, it is quite natural to consider the problem of *minimal surfaces with obstacle*. This has been considered by Nitsche[14] and by Giusti,[15] Giaquinta and Pepe,[16] and for *surfaces with mean curvature fixed*, it has been considered by Mazzone.[17]

7. AN INEQUALITY FOR THE OBSTACLE PROBLEM

In the above proof of the regularity for the obstacle problem, the hypothesis '$A\psi \leq 0$' is much too restrictive. This can be overcome in several ways.
(1) One can introduce more flexibility in the abstract hypothesis of section 4 (see [48]); following Brezis and Stampacchia, one introduces families of operators $B_\varepsilon : V \to X$, $C_\varepsilon : V \to X'$, which are *bounded* as $\varepsilon \to 0$ and such that the equation

$$u_\varepsilon + \varepsilon J(Au_\varepsilon + B_\varepsilon u_\varepsilon) = u + \varepsilon C_\varepsilon u_\varepsilon \tag{54}$$

has a solution $u_\varepsilon \in K$ such that $Au_\varepsilon \in X$; one then obtains the conclusion (same proof) that

$$Au \in X \tag{55}$$

and this allows one to obtain regularity results similar to the above results under the assumption:

$$\begin{aligned} &A\psi \text{ is a measure on } \overline{\Omega} \\ &\sup \{A\psi, 0\} \in L^p(\Omega) \end{aligned} \tag{56}$$

(2) One can use penalty arguments.
(3) One can use an inequality given in Lewy and Stampacchia [9, 10] that we now explain.

Let u be the solution of the obstacle problem. Then one has

$$f \leq Au \leq \max \{A\psi, f\} \tag{57}$$

(14) J.C. Nitsche. 'Vorlesungen uber Minimalflachen'. Grundlehr. Math. Wiss., vol. 199, Springer, Berlin (1975).
(15) E. Giusti. 'Minimal surfaces with obstacles'. *CIME course on Geometric Measure theory and Minimal Surfaces*, Rome, 1973, pp. 119-53.
(16) M. Giaquinta and L. Pepe. 'Esistenza e regolaritá per il problema dell'area minima con ostacoli in *n* variabili'. *Ann. Scu. Norm. Sup., Pisa*, **25** (1971), 481-507.
(17) S. Mazzone. 'Un problema di disequazioni variazionali per superficie di curvatura media assegnata'. *Boll. Unione Mat. Ital.*, **7** (1973), 318-29.

One does not restrict the generality in taking

$$f = 0 \tag{58}$$

(Indeed if ω is defined by $A\omega = f$, $\omega = 0$ on Γ, then it suffices to work on $u - \omega$ instead of u.) Then one has to show that

$$0 \leqslant Au \leqslant \max\{A\psi, 0\} \tag{59}$$

Actually, one can obtain a more precise result [9]. Let us introduce $\theta(s) = 1$ for $s \leqslant 0$, $\theta(s) = 0$ for $s > 0$. Then there exists a unique function u in $W^{2,p}(\Omega)$ such that

$$\begin{aligned} &Au = \max\{A\psi, 0\}\, \theta(u - \psi) \quad \text{in } \Omega \\ &u = 0 \quad \text{in } \Gamma \end{aligned} \tag{60}$$

and u is the solution of the obstacle problem. (Of course (59) follows from (60).)

Proof: The proof of (60) is in two essential steps:

Step 1 One considers an approximation of (60). Let $\theta_n(s)$ be a sequence of Lipschitz continuous functions approximating θ:

$$\theta_n(s) = \begin{cases} 1 & \text{if} \quad s \leqslant 0 \\ 1 - ns & \text{if} \quad 0 \leqslant s \leqslant 1/n \\ 0 & \text{if} \quad s > 1/n \end{cases} \tag{61}$$

One considers the equation

$$Au_n = \max\{A\psi, 0\}\, \theta_n(u_n - \psi), \qquad u_n = 0 \quad \text{on } \Gamma \tag{62}$$

One proves that this equation has a solution by a fixed-point argument: given $\omega \in H_0^1(\Omega)$, one defines $\hat{u}$ as the solution of

$$A\hat{u} = \max\{A\psi, 0\}\theta_n(\omega - \psi) \tag{63}$$

One verifies that $\omega \to \hat{u} = T(\omega)$ maps a suitable ball Σ of $H_0^1(\Omega)$ into itself and that T is continuous. One has also, if $\max\{A\psi, 0\} \in L^p(\Omega)$, and if the coefficients of A are smooth enough, that $\hat{u} \in W^{2,p}(\Omega)$. One then verifies that the mapping T is compact from $\Sigma \to \Sigma$, and hence it has a fixed point, which is a solution u_n of (62).

One has also obtained in this manner that

$$u_n \text{ remains in a bounded set of } W^{2,p}(\Omega) \tag{64}$$

Step 2 The second step in the proof consists of proving that

$$u_n \geqslant \psi \tag{65}$$

Assume that one can find x_0 where $u_n(x_0) < \psi(x_0)$. Then one can find an open set G around x_0 such that

$$\begin{aligned} &u_n < \psi \quad \text{on } G \\ &\text{and} \\ &u_n = \psi \quad \text{on } \partial G \end{aligned} \tag{66}$$

Since on G one has $u_n < \psi$, equation (62) reduces to

$$Au_n = \max\,\{A\psi, 0\} \quad \text{on } G$$

hence

$$\begin{aligned} A(u_n - \psi) &\geqslant 0 \quad \text{on } G \\ u_n - \psi &= 0 \quad \text{on } \partial G \end{aligned} \tag{67}$$

Therefore, by the maximum principle, $u_n - \psi \geqslant 0$ in G, in contradiction to $u_n(x_0) < \psi(x_0)$. Hence (65) follows.

One can then pass to the limit in n, using (64) and (65), and one shows that u is the solution of the VI of the obstacle problem and that u satisfies (60).

Application to the regularity.
The application is obvious, and actually it is already implicitly contained in (64).

Remark A systematic use of inequality (57) or of similar inequalities with other boundary conditions, or with parabolic operators, is made in the work of Mosco, Troianiello, Joly, Hanouzet and others.

8. ELASTO-PLASTIC PROBLEM

Another important VI arises in the theory of elasto-plastic materials; the physical problem corresponds to dimension 2.

One considers the same bilinear form $a(u, v)$ as in section 3 and the convex set

$$K = \{v \mid v \in H_0^1(\Omega), |\nabla v(x)| \leqslant 1 \text{ a.e. in } \Omega\} \tag{68}$$

(a.e. = almost everywhere).

The following result is due to Brezis and Stampacchia [8] : if $f \in L^p(\Omega)$, the solution of the VI corresponding to (68) satisfies

$$Au \in L^p(\Omega) \tag{69}$$

One uses the idea of section 6. One has to consider then the equation

$$u_\varepsilon + \varepsilon J_p A u_\varepsilon = u \tag{70}$$

where u is given in K, $u_\varepsilon = 0$ on Γ and to show that it has a solution u_ε in K such that $Au_\varepsilon \in L^p(\Omega)$. In fact, if $u_\varepsilon \in K$, then it is bounded and (70) implies that $Au_\varepsilon \in L^\infty(\Omega)$. We write (70) in the form

$$Au_\varepsilon = J_{p'}\left(\frac{u - u_\varepsilon}{\varepsilon}\right) \tag{71}$$

and one has to show that there is a solution such that $|\nabla u_\varepsilon(x)| \leqslant 1$ a.e.

More generally, let $\lambda \to \theta(\lambda)$ be a strictly increasing function such that $\theta(0) = 0$ and let us consider the equation:

$$Au = \theta(f - u), \qquad u \in H_0^1(\Omega) \tag{72}$$

Brezis and Stampacchia [8] show that this problem has a solution and that if $|\nabla f(x)| \leqslant 1$ a.e., it follows that $u \in K$.

The proof rests on a comparison lemma and on several technical ideas in order to obtain estimates on ∇u; the authors consider first the case when Ω is convex and then the general case.

This problem of elasto-plasticity has been the object of a large number of interesting works. Two of the main questions are as follows.

(i) Let us consider $\delta(x)$, which is the distance of x to Γ, and let us define

$$K_1 = \{v \mid v \in H_0^1(\Omega), |v(x)| \leqslant \delta(x)\} \tag{73}$$

Let us consider

$$a(u, v) = \int_\Omega \nabla u \nabla v \, dx + \lambda \int_\Omega uv \, dx \tag{74}$$

Let u (and, respectively, $\tilde{u}$) be the solution of

$$\begin{aligned} &a(u, v - u) \geqslant (f, v - u) \quad \forall v \in K \\ &u \in K \end{aligned} \tag{75}$$

and, respectively,

$$\begin{aligned} &a(\tilde{u}, v - \tilde{u}) \geqslant (f, v - \tilde{u}), \quad \forall v \in K_1 \\ &\tilde{u} \in K_1 \end{aligned} \tag{76}$$

Then, under suitable hypotheses on f and λ it has been proven by Brezis and Sibony[18] that

$$\tilde{u} = u \tag{77}$$

Their proof uses, in an essential way, an idea of Hartman and Stampacchia [4].

When Ω is multiply connected, the formulation of the elasto-plastic problem has to be slightly changed with respect to the above – one has to consider functions which are *constant* on the boundaries of the 'holes' of Ω (see Lanchon[19]).

(ii) Another important question is connected with the *regularity of the free boundary*, that is the regularity of the boundary between the elastic region (where $|\nabla u(x)| < 1$) and the plastic region (where $|\nabla u(x)| = 1$). Let us refer the reader to Caffarelli and Friedman[20] and to the bibliography therein.

9. HODOGRAPH METHOD AND VI

Towards the end of the 1960s, as the free-boundary problems solved by the technique of VI were becoming understood (with, of course, still many questions unanswered – in particular, of regularity – at that time), another type of question came into the picture: given a free-boundary problem arising from mathematical physics[21], when can it be formulated (and hopefully solved) by the technique of VI.

This type of question was raised, in particular, in problems of infiltration through porous media. It was observed in 1971 by Baiocchi[22] that by an appropriate transformation of the unknown function, it was possible to reduce, at least in some case, the problem of infiltration through porous media to a VI.

This idea gave rise to a large number of papers. Some of the most interesting among them are those of Brezis and Stampacchia [12, 13].

One considers in the plane x, y the flow of a perfect fluid (assumed to be steady, irrotational and incompressible) around a profile P, which is symmetric with respect to y. The flow is assumed to be uniform at infinity, i.e. if $\mathbf{q} = \{u, v\}$ denotes the velocity

$$\mathbf{q}(x, y) \to \{q_\infty, 0\} \quad \text{as } |x| + |y| \to \infty \tag{78}$$

(18) H. Brezis and M. Sibony. 'Equivalence de deux I.V.' *Arch. Ration. Mech. Anal.*, **41** (1971), 254-65.
(19) H. Lanchon. *J. Mécanique*, **13** (1974), 267-320.
(20) L.A. Caffarelli and A. Friedman. 'The free boundary for elastic plastic torsion problems'. To appear.
(21) We do not speak here of the free-boundary problems arising in the theory of optimal control.
(22) C. Baiocchi. 'Sur un problème a frontière libre traduisant le filtrage de liquides à travers des milieux poreux'. *C. R. Acad. Sci., Paris* **273** (1971), 1215-7.

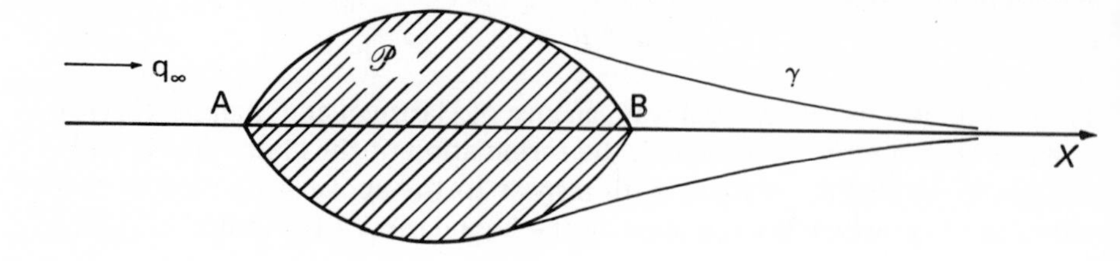

(see the figure), one has

$$\text{div}\, \mathbf{q} = 0, \qquad \text{rot}\, \mathbf{q} = 0 \tag{79}$$

$$\mathbf{q} \text{ is tangential to } P \text{ along } \partial P \tag{80}$$

There is a wake denoted by γ; γ is a free boundary and along γ one has the two conditions

$$\begin{aligned} &\mathbf{q} \text{ is a tangent to } \gamma \\ &|\mathbf{q}| = q_\infty \end{aligned} \tag{81}$$

One introduces the stream function ψ by

$$u = \psi_y, \qquad v = -\psi_x$$

Then

$$\Delta \psi = 0 \tag{82}$$

and by symmetry it suffices to consider the problem for $y > 0$.

Boundary conditions are (see the figure)

$$\psi = 0 \qquad \text{on MAB} \tag{83}$$

$$\psi = 0, \qquad |\nabla_\psi| = q_\infty \qquad \text{on } \gamma \tag{84}$$

One considers the hodograph transformation

$$x, y \to u, v \to \theta, q$$

where

$$\tan \theta = v/u, \qquad q = |\mathbf{q}|$$

One considers ψ as a function of the independent variables

$$\theta \quad \text{and} \quad \sigma = -\log q \tag{85}$$

Then, one has

$$\Delta\psi = \frac{\partial^2 \psi}{\partial \theta^2} + \frac{\partial^2 \psi}{\partial \sigma^2} = 0 \tag{86}$$

In the hodograph transformation, the part of the boundary where one has only *one* condition (i.e. MAB) becomes a free boundary and the part of the boundary where one has *two* conditions (i.e. γ) becomes *known* (in fact, a part of the σ axis) and one ends up with a problem of the following nature.

Let Ω be an open set in $\mathbb{R}^n$; find $D \subset \Omega$ and a function ψ defined in D such that

$$-\Delta\psi = \varphi \quad \text{on } D \quad (\varphi \text{ given in } \Omega) \tag{87}$$

$$\psi \text{ satisfies a standard boundary condition on } \partial D \cap \partial\Omega \tag{88}$$

$$\begin{gathered} \psi = 0 \text{ on } S = \partial D \cap \Omega \\ \partial\psi/\partial\nu = \pi \cdot \nu \text{ on } S \end{gathered} \tag{89}$$

where π is a given vector field in Ω, ν is the normal to S extended to D; S is the free boundary (in the 'hodograph' plane).

By a transformation of an unknown function of the type of that introduced by Baiocchi, one can reduce – at least in some case – this type of problem to a VI (see Brezis and Stampacchia [12,13, 19]).

The analogous problem in a finite strip has been solved by a research student of Stampacchia[23].

10. FOURTH-ORDER VI

In a paper [17] with Brezis, Stampacchia studied the regularity of fourth-order VI; a very simple remark shows that – essentially – one cannot go farther than third-order derivative estimates. Indeed, if one defines

$$K_1 = \{\nu \mid \nu \in H_0^1(\Omega) \cap H^2(\Omega),\ \alpha \leqslant \Delta\nu \leqslant \beta\}$$

where α and β are constants, $\alpha < 0 < \beta$, and if one considers the VI

$$\begin{gathered} (\Delta u, \Delta(\nu - u)) \geqslant (f, \nu - u) \quad \forall\, \nu \in K_1 \\ u \in K_1 \end{gathered} \tag{90}$$

(23) F. Tomarelli. *Graduation Thesis*, Scuola Normale Superiore, Pisa (1978).

then, if $f \in L^2(\Omega)$, one has $u \in W^{3,p}(\Omega)$ for every p finite but 'nothing better'. This is immediate: let us introduce F by

$$\Delta F = f, \qquad F \in H_0^1(\Omega)$$

If we set $\Delta u = \hat{u}$, $\Delta v = \hat{v}$, then (90) becomes equivalent to

$$\begin{aligned} &(\hat{u}, \hat{v} - \hat{u}) \geqslant (F, \hat{v} - \hat{u}) \qquad \forall\, \hat{v} \\ &\alpha \leqslant \hat{v} \leqslant \beta \qquad \text{and} \qquad \alpha \leqslant \hat{u} \leqslant \beta \end{aligned}$$

Then, if $\lambda \to P(\lambda)$ denotes the projection $R \to [\alpha, \beta]$, we have

$$\hat{u} = P(F)$$

so that $u \in W^{1,\infty}(\Omega)$ and u has *exactly* the regularity properties given above.

If one considers instead of K_1 the convex set K_2 given by

$$K_2 = \{v \mid v \in H_0^2(\Omega),\ \alpha \leqslant \Delta v \leqslant \beta\}$$

and if we denote by u the solution of the VI (90) where K_1 is replaced by K_2, then, again one has the same regularity result, namely, $u \in W^{3,p}(\Omega)$ for every p finite and essentially 'nothing better'. The proof consists of showing that there exists $z \in L^1(\Omega)$, such that $\Delta z = 0$ and such that $\hat{u}$ (with the same notations as above) can be represented by

$$\hat{u} = P(F + z)$$

For the 'obstacle problem', i.e. the same problem with K_1 or K_2 replaced by

$$K_3 = \{v \mid v \geqslant \phi,\ v \in H_0^2(\Omega)\}$$

it has been shown by Frehse[24] that the corresponding solution belongs to $H^3_{\text{local}}(\Omega)$ (assuming that ϕ is smooth), a result which has been recently improved by Caffarelli and Friedman[25] (these authors also study the regularity of the free boundary).

Further remarks concerning this problem with the convex set K_2 can be found in Torelli[26].

(24) J. Frehse. *Hamburg Univ. Math. Sem., Abhard.*, **36** (1971), 140-9.

(25) L.A. Caffarelli and A. Friedman. 'The obstacle problem for the biharmonic operator'. To appear.

(26) A. Torelli. 'Some regularity results for a family of variational inequalities,' Ann. Scuola Normale Superiore, Pisa, (IV), **6** (1979). To appear.

11. INFILTRATION IN POROUS MEDIA

We now briefly report on a posthumous work [20] with Brezis and Kinderlehrer on infiltration in porous media. This type of problem has been studied in particular by Baiocchi(27) and by Alt(28). The method introduced in reference [18] is, roughly speaking, as follows. We are given an open set $\Omega \subset \mathbb{R}^2$ with boundary $\partial\Omega$ which consists of three parts: $\partial\Omega = S_1 \cup S_2 \cup S_3$; we want to find $p \in H^1(\Omega)$, $p \geqslant 0$, and $g \in L^\infty(\Omega)$, such that

$$(91) \qquad \begin{array}{ll} g = 1 & \text{if } p > 0 \\ g \in [0,1] & \text{if } p = 0 \end{array}$$

$$(92) \qquad p = \text{given function on } S_2 \cup S_3$$

$$(93) \qquad \int_\Omega \left(\nabla p \, \nabla \zeta + g \frac{\partial \zeta}{\partial y} \right) dx\, dy \leqslant 0, \qquad \forall\, \zeta \in Z$$

where Z is defined as follows:

$$Z = \{\zeta \mid \zeta \in C^1(\overline{\Omega}), \zeta \geqslant 0 \text{ on } S_2, \zeta = 0 \text{ on } S_3\}$$

In order to prove that there exists $p \in W^{1,s}_{\text{loc}}(\Omega)$ $\forall$ finite s, and that there exists $g \in L^\infty(\Omega)$ such that (91), (92) and (93) are true, the authors in [18] introduce the following approximation procedure (of the penalty type). Let $H_\varepsilon(\lambda)$ be defined by

$$(94) \qquad H_\varepsilon(\lambda) = \begin{cases} 1 & \text{if } \lambda \geqslant \varepsilon \\ \lambda\, \varepsilon & \text{if } 0 \leqslant \lambda \leqslant \varepsilon \\ 0 & \text{if } \lambda \leqslant 0 \end{cases}$$

and let us consider the problem of finding $p_\varepsilon \in H^1(\Omega)$ such that

$$(95) \qquad p_\varepsilon = \text{given on } S_2 \cup S_3 \qquad \text{(same values as in (92))}$$

$$(96) \qquad \int_\Omega \left(\nabla p_\varepsilon \, \nabla \zeta + H_\varepsilon(p_\varepsilon) \frac{\partial \zeta}{\partial y} \right) dx\, dy = 0, \qquad \forall\, \zeta \in H^1(\Omega) \text{ such that } \zeta = 0 \text{ on } S_2 \cup S_3$$

The authors prove (i) that (95) and (96) have a unique solution; and (ii) that p_ε converges, as $\varepsilon \to 0$, to a solution of the problem. (The uniqueness

(27) C. Baiocchi. *C. R. Acad. Sci., Paris*, **278** (1974), 1201-4. See also the book of Baiocchi and Capelo. Disequazioni Variazionali e Quasi variazionali. Applicazioni a problemi di frontiera libera. Published by Unione Matematica Italiana, Univ. of Bologna (1978).

(28) H. W. Alt. *Arch. Ration. Mech. Anal.*, **64** (1977), 111-26.

of p is an open question, except in particular cases, solved by Caffarelli and Rivière.(29)

The existence in (95) and (96) follows easily from Schauder's fixed-point theorem.

For the uniqueness, if p_ε and $\hat{p}_\varepsilon$ are two solutions, one introduces

$$q = p_\varepsilon - \hat{p}_\varepsilon$$

and it is enough to prove that $q \leqslant 0$ (since then, by exchanging the roles of p_ε and $\hat{p}_\varepsilon$, $q = 0$).

One verifies that

$$\text{(97)} \qquad \left| \int_\Omega (\text{grad } q)\ (\text{grad } \zeta)\ dx\, dy \right| \leqslant L \int_\Omega |q|\ |\zeta_y|\ dx\, dy$$

where L depends on ε (but ε is fixed for the time being).

Then one chooses, for any $\delta > 0$,

$$\zeta = \frac{1}{q}\ (q - \delta)^+$$

and after some computations, one shows that this implies (97).

12. CONCLUSION

In the above report, we have not spoken of a number of related works. Let us briefly cite some further related work not mentioned in the above report: on the regularity of solution of VI [20, 21]; on the obstacle problem with the obstacles irregular [22] (work with A. Vignoli); or when the boundary conditions are of mixed type [23] (with V. Murthy). Stampacchia was also interested in the numerical aspects of the solution of VI as shown in [24, 25].

For many years, he wanted to write a book, one which would be an introduction to VI and would also carry the reader close to the frontiers of research. Work on this book was undertaken in collaboration with D. Kinderlehrer in July 1976; this book was nearly completed at the time of his death and will be published shortly [26].

Guido Stampacchia has left to us a beautiful example of a mathematician, working with very good taste on strongly motivated problems; he introduced elegant abstract methods but only when necessary and without artificial generality; he introduced, and masterfully used, some techniques which already belong to the classical tools of analysis.

(29) Work of Caffarelli and Rivière. To appear.

REFERENCES

This is *not* a complete bibliography of the works of Guido Stampacchia, for which we refer the reader to the biography written by E. Magenes (*Boll. Unione Mat. Ital.*), but rather a list of his major contributions in the domain of variational inequalities.

[1] 'Formes bilinéaires coercitives sur les ensemble convexes'. *C. R. Acad. Sci., Paris,* **258** (1964), 4413-6.

[2] 'Equations elliptiques du second ordre à coefficients discontinus'. *Sem. J. Leray, College de France* (1963/64).

[3] 'Le problème de Dirichlet pour les équations elliptiques du second ordre à coefficients discontinus'. *Ann. Inst. Fourier,* **15** (1965), 189-258.

[4] with Ph. Hartman. 'On some non linear elliptic differential - functional equations'. *Acta Math.*, **115** (1966), 271-310.

[5] with J. L. Lions. 'Inéquations variationnelles non coercives'. *C. R. Acad. Sci., Paris,* **261** (1965), 25-27.

[6] with J. L. Lions. 'Variational Inequalities'. *Commun. Pure & Appl. Math.*,**20** (1967), 493-519.

[7] with H. Lewy. 'On existence and smoothness of solutions of some non-coercive variational inequalities'. *Arch. Rat. Mech. Anal.*, **41** (1971), 241-53.

[8] with H. Brezis. 'Sur la regularité de la solution d'inéquations elliptiques'. *Bull. Soc. Math. Fr.*, **96** (1968), 153-80.

[9] 'Variational inequalities'. In '*Theory and Applications of Monotone operators*', *Proc. NATO Adv. Study Inst.* Venice, June 1968.

[10] with H. Lewy. 'On the regularity of the solution of a variational inequality'. *Commun. Pure & Appl. Math.*, **22** (1969), 153-88.

[11] with H. Brezis and L. Nirenberg. 'A remark on Ky Fan's minimax principle'. *Boll. Unione Mat. Ital.*, **6** (1972), 293-300.

[12] with H. Brezis. 'Une nouvelle méthode pour l'étude d'écoulements stationnaires'. *C. R. Acad. Sci., Paris,* **276** (1973), 129-32.

[13] with H. Brezis. 'The hodograph method in fluid dynamic in the light of variational inequalities'. *Arch. Ration Mech. Anal.*, **61** (1976), 1-18.

[14] with H. Lewy. 'On the smoothness of superharmonics which solve a minimum problem'. *J. Anal. Math.*, **23** (1970), 227-36.

[15] 'On the filtration of a fluid through a porous medium with variable cross section'. *Russ. Math. Surv.*, **29** (1974), 89-102.

[16] with D. Kinderlehrer. 'A free boundary problem in the plane'. *Ann. Inst. Fourier*, XXV (1975), 323-44.

[17] with H. Brezis. 'Remark on some fourth order variational inequalities'. *Ann. Scu. Norm. Sup., Pisa,* **4** (1977), 363-71.

[18] with H. Brezis and D. Kinderlehrer. 'Sur une nouvelle formulation du problème de l'écoulement à travers une digue'. *C. R. Acad. Sci., Paris,* (1978).

[19] with H. Brezis. 'Problèmes elliptiques avec frontière libre'. *Sem. Goulaouic Schwartz*, December 1972.

[20] with H. Brezis and D. Kinderlehrer. 'On the regularity of solutions of V.I.' *Proc. Int. Conf. on Functional Analysis and Related Topics*, Tokyo, 1969, 285-9.

[21] 'Regularity of solutions of some V.I.'. In *Non linear Functional Analysis*; *Proc. Symp. Pure Math.*, **18**, part I (1970), 271-81.

[22] 'A remark on V.I. for a second order non linear differential operator with non Lipschitz obstacles'. *Boll. Unione Mat. Ital.*, **5** (1972), 123-31.

[23] 'A V.I. with mixed boundary conditions'. *Isr. J. Math.*, **13** (1972), 188-334.

[24] 'On a problem of numerical analysis connected with the theory of V.I.'. *Symp. Math.*, **10** (1972), 281-93.

[25] '*Programmazione convessa e disequazioni variazionali*.' Ist. di Calcolo delle Probabilità, Univ. Roma, (1973).

[26] with D. Kinderlehrer. '*An Introduction to Variational Inequalities and Their Applications*.' Academic Press, New York.

Chapter 2

Variational Inequalities and Free-Boundary Problems

C. Baiocchi

1. THE SIMPLEST ELLIPTIC FREE-BOUNDARY PROBLEM. INTERCONNECTIONS WITH VARIATIONAL INEQUALITIES

Let D be an open subset of $\mathbb{R}^n$; Δ will denote the Laplace operator, say

$$\Delta v = \sum_{i=1}^{n} \frac{\partial^2 v}{\partial x_i^2}$$

A typical free-boundary problem associated with the operator Δ (for which a physical interpretation will be given later) is the following one.

Problem 1.
Given f, ψ real smooth functions on $\overline{D}$, we look for a couple $\{\Omega, u\}$ with [1]:

(1) Ω is an open subset of D; $u : \overline{\Omega} \to \mathbb{R}$

(2) $-\Delta u = f$ in Ω

(3) $u = \psi$ on $\partial\Omega \cap D$

(4) grad u = grad ψ on $\partial\Omega \cap D$

(5) $u = 0$ on $\partial\Omega \cap \partial D$

The set $\partial\Omega \cap \partial D$ is the 'fixed boundary'; the condition (5) on it is 'of usual type' (it could be replaced by a non-homogeneous condition, as well as by a Neumann condition, or some other ...). In contrast, on the set $\partial\Omega \cap D$, which is called a 'free boundary' (in short, FB) we want to impose 'too many conditions' (see conditions (3), (4)).

More generally, we will be concerned with FB problems associated with an operator L of the form:

(1) Ω and u must be 'smooth': we do not want to be precise here about this smoothness, nor about the smoothness of the data.

(6) $$Lv = -(a_{ij}v_{x_i})_{x_j} + b_i v_{x_i} + cv$$

and, instead of (4), we will impose

(7) $$a_{ij}(u-\psi)_{x_i}(u-\psi)_{x_j} + g_j(u-\psi)_{x_j} = 0 \qquad \text{on } \partial\Omega \cap D$$

for a prescribed $\mathbf{g} \equiv (g_j(x))$. (We remark that in Problem 1, we have $a_{ij} = \delta_{ij}$, so that (7) with $\mathbf{g} \equiv \mathbf{0}$ is equivalent to (4). We will often disgregard the conditions (like (5)) on the fixed boundary, and we will consider problems of the following form:

Problem 2
Given f, ψ, a_{ij}, b_i, c, g_i, and L being defined through (6), find a couple $\{\Omega, u\}$ satisfying (1), (3), (7) and

(8) $$Lu = f \qquad \text{in } \Omega$$

Remark 1 Let us point out an 'equivalent' formulation for relation (7); denoting by υ_Ω the unit outward normal vector to $\partial\Omega$, from (3), we have that υ_Ω and grad $(u - \psi)$ are parallel, so that we can rewrite (7) in the form

(9) $$[\mathrm{a} \cdot \operatorname{grad}(u-\psi) - \mathrm{g}] \cdot \upsilon_\Omega = 0 \qquad \text{on } \partial\Omega \cap D$$

where $\mathrm{a} \cdot \operatorname{grad}(u-\psi)$ is the vector $[a_{ij}(u-\psi)_{x_i}]_{j=1,\dots,n}$. The form (7) is, however, 'more general' because it does not require the existence of υ_Ω.

In order to show the interconnection between FB problems and variational inequalities (in short, VI), let us start from the simplest physical problem which can be formulated both as an FB problem and as a VI.

Example 1
Elastic membrane stretched upon an obstacle. We fix an elastic membrane[(2)] at height $z = 0$ on the boundary ∂D of a plane region D; we submit the membrane to a system of vertical forces $f(x, y)$; and we force the membrane to stay above an obstacle $z = \psi(x, y)$; we look for the equilibrium shape $z = u(x, y)$. Ω will be the 'non-contact region', say:

(10) $$\Omega = \{(x, y) \in D : u(x, y) > \psi(x, y)\}$$

From the mathematical point of view, this problem can be formulated in

(2) Which is assumed isotropic, homogeneous, and with unitary elastic coefficient.

many ways; we will set up two of them.

Formulation 1 As Ω is defined through (10), in Ω the obstacle cannot give reactions; writing a balance between external forces f and interior elastic forces, we get (2); (3) holds by definition of Ω; (4) means that the membrane leaves the obstacle in a smooth way[3]; finally (5) means that the membrane is fixed on ∂D at the height $z = 0$, so that this physical problem leads to Problem 1.

Remark 2 Without entering into details about smoothness, let us point out that, across $\partial\Omega$, u cannot be very smooth; e.g. for

$$D = \{(x, y) \in \mathbb{R}^2 : x^2 + y^2 < 4\} \qquad f \equiv 0 \qquad \psi = 1 - x^2 - y^2$$

we have

$$\Delta u = \Delta\psi \equiv -4 \quad \text{in } D\backslash\overline{\Omega}$$

and

$$\Delta u = 0 \quad \text{in } \Omega$$

(see (2)); in particular, also for analytic data D, f, ψ, we will have $u \notin C^2(D)$.

Formulation 2 Let us denote by K the set of all 'admissible configurations', say[4]:

$$K = \{v \in H_0^1(D) : v \geqslant \psi\} \tag{11}$$

and let us define:

$$a(v, w) = \iint_D \text{grad } v \cdot \text{grad } w \, dxdy, \qquad \ell(v) = \iint_D fv \, dxdy \tag{12}$$

Then, by the virtual work principle, we must have:

$$\begin{aligned} &u \in K \\ &a(u, u - v) \leqslant \ell(u - v) \qquad \forall\, v \in K \end{aligned} \tag{13}$$

Remark 3 (13) is a typical VI: K being a closed convex subset of the Hilbert space[5] $H_0^1(D)$ and ℓ being a linear continuous functional on such a space[6], an existence and uniqueness result for u, the solution of (13), follows from the fact that $\{u, v\} \to a(u, v)$ is a scalar product on $H_0^1(D)$ equiv-

(3) It is a natural requirement for ψ smooth; we remark, however, that for conical or pyramidal ψ, relation (4) fails and a different formulation is needed.

(4) $H^1(D)$ is the usual Sobolev space of functions $v \in L^2(D)$ whose first weak derivatives v_x, v_y are in $L^2(D)$ (and $v \in H^1(D)$ means that v has finite elastic energy); $v \in H_0^1(D)$ means $v \in H^1(D)$ and $v|_{\partial D} = 0$.

(5) Moreover, K is non-empty under large assumptions on ψ; e.g. $\psi \in C^0(\overline{D})$ and $\psi|_{\partial D} < 0$, or $\psi \in H^1(D)$ and $\psi|_{\partial D} \leqslant 0$.

(6) For example for $f \in L^p(D)$ for some $p > 1$.

alent to the original one[7].

In order to show that, for u a solution of (13) and Ω defined through (10), the couple $\{\Omega, u\}$ solves Problem 1 (as checked in the first formulation), we need some further smoothness properties of[8] u; to this aim let us assume, for example, that f and second derivatives of ψ are locally in L^p for some $p>2$; then a fundamental theorem of Lewy and Stampacchia[9] implies:

(14) the solution u of (13) verifies $u \in C^1(D)$

Then, denoting by Ω the set defined in (10) (which is open; also ψ is $C^1(D)$ in our hypotheses) and by u also the restriction of u to $\bar{\Omega}$, (1) holds; (3) follows from the definition of Ω; (5) follows from $u \in H_0^1(D)$; (4) holds because $u - \psi$ is a C^1 function which is non-negative ($u \in K$!) so that grad $(u - \psi) = 0$ where $u - \psi = 0$; finally (2), e.g. in the distributions sense in Ω, can be proved on choosing in (13) $v = u + \lambda \phi$ with $\phi \in \mathfrak{D}(D)$, supp $(\phi) \subset \Omega$, and $|\lambda|$ small enough in order to have $u + \lambda \phi \in K$.

The fundamental role played by the regularity result (14) in the interconnections between FB problem and VI has already been pointed out in [12] (where the FB problem solved from (13) was interpreted in a different way). We remark, however, that (13) also implies that FB problems like Problem 2 with $g \not\equiv 0$ *cannot* be solved directly through VI. The next section is devoted to a list of examples of concrete FB problems of the type of Problem 2; we will see in section 3 that it is still possible to associate a VI to such problems.

2. SOME OTHER EXAMPLES OF FB PROBLEMS

Example 2

Steady dam problem. Let D be a porous dam in which some water infiltrates with a steady flow. Denoting by x_1 the vertical axis, by x_2 the horizontal one (for two-dimensional flow; otherwise (x_2, x_3) will be a system of coordinates in the horizontal plane), by Ω the flow region, and by u the piezometric head (see, e.g., [6] for details) we get a problem like[10] (1), (2), (3) with $f \equiv 0$,

(7) More generally, it would be sufficient that $a(u, v)$ is continuous and coercive; it is the well known Stampacchia's theorem [14].

(8) Beside the control of (4), which requires the traces of grad u on $\partial\Omega \cap D$, the same formula (10) defining Ω gives some trouble if u is just in $H^1(D)$!

(9) See [12]. We confine ourselves to stating a 'local' result; under suitable assumptions on ∂D, a 'global' result also holds; see [12] again.

(10) We assume, here, the dam to be isotropic and homogeneous; in general, we will have, instead of $-\Delta$, an operator L like in (6) with $b_i \equiv 0$, $c \equiv 0$, (a_{ij}) symmetric positive definite.

$\psi \equiv x_1$. The second condition on the FB is usually written in the form

$$\frac{\partial u}{\partial \upsilon_\Omega} = 0$$

υ_Ω being the unit normal vector to $\partial\Omega$; but we can also write it in the form (7) (or (9)) on choosing[(11)]:

(15) $$g_1 \equiv -1 \qquad g_i \equiv 0 \qquad \text{for } i \neq 1$$

so that we get Problem 2.

Remark 4 As already done for Problem 2, we do not make explicit the conditions on the fixed boundary, which are unessential for the FB nature of the problem; often, however, when we try to give a variational formulation for the FB problem (see section 3 below) these conditions give rise to serious difficulties.

Example 3

Non-steady dam problem. With a little change in notation, i.e. let D_0 be the dam, $D_0 \subset \mathbb{R}^m$ ($m = 2$ or 3); we set $D = D_0 \times]0,T[\subset \mathbb{R}^n$, $n = m + 1$, $x_n = t$; the problem still has the form (1), (2), (3) with $f \equiv 0$, $\psi \equiv x_1$ (the vertical component) but Δ is just in the space variables; the only derivative in t appears on the second condition on the FB which is usually written in the form $u_t =$ $=$ grad $u \cdot$ grad $(u - x_1)$ (see again [6]; also grad refers only to the space-variables); this relation can be written in the form (7) on choosing:

(16) $$\begin{aligned} &g_1 \equiv -1 \\ &g_i \equiv 0 \qquad \text{for } 1<i<n \\ &g_n \equiv -1 \end{aligned}$$

Example 4

Oxygen diffusion-absorption in a tissue. We have here $D = \{(x, t) \in \mathbb{R}^2 : x > 0, t > 0\}$; for each t the set $\{x : (x, t) \in \Omega\}$ is the part of the tissue which, at the time t, has oxygen to live; see, e.g., [8] for details. Then we get Problem 2 with

$$L = \frac{\partial}{\partial t} - \frac{\partial^2}{\partial x^2}, \qquad f \equiv -1, \qquad \psi \equiv 0$$

The second condition on $\partial\Omega \cap D$ is usually written in the form $u_x = 0$, say (7) with:

(17) $$g \equiv 0$$

(11) Or $g_i = -a_{i1}$ $\forall$ in the general approach of footnote (10).

Example 5

One-phase Stefan problem. Let D_0 be a region of water-ice melting; $D = D_0 \times]\, 0,T\, [$ (notation like in Example 3 also for Δ and grad); ice is supposed at 0 temperature (degrees Celsius); Ω is the set $\{(x, t) \in D :$ at time t, at point x there is water$\}$; u is the temperature. Then (see, e.g., [13] for details) we have Problem 2 with

$$f \equiv \psi \equiv 0, \qquad L = \frac{\partial}{\partial t} - \Delta$$

The second condition on the FB is usually written in the form[12] $\text{grad}\, u \cdot \text{grad}\, \phi = -k$ but (as u and $t - \phi(x)$ are positive in Ω and 0 on $\partial\Omega$ we must have that the vector $(\text{grad}\, u, u_t)$ is a multiple of $(-\text{grad}\, \phi, 1)$; say $\text{grad}\, u = -u_t\, \text{grad}\, \phi$; so that this conditions becomes

$$|\text{grad}\, u\,|^2 = ku_t$$

which has the form (7) with:

$$g_i \equiv 0 \quad \text{for } i < n \qquad (18)$$
$$g_n = -k$$

Example 6

Flow around a profile. The problem, in the physical plane, is a fixed-boundary problem for a bad non-linear differential operator; under suitable assumptions (see, e.g., [7]) in the 'velocity plane' the problem becomes an FB problem which, in logarithmic polar coordinates (σ, θ), takes the form of Problem 2 with

$$f \equiv \psi \equiv 0, \qquad L = -\frac{\partial^2}{\partial \sigma^2} - k(\sigma) \frac{\partial^2}{\partial \theta^2}$$

($k(\sigma)$ given function); the second condition on the FB takes the form

$$u_\theta^2 + u_\sigma^2 + ku_\sigma = 0$$

say (7) with (h given function):

$$g_1 = h, \qquad g_2 = 0 \qquad (19)$$

Until now, have presented some examples (but the list could continue . . . .) of FB problems of 'type 2', say problems in which the second condition on

(12) k being a physical constant, and $t = \phi(x)$ being the equation of $\partial\,\Omega \cap D$; say $\phi(x)$ is the time in which the ice becomes water at point x.

the FB can be written in the form (7). Now let us show an example where this fails.

Example 7

Jet cavities. A large class of FB problems (see, e.g., [9]) leads to (1), (2), (3) with $f \equiv 0$, $\psi \equiv 1$; the second conditions on the FB being:

$$|\text{grad}\, u| = q \qquad \text{on } \partial\Omega \cap D \tag{20}$$

with q a given function with $q \neq 0$; obviously (20) cannot be written in the form (7).

3. VARIATIONAL APPROACH FOR PROBLEM 2

Some years ago I proved an existence uniqueness result for the problem described in Example 2 (under suitable assumptions on the geometry of D; see [1], also, for conditions on the fixed boundary) by setting:

$$p = \begin{cases} u - \psi & \text{in } \overline{\Omega} \\ 0 & \text{in } \overline{D}\backslash\overline{\Omega} \end{cases} \tag{21}$$

(it is a 'continuous extension' because of (3)) and choosing as new unknown function a *primitive* of $\tilde{p}$ along the direction x_1; this new unknown function was the solution of a suitable VI (of type (13)), so that a variational approach was possible also for an FB problem whose solution cannot be the solution of a VI (see the last part of section 1); u was, in fact, a *derivative* of the solution of a VI!

Later on, a lot of FB problems have been solved on the basis of such an idea; for steady dam problem, in more general situations with respect to that studied in [1], the 'good direction of integration' is always the direction x_1, but the conditions on the 'fixed boundary' give rise to more difficult variational (or quasi-variational) problems; we refer the reader to [4], also, for more detailed references(13).

Concerning Example 3, the 'good direction' of integration was the direction $x_1 - t$ (see [15]); in Example 4, u solves directly a VI (of parabolic type; see [5]) so that, as in Example 1, no integration is needed; in Example 5 we need an integration in t (see [10]); in Example 6 we need an integration in σ (see [7]). The examples of FB problems differ from each other, so that the 'strategy' in order to get a variational approach changes. We remark, how-

(13) We confine ourselves to give a few references for each example; for other references see the quoted papers. Other problems, references and results can be found in [3, 4].

ever, that, writing the second condition on the FB in the form (7) (or (9)), the direction of integration was always the direction of **g** (and no integration is needed if $\mathbf{g} \equiv \mathbf{0}$); see (15), (19). In fact, this is a general result, as pointed out in [2] (see also appendix 4 of vol. II of [4]; a general approach for parabolic FB problems is given in [11]): Problem 2, in terms of p defined through (20), becomes[(14)]:

$$(22)\qquad \begin{aligned} &Lp = \chi_\Omega (f - L\psi) - \mathbf{g} \cdot \operatorname{grad} \chi_\Omega \qquad \text{in } \mathcal{D}'(D) \\ &\operatorname{supp}(p) \subset \bar{\Omega} \end{aligned}$$

so that it is an integration along the direction of **g** which can destroy the bad term grad χ_Ω in (22).

(14) χ_Ω denotes the characteristic function of Ω in D, say $\chi_\Omega = 1$ in Ω and $\chi_\Omega = 0$ in $D \backslash \Omega$. Relation (21) is quite simple, starting from (21), (8), and (9); see [2, 4] for more details.

REFERENCES

[1] C. Baiocchi. 'Su un problema di frontiera libera connesso a questioni di idraulica'. *Ann. Mat. Pura Appl.*, **92** (1972), 107-27.
[2] C. Baiocchi. 'Problèmes à frontière libre. et inéquations variationnelles'. *C. R. Acad. Sci., Paris*, **283** (1976), 29-32.
[3] C. Baiocchi. 'Free boundary problems and variational inequalities'. Invited talk to the *SIAM 1978 Spring Meeting* in Madison. Submitted to *SIAM Rev.*
[4] C. Baiocchi, and A. Capelo. '*Disequazioni variazionali e quasivariazionali. Applicazioni a Problemi di Frontiera Libera*', 2 vol., Pitagora, Bologna (1978).
[5] C. Baiocchi, and G. A. Pozzi. 'An evolution variational inequality related to a diffusion-absorption problem'. *Appl. Math. Opt.*, **2** (1976), 304-14.
[6] J. Bear. '*Dynamics of Fluids in Porous Media*.' American Elsevier, New York (1972).
[7] H. Brezis, and G. Stampacchia. 'The Hodograph method in fluid dynamics in the light of variational inequalities'. *Arch. Ration. Mech. Anal.*, **61** (1976), 1-18.
[8] J. Crank.'*The Mathematics of Diffusion*.' Clarendon Press, Oxford (1975).
[9] I. I. Daniljuk. 'Sur une classe de fonctionnelles intégrales à domaine variable d'intégration'. *Actes Congrès Intern. Math., Nice, 1970,* Tome II, Gauthier-Villars, Paris (1971), 703-15.
[10] G. Duvaut. 'Résolution d'un problème de Stefan (Fusion d'un bloc de glace à zero degré)'. *C. R. Acad. Sci., Paris*, **276** (1973), 1461-3.
[11] F. Gastaldi. 'About the possibility of setting Stefan-like problems in variational form'. To apperar on *Boll. Unione Mat. Ital.*
[12] H. Lewy, and G. Stampacchia. 'On the regularity of the solution of a variational inequality'. *Commun. Pure & Appl. Math.*, **22** (1969) 153-88.
[13] L. I. Rubinstein. '*The Stefan Problem*.' *Am. Math. Soc.* Providence, RI (1971).
[14] G. Stampacchia. 'Formes bilinéaires coercitives sur les ensembles convexes'. *C. R. Acad. Sci., Paris,* **258** (1964), 4413-6.
[15] A. Torelli. 'On a free boundary value problem connected with a non steady filtration phenomenon'. *Ann. Scu. Norm. Sup., Pisa,* **4** (1977), 33-59.

Chapter 3

Existence of a Ground State in non-Linear Equations of the Klein-Gordon Type

H. Berestycki and P.L. Lions

1. INTRODUCTION

1.1. Physical motivation

Consider the non-linear Klein-Gordon equation:

$$\Phi_{tt} - \Delta\Phi + a^2\Phi - f(\Phi) = 0 \tag{1}$$

where $\Phi = \Phi(t,x)$, $x \in \mathbb{R}^N$. Suppose that $f(\rho e^{i\theta}) = e^{i\theta} f(\rho)$, $\forall\, \rho \geqslant 0$, $\theta \in [0,2\pi)$, and $f(\rho)$ is real. Then, looking for solitary waves in (1) of the 'standing wave' form, i.e. $\Phi(t,x) = e^{i\omega t}u(x)$, or of the 'travelling wave' form, i.e. $\Phi(t,x) = u(x - ct)$, $c \in \mathbb{R}^N$ being a constant vector, $|c| < 1$, one is led to the semi-linear equation

$$-\Delta u + mu = f(u) \quad \text{in } \mathbb{R}^N \tag{2}$$

where m is a constant. (Actually, for a travelling wave, (2) is obtained after a rotation and stretching of axes.) It is always assumed that $f(0) = 0$, and one looks for non-trivial solutions of (2). Also, for physical reasons, one requires u to be 'small at infinity'. Thus, the problem is the following: given a function g such that $g(0) = 0$, find a function u such that

$$\begin{aligned} -\Delta u &= g(u) \quad \text{in } \mathbb{R}^N \\ u &\in H^1(\mathbb{R}^N) \quad u \not\equiv 0 \end{aligned} \tag{3}$$

The same problem also arises when looking for solitary waves of the standing wave form in non-linear Schrödinger equations. In another context, a solution of (3) can be interpreted as a stationary state for a non-linear heat equation

$$\frac{\partial u}{\partial t} - \Delta u = g(u) \quad \text{in } \mathbb{R}^N \tag{4}$$

Such problems arise, in particular, in certain models of biology (for instance, in the study of the dying out species) [2, 7, 10, 16].

In the following, we study the existence of a positive solution of (3).

Such a solution represents a so-called ground state for (1). In this paper, we restrict ourselves to the model case where [(1)]

$$g(u)=\lambda u^p-\mu u^q-mu$$

(which models different types of behaviour for g near $u=0$ und $u=\infty$). A more general result is stated in [3] (for the proof and more precise details, the reader in referred to [5]). The next theorem gives conditions which are sufficient and 'almost' necessary for the existence of a positive solution in the model case, in dimension $N\geqslant 3$ [(2)].

1.2. The main result

Theorem 1.1.
Suppose $N\geqslant 3$. Let $g(t)=\lambda t^p-\mu t^q-mt$, where $\lambda,m>0$, $\mu\geqslant 0$, and $1<p,q$. Suppose that

$$p<\max\left(q,\frac{N+2}{N-2}\right) \tag{5}$$

$$\text{if } p\geqslant\frac{N+2}{N-2}, \text{ then } \mu>0 \tag{6}$$

$$\exists\ \zeta>0 \text{ such that } G(\zeta)>0, \text{ where } G(s)=\int_0^s g(t)\mathrm{d}t. \tag{7}$$

Then, problem (3) has a solution $u\in C^\infty(\mathbb{R}^N)$ which is strictly positive, spherically symmetric, decreases with $r=|x|$, and is such that

$$\forall\,|\alpha|\in\mathbb{N}\qquad |D^\alpha u(x)|\leqslant C_\alpha\,\mathrm{e}^{-\delta_\alpha r}$$

where $C_\alpha,\delta_\alpha>0$.

Remark 1.1. This theorem generalizes results of several authors.
(*a*) In the case $\mu=0$, we have the result of Pohozaev [13], that is the existence of a non-trivial solution when $p<\ell=(N+2)/(N-2)$. Moreover, it is known [13, 15] that such a solution does not exist when $p\geqslant\ell$.
(*b*) In the case $\mu>0$, this theorem extends and complements the results of Strauss [15], and Coleman, Glazer and Martin [6]. The main new case here is $1<p<q$. For this case, μ and m being fixed, Strauss [15] showed the existence of at least one λ for which (3) has a positive solution. From Theo-

(1) This is for $u\geqslant 0$: in general, for $u\in\mathbb{R}$, we consider $g(u)=\lambda|u|^{q-1}u-mu$.

(2) Actually, in [3,5] we prove, under general hypotheses on g, the existence of infinitely many distinct solutions of (3). The solution of (3) we will construct here can be seen to be that of minimal energy (cf. [6]) and is called 'ground state'. The energy of the other solutions ('bound states') can be arbitrarily large.

rem 1.1 we obtain the existence of λ^*, determined by condition (7), such that for any $\lambda > \lambda^*$, (3) has a positive solution. We remark that λ^* is explicitly given from (7) by

$$\lambda^* = \left(\frac{m}{2}\right)^a \mu^b (q-1)(p+1)(q-p)^{-a}(p-1)^{-b}(q+1)^{-b} \tag{8}$$

where $a = (q-p)/(q-1)$ and $b = (p-1)/(q-1)$. Such a condition had already been observed by Anderson [1] in the special case $p = 3$, $q = 5$ and $\lambda = m = 1$. In agreement with his study, formula (8) shows that a necessary condition for positive solutions to exist is

$$1 > \lambda^* = \mu^{1/2} \times 4 \times 3^{-1/2} \quad \text{i.e. } \mu < 3/16$$

We also show later that for $\lambda < \lambda^*$, (3) has no positive solution. In the next section we prove Theorem 1.1. In section 3, we show the 'almost necessary' character of conditions (5), (6), and (7), and give further details. In section 4, we give some indications on a 'local method' for solving (3) which is valid in any dimension.

2. PROOF OF THEOREM 1.1

2.1. The constrained minimization method

This method was introduced in [6]. Let

$$E = H^1(\mathbb{R}^N) \cap L^{p+1}(\mathbb{R}^N) \cap L^{q+1}(\mathbb{R}^N)$$

(in the special case $\mu=0$, where the term in u^q disappears, define E simply by $E=H^1(\mathbb{R}^N) \cap L^{p+1}(\mathbb{R}^N)$.) On E, define

$$T(u) = \frac{1}{2} \int_{\mathbb{R}^N} |\nabla u(x)|^2 \, dx \quad \text{and} \quad V(u) = \int_{\mathbb{R}^N} G(u(x)) \, dx$$

The energy attached to a solution u of (3) is given by $S(u)=T(u) - V(u)$. This energy is not bounded (neither from above, nor from below) on the space E. Therefore, rather than searching directly for non-trivial critical points of S, we consider the following constrained minimization problem:

$$\text{minimize } \{T(w) : w \in E,\ V(w) = 1\} \tag{9}$$

A solution to (9) leads to a solution of (3). Indeed if u solves (9), then

$u \not\equiv 0$, $u \in H^1(\mathbb{R}^N)$, and (cf section 2.3) there exists a Lagrange multiplier $\theta > 0$ such that

$$(10) \qquad -\Delta u = \theta g(u) \quad \text{in } \mathbb{R}^N$$

But then, operating a scale change $u_\sigma(x) = u(x/\sigma)$, with $\sigma = 1/\sqrt{\theta}$, one obtains a solution of (3). The result concerning (9) is the following.

Theorem 2. 1
Under the hypotheses of Theorem 1.1. the minimization problem (9) has a solution $u \in H^1(\mathbb{R}^N)$, which is positive, spherically symmetric and decreases with $r = |x|$. Furthermore, there exists a Lagrange multiplier $\theta > 0$ such that u satisfies (10).

Proof: The proof is divided into three parts: section 2.2 : existence of a solution to problem (9); section 2.3 : existence of a solution to problem (3); and section 2.4 : further properties of the solution.

2.2. Existence of a solution to the constrained minimization problem

Step 1 The set $\{w \in E : V(w) = +1\}$ is not empty.
This is where hypothesis (7) is being used. Let $\zeta > 0$ satisfy $G(\zeta) > 0$. Set

$$w_R = \begin{cases} \zeta & \text{on } B_R = \{x \in \mathbb{R}^N : |x| \leqslant R\} \\ \zeta\,\{R+1-r\} & \text{on } |x| = r \in [R, R+1] \\ 0 & \text{on } |x| \geqslant R+1 \end{cases}$$

It is easily seen that

$$V(w_R) \geqslant C_N R^N G(\zeta) - C'_N R^{N-1} \max_{0 \leqslant s \leqslant \zeta} |G(s)|$$

whence, for R large enough, $V(w_R) > 0$. Then, operating a scale change on $w_R : w_{R,\sigma}(x) = w_R(x/\sigma)$, we have $V(w_{R,\sigma}) = \sigma^N V(w_R)$. Thus, $V(w_{R,\sigma}) = 1$ for $\sigma = [V(w_R)]^{-1/N}$; it is also obvious that $w_{R,\sigma} \in E$.
Step 2 Construction of an adequate minimizing sequence.
There exists $(u_n) \subset E$, such that $V(u_n) = +1$ and

$$\lim_{n \to \infty} T(u_n) = I \equiv \inf_{\{w \in E:\, V(w) = +1\}} T(w) \geqslant 0$$

Let u_n^* be the Schwarz spherical rearrangement of u_n (cf references [8, 11] for the definition and main properties of the Schwarz symmetrization). Then, we have

$$I \leqslant T(u_n^*) \leqslant T(u_n)$$

and $$V(u_n^*) = V(u_n) = +1, \quad u_n^* \in E$$

Thus, we may suppose henceforth that u_n is spherically symmetric, positive, and decreasing with $r = |x|$.

Step 3 Estimates on (u_n).

We first show that u_n is bounded in E. By construction, $|\nabla u_n|$ is bounded in $L^2(\mathbb{R}^N)$. Hence, by Sobolev's inequality, we know that

$$\|u_n\|_{L^{\ell+1}(\mathbb{R}^N)} \leqslant C \quad \text{where} \quad \ell + 1 = 2^* = \frac{2N}{N-2}$$

(Here and thereafter C designates various positive constants.) We also know that

$$(11) \qquad V(u_n) = \frac{\lambda}{p+1}\int_{\mathbb{R}^N} u_n^{p+1}\,dx - \frac{\mu}{q+1}\int_{\mathbb{R}^N} u_n^{q+1}\,dx - \frac{m}{2}\int_{\mathbb{R}^N} u_n^2\,dx = +1$$

(i) If $\ell < p < q$, we have by Hölder's inequality

$$\int_{\mathbb{R}^N} u_n^{p+1}\,dx = \|u_n\|^{p+1}_{L^{p+1}(\mathbb{R}^N)} \leqslant \|u_n\|^{\alpha(p+1)}_{L^{q+1}(\mathbb{R}^N)}\|u_n\|^{(1-\alpha)(p+1)}_{L^{\ell+1}(\mathbb{R}^N)}$$

where

$$\alpha \in (0,1); \quad \frac{\alpha}{q+1} + \frac{1-\alpha}{\ell+1} = \frac{1}{p+1} \quad \text{and} \quad \alpha(p+1) < q+1$$

Thus

$$(12) \qquad \|u_n\|^{p+1}_{L^{p+1}(\mathbb{R}^N)} \leqslant C\|u_n\|^t_{L^{q+1}(\mathbb{R}^N)}$$

where $0 < t < p+1$. Since from (11) we have

$$\|u_n\|^{q+1}_{L^{q+1}(\mathbb{R}^N)} \leqslant C\|u_n\|^{p+1}_{L^{p+1}(\mathbb{R}^N)}$$

we derive, using (12), that

$$\|u_n\|_{L^{q+1}(\mathbb{R}^N)} \leqslant C$$

Again from (12), and then from (11) we also obtain

$$\|u_n\|_{L^{p+1}(\mathbb{R}^N)} \leqslant C \quad \text{and} \quad \|u_n\|_{L^2(\mathbb{R}^N)} \leqslant C$$

(ii) In the case $p = \ell - 1$, we have directly

$$\|u_n\|_{L^{p+1}(\mathbb{R}^N)} \leq C$$

Hence we have the same bounds on u_n as before using (11) (i.e.

$$\|u_n\|_{L^{q+1}}, \|u_n\|_{L^2} \leq C)$$

(iii) If $p < \ell$, we have by Hölder's inequality

$$\|u_n\|^{p+1}_{L^{p+1}(\mathbb{R}^N)} \leq \|u_n\|^{\beta(p+1)}_{L^2(\mathbb{R}^N)} \|u_n\|^{(1-\beta)(p+1)}_{L^{\ell+1}(\mathbb{R}^N)}$$

where

$$\beta \in (0,1); \quad \frac{\beta}{2} + \frac{1-\beta}{\ell+1} = \frac{1}{p+1} \quad \text{and} \quad \beta(p+1) < 2$$

Thus,

$$\|u_n\|^{p+1}_{L^{p+1}(\mathbb{R}^N)} \leq \|u_n\|^{t}_{L^2(\mathbb{R}^N)}$$

with $t < 2$. Then, as before, using (11) we see that (u_n) is bounded in E.

Thus, in all cases, (u_n) is bounded in E. In particular $\|u_n\|_{L^2(\mathbb{R}^N)} \leq C$. This implies uniform behaviour of u_n at infinity as can be seen in the following lemma.

Lemma 2.1 There exists a constant C_N (depending on N only) such that for any function $u \in H^1(\mathbb{R}^N)$ which is spherically symmetric and decreasing with $r = |x|$, one has

(13) $$|u(r)| \leq C_N\, r^{-N/2} \|u\|_{L^2(\mathbb{R}^N)} \quad \forall r > 0$$

Proof: Notice that (13) makes sense since $u(r)$ coincides almost everywhere with a continuous function of $r>0$ by the radial lemma of Strauss [15]. Now (13) follows from

$$N^{-1} r^N u(r)^2 \leq \int_0^r u(\rho)^2 \rho^{N-1} \mathrm{d}\rho \leq C_N \|u\|^2_{L^2(\mathbb{R}^N)} \quad \forall r>0$$

Step 4 Convergence of a subsequence of (u_n).

Since (u_n) is bounded in E, there exists a subsequence of (u_n), denoted again by (u_n), which converges weakly in the spaces $H^1(\mathbb{R}^N)$, $L^{p+1}(\mathbb{R}^N)$, and $L^{q+1}(\mathbb{R}^N)$, and also almost everywhere in $\mathbb{R}^N$, to $u \in E$. We remark that u is spherically symmetric and decreases with $r = |x|$. We claim that u_n converges strongly to u in $L^{p+1}(\mathbb{R}^N)$. This can be seen to follow from a compactness lemma of Strauss [15]. We give a direct proof. Let $\epsilon > 0$ be given. Since $0 \leq u_n(r), u(r) \leq Cr^{-N/2}$, one has

$$\int_{\{r \geqslant R\}} |u_n - u|^{p+1}\, dx \leqslant \{CR^{-N/2}\}^{p-1} \int_{\{r \geqslant R\}} |u_n - u|^2\, dx \leqslant \epsilon$$

for $R = R_0$ large enough, and any $n \in N$. Since $u_n - u$ is bounded in $L^{q+1}(B_{R_0})$ and $u_n - u \to 0$ almost everywhere, $|u_n - u|^{p+1}$ is uniformly integrable on B_{R_0}, whence

$$\int_{B_{R_0}} |u_n - u|^{p+1}\, dx \to 0 \quad \text{as } n \to +\infty$$

Combining these two facts, we see that $u_n \to u$ in $L^{p+1}(\mathbb{R}^N))$ as $n \to +\infty$.

Then, by Fatou's theorem, we find, that $V(u) \geqslant 1$. Furthermore $T(u) \leqslant I$. Suppose $V(u) > 1$; then $\exists\ \sigma \in (0,1)$ such that $V(u_\sigma) = \sigma^N V(u) = 1$ $(u_\sigma(x) = u(x/\sigma))$. Also, $T(u_\sigma) = \sigma^{N-2} T(u) \leqslant \sigma^{N-2} I$. But by definition $I \leqslant T(u_\sigma)$. This would imply $I = 0 = T(u)$, contradicting $V(u) > 0$. Hence, one has $V(u) = +1$ and $T(u) = I$, that is, u is a solution to the constrained minimization problem (9).

Remark 2.1 The preceding proof shows, in particular, that the subsequence (u_n) converges to u *strongly in E.*

Remark 2.2 In dimensions $N = 1$ and $N = 2$, the method used in step 3 for obtaining bounds on the sequence (u_n) fails. The reason is that in those dimensions a bound on $|\nabla u|$ in $L^2(\mathbb{R}^N)$ alone no longer provides a bound for u in an $L^{\ell+1}(\mathbb{R}^N)$ space. Actually, the constrained minimization problem (9) has *no* solutions when $N = 1$ or $N = 2$. Indeed, let us examine separately the case of each of the two dimensions.

Case (i) : $N = 2$. Under a scale change $u_\sigma(x) = u(x/\sigma)$, one has the following relations:

$$T(u_\sigma) = T(u) \quad \text{and} \quad V(u_\sigma) = \sigma^2 V(u)$$

Then, it is easily seen that

$$\inf_{\{V(u) = 1\}} T(u) = \inf_{\{V(u) > 0\}} T(u)$$

Now, if u_0 is a solution of (9), one has $V(u_0) = 1$ and

$$T(u_0) = \min_{\{V(u) > 0\}} T(u)$$

Hence, u_0 is a 'interior min' for $T(u)$, and thus $T'(u_0) = 0$, hence $u_0 = 0$ which contradicts $V(u_0) = 1$.

Case (ii) : $N = 1$. Under a scale change, one has the relations

$$T(u_\sigma) = \sigma^{-1} T(u) \quad \text{and} \quad V(u_\sigma) = \sigma V(u)$$

Choose a $w \in E$ such that $V(w) = 1$. Recalling that

$$\lim_{s \to 0^+} g(s)/s = -m < 0$$

we see that there exists $\theta_0 \in (0,1)$ such that $V(\theta_0 w) = 0$ and $V(\theta w) > 0$ for $\theta_0 < \theta \leqslant 1$. Clearly, $V(\theta w) \to 0^+$ as $\theta \to \theta_0^+$. Let $\sigma(\theta) = V(\theta w)^{-1}$; $V(\theta w_{\sigma(\theta)}) = \sigma(\theta) V(\theta w) = 1$. Now,

$$T(\theta w_{\sigma(\theta)}) = \sigma(\theta)^{-1} T(\theta w) = \theta^2 V(\theta w) T(w)$$

This obviously implies that

$$\inf_{\{V(u)=1\}} T(u) = 0$$

and hence equation (9) has no solutions.

The case of dimension $N=1$ is treated in the appendix of this chapter, where a very general existence result is proved by means of simple methods of ordinary differential equations. For the case of dimension $N=2$, one has, under suitable assumptions, the existence result obtained by the 'local method' of which some indications are presented in section 3.3, and which is developped in [4]. However, for the moment, the existence results in dimension $N=2$ have not attained the same generality as they have for the other dimensions.

2.3. Existence of a solution to problem (3)

Consider a solution u of the constrained minimization problem (9), which is positive, spherically symmetric, and decreases with $r = |x|$. We denote by E_r the space of functions in E which are radial (i.e. spherically symmetric). It is easily verified that T and V are C^1 functionals on E_r. Hence, there exists a Lagrange multiplier θ such that $T'(u) = \theta V'(u)$, that is

$$(14) \qquad -\Delta u = \theta g(u) \quad \text{in } \mathbb{R}^N$$

in the sense of distributions. Observe that since $V(u) = 1$, necessarily $\theta \neq 0$. Since $u \in E_r$, by a radial lemma of Strauss [15], u is continuous on $\mathbb{R}^N - \{0\}$. Furthermore, using the fact that u decreases with $r = |x|$, one sees that $u \in W^{2,p}_{\mathrm{Loc}}(\mathbb{R}^N - \{0\})$ for any p, and thus $u \in C^{1,\alpha}(\mathbb{R}^N - \{0\})$, $\alpha \in (0,1)$. By the uniqueness in the initial value problem for the ordinary differential equation derived from (14), one sees that u cannot vanish on $\mathbb{R}^N$. (Indeed if $u(r_0) = 0$, then $u(r) = 0$, $\forall\ r \geqslant r_0$ and $u'(r_0) = 0$.) Then by a classical reiteration of the regularity argument, one obtains that $u \in C^\infty(\mathbb{R}^N - \{0\})$.

Let us show that $\theta > 0$. Suppose $\theta < 0$; denoting $u'(r) = \partial u/\partial r$, one has

$$-u'' - \frac{N-1}{r} u' = \theta g(u), \qquad \forall r > 0 \tag{15}$$

Since, by Lemma 2.1, $0 < u(r) \leqslant C r^{-N/2}$, one has for $r \geqslant r_0 > 0$

$$\theta g(u) \geqslant |\theta| \frac{m}{2} u(r)$$

Hence,

$$-r^{-N+1} \frac{d}{dr}(r^{N-1} u') \geqslant |\theta| \frac{m}{2} u(r) \qquad \forall r \geqslant r_0 > 0$$

Thus,

$$r_0^{N-1} u'(r_0) - R^{N-1} u'(R) \geqslant |\theta| \frac{m}{2} \int_{r_0}^{R} r^{N-1} u(r) dr$$

$$\geqslant |\theta| \frac{m}{2} u(R) \frac{1}{N} (R^N - r_0^N)$$

Therefore, there exists $r_1 > r_0$ and $C > 0$ such that

$$u(r) \leqslant C r^{-1} [-u'(r)], \quad \forall r \geqslant r_1$$

Whence,

$$\frac{d}{dr}[u(r) e^{r^2/2C}] \leqslant 0$$

and

$$u(r) \leqslant e^{-r^2/2C} u(r_1) e^{r_1^2/2C}$$

But then we may write

$$-\Delta u + q(x) u = |\theta| m u, \quad u \not\equiv 0, \quad u \in H_r^1 \tag{16}$$

where $|q(x)| = o(|x|^{-1})$ as $|x| \to = \infty$, and this is impossible by a result of Kato [9].

Therefore, letting $u_{\sqrt{\theta}}(x) = u(x/\sqrt{\theta})$, we obtain a solution of problem (3), that will be denoted again by $u(= u_{\sqrt{\theta}})$ in the following:

$$\begin{aligned} -\Delta u &= g(u) && \text{on } \mathbb{R}^N \\ u &> 0 && \text{on } \mathbb{R}^N \\ u \in C^\infty(\mathbb{R}^N - |0|) & \text{ and} && u \text{ decreases with } r = |x|: \end{aligned}$$

2.4. Further properties of the solution

(1) First, we show that $u \in C^\infty(\mathbb{R}^N)$. Clearly, it suffices to prove that $u \in L^\infty(\mathbb{R}^N)$, because then $g(u) \in L^\infty(\mathbb{R}^N)$, hence $u \in W^{2,p}_{loc}(\mathbb{R}^N)$ for all $p > 1$, and one can reiterate the regularity by a standard argument. We examine two cases.
(i) $p < q$ and $\mu > 0$: then $g(u) \leqslant C$, and on B_1 (the unit ball of $\mathbb{R}^N$), one has

$$-\Delta u \leqslant C \quad \text{on } B_1$$
$$u\Big|_{\partial B_1} = u\,(|x| = 1)$$

It follows that $u \leqslant u(1) + C'$ on B_1, whence $u \in L^\infty(B_1)$ and obviously $u \in L^\infty(\mathbb{R}^N)$.
(ii) $p,\ q < (N+2)/(N-2) = \ell$ then, since $u \in H^1(B_1)$, $u \in L^{\ell+1}(B_1)$ and hence $g(u) \in L^\alpha(B_1)$, where

$$\alpha = \frac{\ell + 1}{\max(p, q)} > \frac{\ell + 1}{\ell}$$

This implies $u \in W^{2,\alpha}(B_1)$, and $u \in L^{\alpha^{**}}(B_1)$, where

$$\frac{1}{\alpha^{**}} = \frac{1}{\alpha} - \frac{2}{N}$$

One can check that $\alpha^{**} > \ell + 1$, and that one can reiterate the regularity to obtain $u \in L^\infty(B_1)$, hence $u \in L^\infty(\mathbb{R}^N)$.

Thus, is all cases, $u \in C^\infty(\mathbb{R}^N)$.
(2) To show that, $\forall j \in \mathbb{N}$, $|D^j u(r)| \leqslant C_j\, e^{-\delta_j r}$, with C_j and δ_j positive constants, it suffices to prove it in the case $j = 0$ and $j = 1$. Indeed, the other derivatives can be estimated by iteration, using the equation, from the estimates on u and u'. The fact that u decreases exponentially is proved in Strauss [15]. For u', we just observe that for r large enough, say $r \geqslant r_0$, $g(u) < 0$. Hence,

$$-u'' \leqslant -u'' - \frac{N-1}{r} u' = g(u) < 0, \qquad \text{for } r \geqslant r_0$$

Then, for $r > r_0 + 1$, one has

$$0 \geqslant u'(r) \geqslant \int_{r-1}^{r} u'(s)\,ds = u(r) - u(r-1) \geqslant -C'\, e^{-\delta_0 (r-1)}$$

This concludes the proof of Theorem 2.1.

3. DISCUSSION ON THE HYPOTHESES AND INDICATIONS ON OTHER METHODS

3.1. 'Almost necessary' character of hypotheses (5) - (7)

The proof given in section 2.4 shows that if $u \in H_r^1$ is any positive solution of equation (3), then necessarily $u \in C^\infty(\mathbb{R}^N)$, and u together with all its derivatives tend to zero exponentially as $r \to +\infty$. Then, for such a solution the following identity is verified (cf [13, 15]):

$$(17) \qquad \int_{\mathbb{R}^N} |\nabla u|^2 \, \mathrm{d}x = \int_{\mathbb{R}^N} ug(u)\mathrm{d}x = \frac{2N}{N-2} \int_{\mathbb{R}^N} G(u)\mathrm{d}x$$

This identity can be obtained by writing that u is a critical point of the energy $S = T - V$. Indeed, under a scale change u_σ, one has

$$S(u_\sigma) = \sigma^{N-2}\, T(u) - \sigma^N\, V(u)$$

and if u is a critical point of S, $\mathrm{d}S(u_\sigma)/\mathrm{d}\sigma|_{\sigma=1} = 0$ which yieds precisely equation (17).

Another way to obtain (17) is to multiply the differential equation

$$(18) \qquad -u''(r) - \frac{N-1}{r}\, u'(r) = g(u)$$

by $u'r^N$. Integrating from 0 to R yields

$$-\frac{1}{2} \int_0^R \frac{\mathrm{d}}{\mathrm{d}r}(u'^2)r^N \mathrm{d}r - (N-1)\int_0^R u'^2 r^{N-1}\mathrm{d}r = \int_0^R \frac{\mathrm{d}}{\mathrm{d}r}[G(u)]\, r^N \mathrm{d}r$$

After integration by parts one obtains

$$(19) \quad -\frac{1}{2}u'(R)^2 R^N - \frac{N-2}{2}\int_0^R u'^2 r^{N-1}\mathrm{d}r = G(u(R))R^N - N\int_0^R G(u) r^{N-1}\mathrm{d}r$$

Thus, if u and u' are exponentially small at infinity, letting $R \to +\infty$ in equation (19) we have identity (17).

From (17) we see that if $u \in H_r^1$ (3) is a non-trivial positive solution of (3), one necessarily has

(3) H_r^1 is the subspace of $H^1(\mathbb{R}^N)$ formed by the functions which are radial (or spherically symmetric).

$$\int_{\mathbb{R}^N} G(u(x))\,\mathrm{d}x > 0 \qquad \text{(in dimension } N \geq 3)$$

More precisely, if condition (7) is not satisfied, then, there does *not* exist any non-trivial solution of (3) in H_r^1, for $N \geq 2$.

Remark 3.1. One can also show this fact by multiplying (18) by u' and integrating:

$$-\frac{1}{2}u'(R)^2 - (N-1)\int_0^R \frac{u'^2(s)}{s}\,\mathrm{d}s = G[u(R)] - G[u(0)]$$

whence

$$G(u(0)) = (N\text{-}1)\int_0^\infty \frac{u'^2(s)}{s}\,\mathrm{d}s$$

Thus, for $N \geq 2$, this shows the necessary character of condition (7), and 'almost' necessary when $N = 1$ (i.e. $\exists\, \zeta > 0$ such that $G(\zeta) \geq 0$ is then necessary). In the appendix, a necessary and sufficient condition is given for dimension $N = 1$.

We now examine conditions (5) and (6).

Case 1: $\mu = 0$. Then, it is known (cf [13, 15]) that condition (5) (or (6) to which (5) reduces in this case) is necessary for the existence of a non-trivial solution of (3). This can readily be seen by writing down explicitly g and G in identity (17).

Case 2 : $\mu > 0$. Then, if one does not assume (5) any longer, one has: (i) either $p \geq (N+2)/(N-2) \geq q$, in which case it is known (cf. [15]) that there do *not* exist non-trivial solutions of (3) (this readily follows again from (17));
(ii) or $p > q > (N+2)/(N-2)$. In this case, we do not know whether or not there may exist a positive solution of (3). Nevertheless, we can show that in this case, the constrained minimization problem (9) does *not* posses any solution. Indeed, since the Sobolev injection $H^1(\mathbb{R}^N) \subset L^{\ell+1}(\mathbb{R}^N)$ is the 'best possible' (cf. [12]), there exists a sequence $(u_n) \subset H^1(\mathbb{R}^N)$ such that

$$\|\nabla u_n\|_{L^2(\mathbb{R}^N)} \leq 1, \quad \|u_n\|_{L^2(\mathbb{R}^N)} \leq 1 \quad \forall\, \mathrm{n}$$

and

$$\|u_n\|_{L^{q+1}(\mathbb{R}^N)} \to +\infty \quad \text{as } n \to +\infty$$

(For instance, one may choose $u_n(r) = Cn^{(N/2)-1}\,\mathrm{e}^{-nr}$.) It follows that

$$\|u_n\|_{L^{p+1}(\mathbb{R}^N)} \to +\infty \quad \text{as } n \to +\infty$$

Indeed, using Hölder's inequality and the same methods as in section 2.2, step 3, one shows

(20) $$\|u_n\|^{q+1}_{L^{q+1}(\mathbb{R}^N)} \leq C\,\|u_n\|^{t}_{L^{p+1}(\mathbb{R}^N)}$$

where $0 < t < p+1$. From (20) it follows that

$$V(u_n) \geq C\,\|u_n\|^{p+1}_{L^{p+1}(\mathbb{R}^N)} - C'\,\|u_n\|^{t}_{L^{p+1}(\mathbb{R}^N)} - C''$$

with C, C' and C'' being positive constants. Hence, $\lim_{n\to+\infty} V(u_n) = +\infty$. We define v_n from u_n by the scale change $v_n(x) = u_n(x/\sigma_n)$, where $\sigma_n = V(u_n)^{-1/N}$.

Then, $(v_n) \subset E$ satisfies

$$V(v_n) = \sigma_n^N\, V(u_n) = +\,1$$

and

$$T(v_n) = \sigma_n^{N-2}\, T(u_n) \leq \tfrac{1}{2}\, V(u_n)^{-(N-2)/N}$$

Thus $\lim_{n\to+\infty} T(v_n) = 0$; this shows that

$$\inf_{\{w\in E:\, V(w)=+1\}} T(w) = 0$$

3.2. A dual variational method

The dual formulation of (9) is the following problem:

(21) $$\text{maximize } \{V(w):\, w \in E,\, T(w) = 1\}$$

For this problem, one has the following result which yields Theorem 1.1, by the same method as in section 2.3.

Theorem 3.1.
Under the same hypotheses as in Theorem 1.1, there exists a solution $u \in E$ of (21), which is positive, spherically symmetric and decreases with $r = |x|$.

Proof: Analogous to that of Theorem 2.1.

3.3. Indications on a 'local method'

A natural approach to problem (3) is to try to approximate the solution on $\mathbb{R}^N$ by solving the equation on bounded domains. In fact, in the case

$1 < p < q$, and assuming (7), one can construct a solution u_R of

$$\text{(22)} \qquad \begin{aligned} &-\Delta u_R = g(u_R) \quad \text{in } B_R = \{x \in \mathbb{R}^N : |x| < R\} \\ &u_R = 0 \quad \text{on } \partial B_R \end{aligned}$$

such that u_R converges to a solution u of (3) when $R \to +\infty$. Moreover one can thus obtain a solution u having the properties stated in Theorem 1.1. In this approach, however, one encounters a certain number of technical difficulties. To illustrate this point, we mention in particular that (22) possesses a maximum positive solution $\bar{u}_R$. This solution $\bar{u}_R$ increases with R, and it is possible to prove that $\bar{u}_R$ converges everywhere, as $R \to +\infty$, to a constant $\beta > 0$, where β is the maximum solution of $g(\beta) = 0$ (recall that $1 < p < q$). Therefore, one needs to select adequately the non-trivial solution u_R of (12). This is done by a method of super- and sub-solution and a topological degree argument inspired from Rabinowitz [14]. The precise results obtained by this method and the detailed proofs will be given in [4].

APPENDIX: A NECESSARY AND SUFFICIENT CONDITION FOR THE EXISTENCE OF A SOLUTION IN THE CASE OF DIMENSION $N=1$

In this appendix, we consider a general function $f \in C(\mathbb{R}, \mathbb{R})$. We assume that f is locally Lipschitz continuous and satisfies $f(0) = 0$. We will denote

$$F(z) = \int_0^z f(s)\, ds$$

Consider the problem:

$$\text{(A.1)} \qquad \begin{aligned} &-u'' = f(u) \quad \text{on } \mathbb{R} \\ &u \in C^2(\mathbb{R}) \quad \lim_{x \to +\infty} u(x) = 0 \\ &\text{and} \\ &u \text{ is symmetric } (u(x) = u(-x)) \end{aligned}$$

Theorem A

A necessary and sufficient condition for the existence of a solution u to problem (A.1) such that $u(0) > 0$ is

(A.2) $\quad \zeta_0 = \inf\{\zeta > 0 : F(\zeta) = 0\}$ exists, is positive and is such that $f(\zeta_0) > 0$

Moreover, if (A.2) is satisfied there is a unique solution u to (A.1), and u has the following properties: u is positive, $u(0)=\zeta_0$, and $u'(x)<0$ if $x>0$.

Remark A.1 Under the assumption (A.2), the solution u of (A.1) is thus given by the solution to the following initial value problem (IVP):

$$-u''=f(u) \quad \text{on } \mathbb{R}^+ \quad (u(-x)=u(x) \text{ on } \mathbb{R})$$
$$u(0)=\zeta_0, \quad u'(0)=0 .$$

Remark A.2 It is readily seen that (A.2) implies:

(i) $\exists\, \zeta>0$ such that $F(\zeta)>0$.

(ii) If f is differentiable at 0, then $f'(0)\leqslant 0$.

Remark A.3 If, in addition to (A.2), we assume that

$$\lim_{x\to 0^+} \sup f(x)/x \leqslant -m<0$$

then there exist constants $C, \delta>0$ such that

$$0 \leqslant u(x), -u'(x)$$
$$|u''(x)| \leqslant C\, e^{-\delta x} \qquad \forall\, x \geqslant 0$$

Proof of Theorem A: First we show that (A.2) is a *sufficient* condition for the existence of a solution u to (A.1), with the properties stated in the theorem. Let u be the solution of the following IVP

$$\text{(A.3)} \qquad \begin{aligned} -u'' &= f(u) \quad \text{on } \mathbb{R}^+ \\ u(0) &= \zeta_0\,, \quad u'(0)=0 \end{aligned}$$

Multipying (A.3) by u' and integrating between 0 and r_0 yields

$$-\frac{1}{2}\, u'(r_0)^2 + \frac{1}{2}\, u'(0)^2 = F(u(r_0)) - F(u(0))$$

and hence

$$\text{(A.4)} \qquad -\frac{1}{2}\, u'(r_0)^2 = F(u(r_0))$$

(i) $u(r)>0,\ \forall\, r\geqslant 0$. For if not, there exists $r_0>0$ such that $u(r_0)=0$ and $u(r)>0\ \forall\, r\in(0,r_0)$. But this is impossible since (A.4) would imply $u'(r_0)=$ $=0$, whence $u=0$ since $u\equiv 0$ is unique solution of the IVP

$$-u''=f(u) \quad \text{on } \mathbb{R}^+$$
$$u(r_0)=u'(r_0)=0$$

(ii) $u(r)<\zeta_0,\ \forall\, r>0$. Observe first that $u(r)<\zeta_0$, for $r>0$ sufficiently small,

for $u'(0)=0$, $u''(0)=-f(\zeta_0)<0$. Now if $u(r)\geqslant\zeta_0$ for some r, there must exist an $r_1>0$ such that $u(r_1)=\zeta_0$ and $u(r)<\zeta_0$ in $(0, r_1)$. Hence, there exist $r_0>0$ such that $u(r_0)<\zeta_0$ and $u'(r_0)=0$. From (A.4) one now derives that $F(u(r_0))=0$ which together with $0<u(r_0)<\zeta_0$ is impossible. Hence $0<$ $<u(r)<\zeta_0$ $\forall r$ (which also shows that u exists everywhere).
(iii) $u'(r)<0$, $\forall r>0$. Indeed, this is true for $r>0$ sufficiently small, and from (A.4) we see that $u'(r)=0$ would imply $F(u(r))=0$ which is impossible since $0<u(r)<\zeta_0$.
(iv) $\lim_{x\to+\infty} u(x)=0$. We define $L=\lim_{x\to+\infty} u(x)$, ($u$ decreases); $0\leqslant L<\zeta_0$. Again by (A.4) we see that $-\frac{1}{2}u'(r)^2$ converges to $F(L)$, which implies $F(L)=0$ and $L=0$.
(v) The solution u to (A.1) is unique, Suppose ν is a solution of (A.1) with $\nu(0)>0$. Then, one has $-\frac{1}{2}\nu'(r)^2=F(\nu(r))-F(\nu(0))$ $\forall r\geqslant 0$.
Therefore $\lim_{r\to+\infty}\frac{1}{2}\nu'(r)^2=F(\nu(0))$. This obviously implies $F(\nu(0))=0$. Suppose $\nu(0)>\zeta_0$, then there exists $r_0>0$ such that $\nu(r_0)=\zeta_0$ $\forall r\in(0,r_0)$. From (A.4) again, one has $\nu'(r_0)=0$ and $\nu''(r_0)=-f(\zeta_0)<0$, and this cannot happen since $\nu(r)>\nu(r_0)$ for $r<r_0$. Hence the solution is unique[(4)].

We now prove that (A.2) is *necessary* condition. Suppose (A.2) is violated and there exists a solution of (A.1) such that $\nu(0)>0$. The same argument as given in part (v) above shows that $F(\nu(0))=0$, Thus, necessarily ζ_0 exists. There are two possibilities for (A.2) to be violated.
(i) We assume $\zeta_0>0$ but $f(\zeta_0)\leqslant 0$. Since $\nu(0)\geqslant\zeta_0$, there exists $r_0\geqslant 0$ such that $\nu(r_0)=\zeta_0$. At this point, one has from (A.4) that $\nu'(r_0)=0$ and also $\nu''(r_0)=-f(\zeta_0)\geqslant 0$. This shows that $\nu(r)$ can never go below ζ_0 (notice that if $f(\zeta_0)=0$, then by the uniqueness to the IVP, $\nu\equiv\zeta_0$).
(ii) We assume $\zeta_0=0$. Denote $\nu(0)=\zeta_1>0$. Then, $F(\zeta_1)=0$, and also $f(\zeta_1)>0$, for if $f(\zeta_1)\leqslant 0$, then $\nu(r)$ could never go below ζ_1 by the same argument as above. Since $\zeta_0=0$, there exists ζ_2, $0<\zeta_2<\zeta_1$ such that $F(\zeta_2)=0$. There exists $r_2>0$ such that $\nu>\zeta_2$ on $(0, r_2)$ and $\nu(r_2)=\zeta_2$. From (A.4) it is seen that $\nu'(r_2)=0$. Then, one has necessarily $\nu''(r_2)\geqslant 0$, that is $f(\zeta_2)\leqslant 0$. Now, $f(\zeta_2)=0$ would imply $\nu\equiv\zeta_2$ (uniqueness in the IVP). On the other hand, $f(\zeta_2)<0$ would imply that ν can never go below ζ_2. This contradiction concludes the proof.

(4) It is easily seen that any solution to the problem

$$-\nu''=f(\nu) \text{ on } \mathbb{R}$$

$$\lim_{x\to\pm\infty}\nu(x)\times 0$$

$$\exists\, x\in\mathbb{R},\quad \nu(x)>0$$

is derived from the unique solution u to problem (A.1) by a translation, that is: $\exists\ C\in\mathbb{R}$, $\nu(x)=u(x+C)$.

REFERENCES

[1] D. Anderson. 'Stability of time-dependent particle-like solutions in nonlinear field theories'. *J. Math. Phys.*, **12** (1971), 945-52.

[2] D. G. Aronson and H. F. Weinberger. 'Nonlinear diffusion in population genetics'. *Lecture Notes in Math.*, vol. 446, Springer, Berlin (1975).

[3] H. Berestycki and P. L. Lions. 'Existence d'ondes solitaires dans des problèmes non linéaires du type Klein-Gordon'. *Notes aux Comptes Rendus Acad. Sci. Paris,* série A, **287** (1978), 503-6, and **288** (1979), 395-8.

[4] H. Berestycki and P. L. Lions. 'A local method for the existence of a ground state in nonlinear problems of the type Klein-Gordon'. To appear.

[5] H. Berestycki and P. L. Lions. *'Existence of solitary waves in non-linear Klein-Gordon type equations. Part 1: The ground state. Part 2: Existence of infinitely many bound states.'* To appear.

[6] S. Coleman, V. Glazer, and A. Martin. 'Action minima among solutions to a class of Euclidean scalar fields equations'. *Commun. Math. Phys.*, **58** (1978), 211-21.

[7] P.C. Fife. 'Asymptotic states for equations of reaction and diffusion'. *Bull. Am. Math. Soc.*, **84** (1978). 693-726.

[8] G.H. Hardy, J.E. Littlewood, and G. Polya. *'Inequalities.'* Cambridge Univ. Press., London and New York (1952).

[9] T. Kato. 'Growth properties of solutions of the reduced wave equation with a variable coefficient'. *Commun. Pure Appl. Math.*, **12** (1959), 403-25.

[10] A. Kolmogorov, I. Petrovsky, and N. Piskounov. 'Etude de l'équation de la diffusion avec croissance de la quantité de matiére et son application à un problème biologique'. *Bull. Univ. Moscou,* série Intern. section A.1, no. 6 (1937), 1-37.

[11] E.H. Lieb. 'Existence and uniqueness of the minimizing solution of Choquard's nonlinear equation', *Stud. Appl. Math.*, **57** *(1977), 93.*

[12] J.L. Lions. *'Problèmes aux limites dans les équations aux dérivées partielles.'* Presses de l'Université de Montréal (1962).

[13] S.I. Pohozaev. 'Eigenfunctions of the equation $\Delta u + \lambda f(u) = 0$'. *Sov. Math. Dolk.*, 5 (1965), 1408-11.

[14] P.H. Rabinowitz. 'Pairs of positive solutions of nonlinear elliptic partial differential equations'. *Indiana Univ. Math. J.*, **23** (1973/74), 173-86.

[15] W.A. Strauss. 'Existence of solitary waves in higher dimensions', *Commun. Math. Phys.*, **55** (1977), 149-62.

[16] M.F. Weinberger. 'Asymptotic behaviour of a model in population genetics'. *Lecture Notes in Math.*, Springer, Berlin (Indiana Univ. Seminar in Applied Math., ed J. Chadam). To appear.

Chapter 4

Some Variational Problems of the Thomas-Fermi Type

H. Brézis

1. INTRODUCTION

Let $\rho(x)$ be a function defined on $\mathbb{R}^3$ with values in $[0, +\infty)$. Consider the functional

$$\mathcal{E}(\rho) = \int_{\mathbb{R}^3} \rho^{5/3}(x)\,dx - \int_{\mathbb{R}^3} V(x)\,\rho(x)\,dx + \frac{1}{2}\int_{\mathbb{R}^3}\int_{\mathbb{R}^3} \frac{\rho(x)\,\rho(y)}{|x-y|}\,dx\,dy$$

where $V(x)$ is a given function – a special case of importance is the Coulomb potential:

$$V(x) = \sum_{i=1}^{k} \frac{m_i}{|x - a_i|}$$

where $m_i > 0$ and $a_i \in \mathbb{R}^3$ are fixed.

The following minimization problem arises in quantum mechanics:

$$\min_{\rho \in K} \mathcal{E}(\rho) \tag{1}$$

where K denotes the convex set

$$K = \left\{ \rho \in L^1(\mathbb{R}^3) : \rho \geq 0 \text{ a.e. and } \int_{\mathbb{R}^3} \rho(x)\,dx = I \right\}$$

(a.e. = almost everywhere) and $I > 0$ is given. $\mathcal{E}(\rho)$ is called the Thomas-Fermi functional.

The function ρ–to be determined – represents a (probability) density of electrons (or rather Fermions). The system consists of k positive nuclei of charge m_i placed at the points a_i in space surrounded by a 'cloud' of Fermions.

The functional $\mathcal{E}$ has three terms corresponding, respectively, to the kinetic energy, the attractive potential energy (interaction between Fermions and positive nuclei), and the repulsive potential energy (interaction between Fermions).

It is a surprising fact that problem (1), which looks like a reasonable convex minimization problem, need not have a solution. To illustrate this fact, suppose, for example, that $V \in L^\infty \cap L^1$, so that $\inf \mathcal{E} > -\infty$ and let ρ_n be a minimizing sequence.

Clearly, ρ_n remains bounded in L^p for each $1 \leqslant p \leqslant 5/3$ and therefore (after choosing a subsequence) we have $\rho_n \rightharpoonup \bar{\rho}$ weakly in L^p for $1 < p \leqslant 5/3$.

Therefore, we have

$$\underline{\lim} \int \rho_n^{5/3} \geqslant \int \bar{\rho}^{\,5/3}$$

$$\lim \int V\rho_n = \int V\bar{\rho}$$

$$\underline{\lim} \iint \frac{\rho_n(x)\,\rho_n(y)}{|x-y|}\,dx\,dy \geqslant \lim \iint \frac{\bar{\rho}(x)\,\bar{\rho}(y)}{|x-y|}\,dx\,dy$$

(The last assertion follows from the fact that the function

$$\rho \mapsto \iint \frac{\rho(x)\,\rho(y)}{|x-y|}\,dx\,dy$$

is convex and continuous on $L^{6/5}$).

Therefore, we have

$$\mathcal{E}(\bar{\rho}) \leqslant \inf_K \mathcal{E}$$

However, in general, $\bar{\rho} \notin K$; one can only assert that

$$\int_{\mathbb{R}^3} \bar{\rho}(x)\,dx \leqslant I$$

Indeed, let $A \subset \mathbb{R}^3$ be an arbitrary set of finite measure. Then

$$\int_A \rho_n \leqslant I$$

and hence

$$\int_A \bar{\rho} \leqslant I$$

The argument above shows that, if in (1) we replace K by

$$\widetilde{K} = \left\{ \rho \in L^1(\mathbb{R}^3) : \rho \geqslant 0 \quad \text{a.e. and} \quad \int \rho \leqslant I \right\}$$

then the problem

$(\tilde{1})$ $$\min_{\rho \in K} \mathcal{E}(\rho)$$

is easy and has a solution (if V is reasonable).

We recall first an important result of Lieb and Simon [4].

Theorem 1

Assume

$$V(x) = \sum_{i=1}^{k} \frac{m_i}{|x - a_i|}$$

and set

$$I_0 = \sum_{i=1}^{k} m_i$$

Then

(i) if $0 < I \leqslant I_0$, the problem (1) has a unique solution $\bar{\rho}$;
(ii) if $I > I_0$, the problem (1) has no solution;
(iii) if $I < I_0$, the solution $\bar{\rho}$ of (1) has a compact support.

Proof: The proof of theorem 1 given in [4] is indirect: first solve $(\tilde{1})$ and then show that for $0 < I \leqslant I_0$ the solution of $(\tilde{1})$ in fact lies in K.

Our purpose is to describe a joint work with Benilan [2], which extends Theorem 1 in several directions:

(a) $\rho^{5/3}$ is replaced by $j(\rho)$, where j is a C^1 convex function. Various functions occur in physical problems. For example (E. Lieb, personal communication), the Thomas-Fermi theory taking into account relativistic effects leads to the function

$$j(r) = \int_0^{r^{1/3}} [(1 + t^2)^{1/2} - 1]\, t^2 \, dt$$

(so that $j(r) \sim r^{5/3}$ for small r and $j(r) \sim r^{4/3}$ for large r).

(b) The Coulomb potential is replaced by a general function $V(x)$. Of course, we have to elucidate the role of I_0 and its connection to V.

(c) Our approach to solving (1) is entirely different from that in [4]. We do not try to solve the variational problem, but rather the Euler-Lagrange equation arising from (1). This is a non-linear partial differential equation which we solve by a direct argument. One advantage is the following. Consider, for example, the case where

$$j(r) = r^p \quad \text{and} \quad V(x) = \sum_{i=1}^{k} \frac{m_i}{|x - a_i|}$$

(i) If $p > \frac{3}{2}$, the variational approach of [4] works nicely (note that $p = \frac{5}{3}$ is good!).

(ii) If $\frac{4}{3} < p \leqslant \frac{3}{2}$, the *variational approach fails* and, in fact,

$$\inf_K \mathcal{E} = -\infty$$

On the other hand, the *Euler-Lagrange equation has a solution* as we shall see in section 3.

A similar situation occurs already in linear equations. For example, if we try to solve

$$(*) \qquad \begin{aligned} \Delta u &= f \quad \text{on } \Omega \\ u &= 0 \quad \text{on } \partial\Omega \end{aligned}$$

with $f \in L^1(\Omega)$ by introducing the Dirichlet integral

$$E(\psi) = \frac{1}{2} \int |\Delta \psi|^2 + \int f \psi$$

it may happen that

$$\inf_\psi E(\psi) = -\infty$$

On the other hand, problem (*) has a unique solution (which can be obtained, for example, by a simple duality argument, see, for example [5], Theorem 4.5).

In Section 2 we derive the Euler-Lagrange equation of (1) and we indicate some assumptions (on j and V) which guarantee the equivalence of the two formulations.

In Section 3 we state our main existence-uniqueness theorem for the Euler-Lagrange equation. We state an estimate for I_0 and we prove (under some assumptions) that $\bar{\rho}$ has a compact support.

In Section 4 we discuss a min-max principle for the Lagrange multiplier occurring in (1). We also answer a question raised in [4].

2. THE VARIATIONAL FORMULATION AND THE EULER-LAGRANGE FORMULATION. CONDITIONS FOR EQUIVALENCE

Let $j : \mathbb{R}_+ \to \mathbb{R}_+$ be a convex C^1 function such that $j(0) = j'(0) = 0$. We consider the conjugate convex function defined for $t \in \mathbb{R}$ by

$$j^*(t) = \sup_{s \geqslant 0} (ts - j(s))$$

Let $V(x)$ be a measurable function on $\mathbb{R}^3$. Set

$$J(\rho) = \int_{\mathbb{R}^3} [j(\rho(x)) - V(x)\rho(x)]\, dx$$

$$D(J) = \{\rho \in L^1(\mathbb{R}^3) : \rho \geq 0 \text{ a.e. and } j(\rho) - V\rho \in L^1(\mathbb{R}^3)\}$$

$$H(\rho) = \frac{1}{8\pi} \int_{\mathbb{R}^3} \int_{\mathbb{R}^3} \frac{\rho(x)\,\rho(y)}{|x-y|}\, dx\, dy$$

$$D(H) = \{\rho \in L^1(\mathbb{R}^3) : \rho \geq 0 \text{ a.e. and } \frac{\rho(x)\,\rho(y)}{|x-y|} \in L^1(\mathbb{R}^3 \times \mathbb{R}^3)\}$$

$$\mathcal{E}(\rho) = J(\rho) + H(\rho), \quad D(\mathcal{E}) = D(J) \cap D(H)$$

Moreover, for $\rho \in L^1(\mathbb{R}^3)$, set

$$B\rho = \frac{1}{4\pi} \frac{1}{|\cdot|} * \rho = \frac{1}{4\pi} \int_{\mathbb{R}^3} \frac{\rho(y)}{|x-y|}\, dy$$

We recall the definition of the Marcinkiewicz space M^p (or weak L^p space) for $1 < p < \infty$:

$$M^p = \left\{ f : \mathbb{R}^3 \to \mathbb{R},\ f \text{ measurable and } \sup_{\substack{A \subset \mathbb{R}^3 \\ 0 < |A| < \infty}} |A|^{-1/p'} \int_A |f(x)|\, dx < \infty \right\}$$

with the norm

$$\|f\|_{M^p} = \sup_{\substack{A \subset \mathbb{R}^3 \\ 0 < |A| < \infty}} |A|^{-1/p'} \int_A |f(x)|\, dx$$

Since

$$\frac{1}{|\cdot|} \in M^3$$

it follows from convolution inequalities (see, e.g., the Appendix of reference [3]) that $B: L^1 \to M^3$ is a bounded operator. Also, it is well known that

$$-\Delta(B\rho) = \rho \quad \text{in } \mathcal{D}'$$

It is not difficult to check (using the positivity of the kernel $1/|x|$) that, if φ, $\psi \in D(H)$, then $\varphi B \psi$ and $\psi B \varphi \in L^1$,

$$\int \varphi B \psi = \int \psi B \varphi \leq H(\varphi) + H(\psi)$$

In particular H is convex.

Theorem 2

Assume $\bar{\rho} \in D(\mathcal{E}) \cap K$ and satisfies

(2) $$\mathcal{E}(\bar{\rho}) \leqslant \mathcal{E}(\rho), \qquad \forall \rho \in D(\mathcal{E}) \cap K$$

Then, there exists a constant λ, such that

(3) $$\begin{aligned} j'(\bar{\rho}) - V + B\bar{\rho} &= -\lambda \quad \text{a.e.} \qquad \text{on } [\bar{\rho} > 0] \\ -V + B\bar{\rho} &\geqslant -\lambda \quad \text{a.e.} \qquad \text{on } [\bar{\rho} = 0] \end{aligned}$$

Conversely, assume the following condition holds:

(4) $$\begin{cases} \text{there exists a constant } C, \text{ such that} \\ j^*(V(x) + C) \in L^1(\mathbb{R}^3) \end{cases}$$

and let $\bar{\rho} \in K$, then (3) implies (2).

Remark 1 Theorem 2 shows that a variational solution is *always* a solution of the Euler-Lagrange equation (3). The converse is true only under some restrictive assumption. Assumption (4) is made up essentially to guarantee that J (and thus $\mathcal{E}$) is bounded below on K.

Remark 2 Let us check the assumption (4) when

$$j(r) = \frac{1}{p} r^p \quad \text{and} \quad V(x) = \sum_i \frac{m_i}{|x - a_i|}$$

Here

$$j^*(r) = \frac{1}{p'} (r^+)^{p'}$$

taking $C = -1$, for example, we see that (4) is equivalent to: $1/|x|^{p'}$ is integrable near $x = 0$, i.e. $p > \frac{3}{2}$.

Remark 3 The constant λ which appears in (3) is a Lagrange multiplier corresponding to the constraint

$$\int \rho(x)\,dx = I$$

$-\lambda$ is called the chemical potential.

Sketch of the proof of Theorem 2:

(3) *and* (4) *imply* (2):

Let $\rho \in L^1$, $\rho \geqslant 0$ a.e. We claim that a.e. on $\mathbb{R}^3$

(5) $$j(\rho) - j(\bar{\rho}) \geqslant (V - B\bar{\rho} - \lambda)(\rho - \bar{\rho})$$

Indeed, on the set $[\bar{\rho} > 0]$ we have, using the convexity of j, and (3),

$$j(\rho) - j(\bar{\rho}) \geqslant j'(\bar{\rho})(\rho - \bar{\rho}) = (V - B\bar{\rho} - \lambda)(\rho - \bar{\rho})$$

While, on the set $[\bar{\rho} = 0]$ we have

$$j(\rho) - j(\bar{\rho}) = j(\rho) \geqslant 0 \geqslant (V - B\bar{\rho} - \lambda)(\rho - \bar{\rho})$$

since

$$V - B\bar{\rho} - \lambda \leqslant 0 \quad \text{and} \quad \rho - \bar{\rho} = \rho \geqslant 0$$

It follows from (5) that

$$(6) \quad [(j(\rho) - V\rho) + \tfrac{1}{2}\rho B\rho] - [(j(\bar{\rho}) - V\bar{\rho}) + \tfrac{1}{2}\bar{\rho}B\bar{\rho}] \geqslant [\tfrac{1}{2}\rho B\rho + \tfrac{1}{2}\bar{\rho}B\bar{\rho} - \rho B\rho] - \lambda(\rho - \bar{\rho})$$

Choosing, in particular, $\rho = 0$ we see that

$$(j(\bar{\rho}) - V\bar{\rho}) + \tfrac{1}{2}\bar{\rho}B\bar{\rho} \leqslant -\lambda\bar{\rho}$$

On the other hand,

$$j(\bar{\rho}) - V\bar{\rho} = j(\bar{\rho}) - (V + C)\bar{\rho} + C\bar{\rho} \geqslant -j^*(V + C) + C\bar{\rho} \in L^1$$

We conclude that $j(\bar{\rho}) - V\bar{\rho} \in L^1$ and $\bar{\rho}B\bar{\rho} \in L^1$, i.e.

$$\bar{\rho} \in D(\mathcal{E})$$

Choosing now $\rho \in D(\mathcal{E}) \cap K$ and integrating (6), we find

$$\mathcal{E}(\rho) \geqslant \mathcal{E}(\bar{\rho})$$

(2) *implies* (3):

We just indicate a very formal argument. Let $\zeta \in D(\mathcal{E})$ and let $0 < t < 1$. Choosing $\rho = (1-t)\bar{\rho} + t\zeta$ in (2) and letting $t \to 0$ we find, after some manipulation

$$(7) \quad \int f(\zeta - \bar{\rho}) \geqslant 0$$

where

$$f = j'(\bar{\rho}) - V + B\bar{\rho}$$

We claim now that there is some constant λ such that

$$f = -\lambda \quad \text{on} \quad [\bar{\rho} > 0]$$
$$f \geqslant -\lambda \quad \text{on} \quad [\bar{\rho} = 0]$$

Indeed, for each $\mu \in \mathbb{R}$, let

$$\rho_\mu = (-f + \bar{\rho} - \mu)^+$$

As $\mu \to -\infty$,

$$\rho_\mu \to +\infty \quad \text{and} \quad \int \rho_\mu \to +\infty$$

As $\mu \to +\infty$,

$$\rho_\mu \to 0 \quad \text{and} \quad \int \rho_\mu \to 0$$

Hence, there is some λ such that

$$\int \rho_\lambda = I$$

Choosing $\zeta = \rho_\lambda$ in (7), we obtain

$$\int (f + \lambda)(\rho_\lambda - \bar{\rho}) \geqslant 0$$

(since $\int (\rho_\lambda - \bar{\rho}) = 0$). Therefore,

$$\text{(8)} \quad \int_{[\rho_\lambda > 0]} (f + \lambda)(\rho_\lambda - \bar{\rho}) - \int_{[\rho_\lambda = 0]} (f + \lambda)\bar{\rho} \geqslant 0$$

But on the set $[\rho_\lambda > 0]$ we have

$$\rho_\lambda = -f + \bar{\rho} - \lambda$$

while on the set $[\rho_\lambda = 0]$ we have

$$-f + \bar{\rho} - \lambda \leqslant 0$$

It follows from (8) that

$$\int_{[\rho_\lambda > 0]} |\rho_\lambda - \bar{\rho}|^2 + \int_{[\rho_\lambda = 0]} (f + \lambda)\bar{\rho} \leqslant 0$$

and consequently

$$\bar{\rho} = \rho_\lambda \quad \text{on} \quad [\rho_\lambda > 0]$$

$$\bar{\rho} = 0 \quad \text{on} \quad [\rho_\lambda = 0]$$

Hence, $\rho_\lambda = \bar{\rho}$ everywhere and

$$\bar{\rho} = (-f + \bar{\rho} - \lambda)^+$$

i.e.

$$f = -\lambda \qquad \text{on } [\bar{\rho} > 0]$$
$$f \geqslant -\lambda \qquad \text{on } [\rho = 0]$$

3. A DIRECT APPROACH FOR SOLVING THE EULER-LAGRANGE EQUATION

We are looking now for a solution of the Euler-Lagrange equation (3). Our main result is the following

Theorem 3
Assume

(9) $$V \in \frac{1}{|\cdot|} * L^1$$

(i.e. $\Delta V \in L^1$ and $V(\infty) = 0$ in some weak sense), and

(10) $$V > 0 \qquad \text{somewhere}$$

i.e. on a set of positive measure. Then,

(A) there is some $0 < I_0 \leqslant \infty$ (depending on j and V), such that:
  (a) if $0 < I \leqslant I_0$ there exists a unique $\bar{\rho}$ and λ solution of (3);
  (b) if $I > I_0$ there exists no solution of (3);

(B) if $I < I_0$ and if $V(x) \to 0$ as $|x| \to \infty$, then the solution $\bar{\rho}$ of (3) has a compact support;

(C) assume $j(r) = r^p$ for small r, with $p \geqslant 4/3$ (this is just an example – a general assumption appears in the proof), then

$$\int (-\Delta V) \leqslant I_0 \leqslant \int (-\Delta V)^+$$

In particular

$$I_0 = \int (-\Delta V) \quad \text{when} \quad -\Delta V \geqslant 0$$

(D) instead of (9) assume now

(11) $$V \in \frac{1}{|\cdot|} * \mathscr{M}$$

where $\mathscr{M}$ denotes the space of bounded measures on $\mathbb{R}^3$ (i.e. $\Delta V \in \mathscr{M}$

and $V(\infty) = 0$). Assume $j(r) = r^p$ for large r, with $p > 4/3$ (again, a general assumption appears in the proof), then (A), (B) and (C) hold.

Remark 1 We emphasize the fact that we make no assumption on j in parts (A) and (B).

Remark 2 In the case

$$V(x) = \sum_i \frac{m_i}{|x - a_i|}$$

we have

$$-\frac{1}{4\pi} \Delta V = \sum_i m_i \, \delta(x - a_i)$$

so we must use part (D). If $j(r) = r^{4/3}$, or in the relativistic Thomas-Fermi problem, part (D) does not apply. In fact, one can prove that *no* solution of the Euler-Lagrange equation exists (see the discussion in the proof of part (D)). In other words, the relativistic Thomas-Fermi equation has no solution for Coulomb potentials. However, if $V(x)$ is smoothed 'a little bit' (in such a way that $\Delta V \in L^1$), then we may apply parts (A), (B), and (C).

Part (A)
First we claim that necessarily $\lambda \geqslant 0$. Indeed by (3) we have

$$j'(\bar{\rho}) - V + B\bar{\rho} \geqslant -\lambda \qquad \text{a.e. on } \mathbb{R}^3$$

Letting $|x| \to \infty$ we see that the left-hand side goes to zero (in a weak sense) and we must have $\lambda \geqslant 0$. (Note that V and $B\bar{\rho}$ lie in M^3 and thus for every $\delta > 0$ we have $|V| < \delta$ and $|B\bar{\rho}| < \delta$ *except* on some set of finite measure).

Next we transform (3) into a non-linear partial differential equation by introducing the new unknown

$$u = V - B\bar{\rho}$$

so that $u \in M^3$ and

$$-\Delta u + \bar{\rho} = -\Delta V$$

On the other hand, we have

$$u - \lambda = j'(\bar{\rho}) \qquad \text{a.e. on } [\bar{\rho} > 0]$$

$$u - \lambda \leqslant 0 \qquad \text{a.e. on } [\bar{\rho} = 0]$$

We may summarize this information in one single equation:

$$\overline{\rho} = \gamma(u - \lambda)$$

where

$$\gamma(t) = \begin{cases} (j')^{-1}(t) & \text{for } t > 0 \\ 0 & \text{for } t \leqslant 0 \end{cases} \tag{12}$$

(for simplicity, we assume that j' is strictly increasing on $[0, \infty)$).

Hence, problem (3) is equivalent to finding $u \in M^3$ and $\lambda \geqslant 0$ such that

$$\begin{aligned} -\Delta u + \gamma(u - \lambda) &= -\Delta V \\ \int \gamma(u - \lambda) &= I \end{aligned} \tag{13}$$

Once problem (13) has been solved, we set $\overline{\rho} = \gamma(u - \lambda)$ and we obtain a solution of (3). Our strategy for solving (13) is the following.

Fix $\lambda \geqslant 0$ and solve the problem

$$\begin{cases} u_\lambda \in M^3 \\ -\Delta u_\lambda + \gamma(u_\lambda - \lambda) = -\Delta V \end{cases} \tag{14}$$

Then, consider

$$I(\lambda) = \int \gamma(u_\lambda - \lambda)$$

and try to determine λ in such a way that

$$I(\lambda) = I \quad (I \text{ is given})$$

The existence and uniqueness of u_λ follows from the next Lemma.

Lemma 1 (see [3]). Let $\beta : \mathbb{R} \to \mathbb{R}$ be a monotonic non-decreasing continuous function with $\beta(0) = 0$. For every $f \in L^1$, there exists a unique $u \in M^3$ solution of

$$-\Delta u + \beta(u) = f \quad \text{in } \mathbb{R}^3$$

In addition

$$\beta(u) \in L^1$$

We apply Lemma 1 for each $\lambda \geqslant 0$ to the function $\beta(r) = \gamma(r - \lambda)$ (note that $\beta(0) = \gamma(-\lambda) = 0$) and to $f = -\Delta V$. Therefore, we obtain a solution $u_\lambda \in M^3$ of (14).

Set

$$I(\lambda) = \quad \gamma(u_\lambda - \lambda)$$

In our next lemma, we summarize the main properties of $I(\lambda)$.

Lemma 2 The function $I(\lambda) : [0, \infty) \to [0, +\infty)$ is non-increasing, continuous and

$$\lim_{\lambda \to +\infty} I(\lambda) = 0$$

In addition $I(\lambda)$ is strictly decreasing on the set $\{\lambda : I(\lambda) > 0\}$, and also $I(0) > 0$.

Part (A) of Theorem 3 is a direct consequence of Lemma 2 with I_0 determined as follows: find $u_0 \in M^3$ unique solution of

$$-\Delta u_0 + \gamma(u_0) = -\Delta V$$

and set

$$I_0 = \int \gamma(u_0)$$

Sketch of the proof of Lemma 2:

* The function $I(\lambda)$ is non-increasing:

Let $\lambda < \mu$ and set $\tilde{u}_\lambda = \tilde{u}_\lambda - \lambda$, $u_\mu = u_\mu - \mu$. Thus, we have

$$\begin{aligned} -\Delta \tilde{u}_\lambda + \gamma(\tilde{u}_\lambda) &= -\Delta V \\ \tilde{u}_\lambda(\infty) &= -\lambda \\ -\Delta \tilde{u}_\mu + \gamma(\tilde{u}_\mu) &= -\Delta V \\ \tilde{u}_\mu(\infty) &= -\mu \end{aligned}$$

Since $-\mu \leqslant -\lambda$ and since γ is non-decreasing, it follows, from an appropriate version of the maximum principle, that

$$\tilde{u}_\mu \leqslant \tilde{u}_\lambda$$

Therefore, $\gamma(\tilde{u}_\mu) \leqslant \gamma(\tilde{u}_\lambda)$ and so $I(\mu) \leqslant I(\lambda)$

* $\lim_{\lambda \to +\infty} I(\lambda) = 0$:

Indeed, $-\Delta(u_\lambda - V) = -\gamma(u_\lambda - \lambda) \leqslant 0$ and since $(u_\lambda - V)(\infty) = 0$, we deduce again from the maximum principle that $u_\mu \leqslant V$.

In particular $\gamma(u_\lambda - \lambda) \leqslant \gamma(V - \lambda)$ and so $\gamma(u_\lambda - \lambda) \to 0$ a.e. as $\lambda \to +\infty$. It follows from the monotonic convergence theorem that

$$\lim_{\lambda \to +\infty} I(\lambda) = 0$$

$*\ I(0) > 0$:

Indeed, suppose by contradiction that $I(0) = 0$, i.e. $\gamma(u_0) = 0$ and so $u_0 \leqslant 0$ a.e.

On the other hand, $-\Delta u_0 + \gamma(u_0) = -\Delta V$ leads to $u_0 = V$. Therefore, $V \leqslant 0$, a contradiction to (10) (in fact, this is the only place where (10) is used!).

$*\ I(\lambda)$ is strictly decreasing on $\{\lambda : I(\lambda) > 0\}$:

Indeed, suppose $\lambda < \mu$ and $I(\lambda) = I(\mu)$, with $I(\lambda) > 0$. Since $\gamma(u_\mu - \mu) \leqslant \gamma(u_\lambda - \lambda)$, we have in fact $\gamma(u_\mu - \mu) = \gamma(u_\lambda - \lambda)$ and also $\Delta u_\lambda = \Delta u_\mu$.

Thus $u_\lambda = u_\mu$. Since $I(\lambda) > 0$, the set $[\gamma(u_\lambda - \lambda) > 0]$ is not empty and, therefore, we have $u_\lambda - \lambda = u_\mu - \mu$ (since γ is strictly increasing on $[0, \infty)$). Hence $\lambda = \mu$.

Part (B)

In the case $0 < I < I_0$, the corresponding λ satisfies $\lambda > 0$. On the other hand (see part (*A*)), $u_\mu - \lambda \leqslant V - \lambda$ and, therefore $\bar{\rho} = \gamma(u_\lambda - \lambda) \leqslant \gamma(V - \lambda)$. Since $V(x) \to 0$ as $|x| \to$ we see that $\bar{\rho}(x) = 0$ for $|x|$ sufficiently large.

Part (C)

Recall that

$$I_0 = \int \gamma(u_0)$$

where $u_0 \in M^3$ is the solution of

$$-\Delta u_0 + \gamma(u_0) = -\Delta V$$

It is a general fact (see [3]) that, under the assumptions of Lemma 1, we have

$$\int \beta(u_0)^+ \leqslant \int f^+$$

Hence,

$$I_0 = \int \gamma(u_0) \leqslant \int (-\Delta V)^+$$

Next we wish to prove that

$$I_0 \geqslant \int (-\Delta V)$$

or, equivalently, that

$$\int \Delta u_0 \geqslant 0$$

Recall that

$$u_0 = \frac{1}{4\pi\,|\cdot|} * (-\Delta u_0)$$

and hence *formally*

$$u_0(x) \sim \frac{C}{|x|} \quad \text{as} \quad |x| \to \infty$$

where

$$C = \frac{1}{4\pi} \int (-\Delta u_0)$$

Assume, by contradiction, that

$$\int \Delta u_0 < 0$$

and so $C > 0$. On the other hand, $\gamma(u_0) \in L^1$ and we would find

$$\int_{|x|>1} \gamma\left(\frac{C}{|x|}\right) \mathrm{d}x < \infty$$

But this is false, for example, when $j(r) = r^p$ and $p \geqslant 4/3$ since

$$\gamma(t) = \left(\frac{t^+}{p}\right)^{1/(p-1)}$$

The general condition, which guarantees that the conclusion of part (C) holds, is

$$\int_{|x|>1} \gamma\left(\frac{1}{|x|}\right) \mathrm{d}x = \infty$$

Part (D)

We need first to extend Lemma 1 to the case where $f \in \mathscr{M}$ (instead of L^1).

Surprisingly, this is possible only under some restrictive assumptions on β (a fact pointed out independently in [1]). Consider, for example, the equation

(15) $$-\Delta u + u^3 = \delta_0 \quad \text{(a Dirac mass at 0).}$$

Then, (15) has no solution $u \in L^3_{\text{loc}}$. Indeed, suppose u is a solution of (15) with $u \in L^3_{\text{loc}}$, then, near $x = 0$, u^3 is 'negligible' in comparison with δ_0 and $u(x)$ tends to 'resemble' $1/4\pi|x|$. On the other hand, $1/|x| \notin L^3$ near $x = 0$. More generally, if u is a solution of

$$-\Delta u + \beta(u) = \pm\,\delta_0$$

such that $\beta(u) \in L^1_{\text{loc}}$, then *necessarily*

$$\beta\left(\pm\frac{1}{|x|}\right) \in L^1 \quad \text{near} \quad x = 0$$

Our next lemma asserts that the converse holds.

Lemma 3 Assume β is non-decreasing, continuous, $\beta(0) = 0$ and

(16) $$\beta\left(\pm\frac{1}{|x|}\right) \in L^1 \quad \text{near} \quad x = 0$$

Then, for every $f \in \mathcal{M}$, there exists a unique $u \in M^3$ solution of

$$-\Delta u + \beta(u) = f$$

and, in addition $\beta(u) \in L^1$.

After this stage, the proofs of parts (A), (B), and (C) are easily modified to handle the case where $\Delta V \in \mathcal{M}$. Note that, if $j(r) = r^p$, then

$$\gamma(t) = \left(\frac{t^+}{p}\right)^{1/(p-1)}$$

and

$$\gamma\left(\frac{1}{|x|}\right) \in L^1 \quad \text{near} \quad x = 0$$

holds provided $p > 4/3$.

Sketch of the proof of Lemma 3: Let $f_n \in L^1$ be a sequence such that $\|f_n\|_{L^1} \leq \|f\|_{\mathcal{M}}$ and $f_n \to f$ in $\mathcal{D}'$. Let $u_n \in M^3$ be the solution of

$$-\Delta u_n + \beta(u_n) = f_n$$

We have

$$\|\beta(u_n)\|_{L^1} \leq \|f_n\|_{L^1} \leq \|f\|_{\mathcal{M}}$$

$$\|\Delta u_n\|_{L^1} \leq 2\|f_n\|_{L^1} \leq 2\|f\|_{\mathcal{M}}$$

$$\|u_n\|_{M^3} \leq C\|f\|_{\mathcal{M}}$$

$$\|\nabla u_n\|_{M^{3/2}} \leq C\|f\|_{\mathcal{M}}$$

We prove now that $\{\beta(u_n)\}$ is relatively compact in $L^1(S)$ for every bounded set $S \subset \mathbb{R}^3$.

It follows then easily that $u_{n_k} \to u$ in L^1_{loc}, $u \in M^3$ and

$$-\Delta u + \beta(u) = f \quad \text{in } \mathcal{D}'$$

with $\beta(u) \in L^1$.

In order to show that $\{\beta(u_n)\}$ is relatively compact in $L^1(S)$, we use a theorem of Vitali. It suffices to check that $\forall \varepsilon > 0, \exists \delta > 0$ such that $|K| < \delta$ implies

$$\int_K |\beta(u_n)| < \epsilon, \quad \forall n$$

Set

$$\overline{\beta}(t) = \beta(t) + |\beta(-t)| \quad \text{for } t \geq 0$$

We have

$$\int_K |\beta(u_n)| = \int_{\substack{K \\ [|u_n| \leq R]}} |\beta(u_n)| + \int_{\substack{K \\ [|u_n| > R]}} |\beta(u_n)|$$

$$\leq |K|\overline{\beta}(R) - \int_R^\infty \overline{\beta}(\lambda)\, d\alpha_n(\lambda)$$

where

$$\alpha_n(\lambda) = \text{meas}\,[|u_n| > \lambda] \leq C/\lambda^3$$

since $\|u_n\|_{M^3} \leq C$. On the other hand,

$$- \int_R^\infty \overline{\beta}(\lambda)\, \mathrm{d}\alpha_n(\lambda) = \overline{\beta}(R)\alpha_n(R) + \int_R^\infty \alpha_n(\lambda)\, \mathrm{d}\overline{\beta}(\lambda)$$

$$\leqslant \frac{C\overline{\beta}(R)}{R^3} + C \int_R^\infty \frac{1}{\lambda^3}\, \mathrm{d}\overline{\beta}(\lambda)$$

$$= 3C \int_R^\infty \frac{\overline{\beta}(\lambda)}{\lambda^4}\, \mathrm{d}\lambda$$

Therefore,

$$\int_K |\beta(u_n)| \leqslant |K|\overline{\beta}(R) + 3C \int_R^\infty \frac{\overline{\beta}(\lambda)}{\lambda^4}\, \mathrm{d}\lambda \equiv X + Y$$

We first fix R large enough so that $Y < \varepsilon/2$ (this is possible since $\overline{\beta}(\lambda)/\lambda^4 \in L^1(1,\infty)$ by (16)) and then we choose $|K|$ so small that $X < \varepsilon/2$.

4. A MIN-MAX PRINCIPLE FOR THE LAGRANGE MULTIPLIER

We use the same assumptions as in parts (A), and (B) (or (D) respectively) of Theorem 3. We assume in addition

(17)
$$\gamma \text{ is convex (i.e. } j' \text{ is concave)}$$
$$\gamma'(V) \in L^1_{\mathrm{loc}}$$

Theorem 4
Let $0 < I < I_0$ and let $\overline{\rho}, \lambda$ be the solution of (3). Then,

(18) $$\lambda = \max_{\rho \in K} \inf_{\substack{x \in \mathbb{R}^3 \\ [\rho > 0]}} \{ V - B\rho - j'(\rho) \}$$

(19) $$\lambda = \min_{\rho \in K} \sup_{x \in \mathbb{R}^3} \{ V - B\rho - j'(\rho) \}$$

In addition, the max in (18) and the min in (19) are achieved only when $\rho = \overline{\rho}$.

When $I = I_0$, then (18) and (19) still hold with $\lambda = 0$, but the uniqueness property fails in general.

Remark Formulae (18) and (19) were proved in [4] for the case where $j(r) = r^{5/3}$ and $V(x)$ is a Coulomb potential.
The uniqueness property was conjectured in [4] (Problem 4).

Sketch of the proof of Theorem 4: We first list some technical tools.
Tool 1 Assume $u \in M^3$ and $\Delta u \in L^1$ with $u(x) \geq 0$ a.e. for $|x|$ large, then

$$\int (-\Delta u) \geq 0$$

Proof: Recall that $u(x) \sim C/|x|$ as $|x| \to \infty$ where

$$C = \frac{1}{4\pi} \int (-\Delta u)$$

Therefore, if $\int(-\Delta u) < 0$ we get a contradiction.
Tool 2 Assume $u \in M^3$ and $\Delta u \in L^1$, $u(x) \geq 0$ a.e. for $|x| > R$, $-\Delta u \geq 0$ a.e. for $|x| > R$ and

$$\int_{\mathbb{R}^3} \Delta u = 0$$

Then, $u(x) = 0$ a.e. for $|x| > R$.
Proof: It follows from the strong maximum principle that either $u(x) = 0$ for $|x| > R$, or $u(x) \geq \varepsilon > 0$ for $|x| = R + 1$ and so, in fact,

$$u(x) \geq \frac{\varepsilon(R+1)}{|x|} \quad \text{for} \quad |x| > R + 1$$

(since u is superharmonic). By tool 1 we would have

$$\int_{\mathbb{R}^3} -\Delta \left(u - \frac{\varepsilon(R+1)}{|x|}\right) \geq 0$$

which is a contradiction since

$$\int_{\mathbb{R}^3} \Delta u = 0$$

Tool 3 (R. Jensen). Assume $u \in L^1_{\text{loc}}$, $u \geq 0$ a.e. $\Delta u \in L^1_{\text{loc}}$ is such that

$$-\Delta u + au \geq 0 \quad \text{a.e. on } \mathbb{R}^3$$

with

$$a \in L^1_{\text{loc}} \qquad a \geq 0 \quad \text{a.e.}$$

If u has compact support, then $u \equiv 0$.

In the case $a \in L^{\infty}_{\text{loc}}$, the conclusion follows from the strong maximum principle. If $a \in L^p_{\text{loc}}$ with $p > 3/2 = (n/2)$, the conclusion follows from a variant of the strong maximum principle (see [5]). We omit the proof in the general case. The following lemma plays a crucial role.

Lemma 4 Let $\tilde{f}, \hat{f}$ be measurable functions such that $\tilde{f} \leq \hat{f}$ a.e. Let $\tilde{\rho}, \hat{\rho} \in K$ be such that

$$\tilde{\rho} = \gamma(\tilde{f} - B\tilde{\rho})$$
$$\hat{\rho} = \gamma(\hat{f} - B\hat{\rho})$$

where γ is C^1 convex, $\gamma(t) = 0$ for $t \leq 0$ and $\gamma'(\tilde{f}) \in L^1_{\text{loc}}$.
Assume $\tilde{f}(x) \leq 0$ for $|x| > R$, then

$$\tilde{\rho} = \hat{\rho}$$

Sketch of the proof of Lemma 4: We first claim that $B\tilde{\rho} \leq B\hat{\rho}$. Indeed, let $\tilde{u} = B\tilde{\rho}$, $\hat{u} = B\hat{\rho}$ so that

$$-\Delta\tilde{u} = \gamma(\tilde{f} - \tilde{u}), \quad -\Delta\hat{u} = \gamma(\hat{f} - \hat{u})$$

It follows from Kato's inequality that

$$\begin{aligned}\Delta(\tilde{u} - \hat{u})^+ &\geq \Delta(\tilde{u} - \hat{u}) \ \text{sgn}^+ \ (\tilde{u} - \hat{u})\\ &= [-\gamma(\tilde{f} - \tilde{u}) + \gamma(\hat{f} - \hat{u})] \ \text{sgn}^+ (\tilde{u} - \hat{u})\\ &\geq 0 \ \text{ since } \tilde{f} \leq \hat{f}\end{aligned}$$

Therefore, $(\tilde{u} - \hat{u})^+ = 0$ and $\tilde{u} \leq \hat{u}$.

Next, set $v = \hat{u} - \tilde{u}$ so that $v \geq 0$ a.e. and

$$-\Delta v = \hat{\rho} - \tilde{\rho} = \hat{\rho} \geq 0 \quad \text{for} \quad |x| > R$$

(since $\tilde{\rho} = \gamma(\tilde{f} - B\tilde{\rho}) \leq \gamma(\tilde{f}) = 0$ for $|x| > R$). Also we have

$$\int_{\mathbb{R}^3} \Delta v = 0$$

and by tool 2 we find $v(x) = 0$ for $|x| > R$. Finally, it is easy to check that

$$-\Delta v + av \geq 0 \qquad \text{a.e.}$$

where a is defined by

$$a = \begin{cases} 1/v\,[\gamma(\tilde{f} - B\tilde{\rho}) - \gamma(\tilde{f} - B\hat{\rho})] & \text{on} \quad [v > 0] \\ 0 & \text{on} \quad [v = 0] \end{cases}$$

(so that $a \geqslant 0$ a.e. and $a \in L^1_{\mathrm{loc}}$ since $a \leqslant \gamma'(\widetilde{f})$). We conclude, using tool 3, that $V \equiv 0$ and therefore $\widetilde{\rho} = \hat{\rho}$.

Proof of Theorem 4

We shall prove (18) as well as the uniqueness property (the proof of (19) is similar).

Choosing $\rho = \overline{\rho}$ we clearly have

$$\max_{\rho \in K} \inf_{[\rho>0]} \{V - B\rho - j'(\rho)\} \geqslant \lambda$$

Assume now that for some $\rho \in K$ we have

$$\inf_{[\rho>0]} \{V - B\rho - j'(\rho)\} \geqslant \lambda \tag{20}$$

We claim that $\rho = \overline{\rho}$ (this will conclude the proof).

Indeed, set $\hat{f} = V - \lambda$, $\hat{\rho} = \overline{\rho}$; since $\overline{\rho}$ is a solution of the Thomas-Fermi equation we have

$$\hat{\rho} = \gamma(\hat{f} - B\hat{\rho})$$

Next set

$$\widetilde{f} = \begin{cases} j'(\rho) + B\rho & \text{on } [\rho > 0] \\ \min\{B\rho, V - \lambda\} & \text{on } [\rho = 0] \end{cases}$$

Set $\widetilde{\rho} = \rho$ and find

$$\widetilde{\rho} = \gamma(\widetilde{f} - B\widetilde{\rho})$$

We have (by (20)) $\widetilde{f} \leqslant \hat{f}$, $\hat{f}(x) \leqslant 0$ for $|x| > R$ (since $\lambda > 0$ and $V(x) \to 0$ as $|x| \to \infty$) and finally

$$\gamma'(\widetilde{f}) \leqslant \gamma'(\hat{f}) = \gamma'(V - \lambda) \leqslant \gamma'(V) \in L^1_{\mathrm{loc}}$$

Therefore, we may apply Lemma 4 and conclude that $\widetilde{\rho} = \hat{\rho}$, i.e. $\rho = \overline{\rho}$. In order to construct an example of non-uniqueness when $I = I_0$ we proceed as follows. We use the assumptions introduced in parts (*A*), (*B*), and (*C*) (or (*D*) respectively).

Assume also that $-\Delta V \geqslant 0$, and let u be the solution of $u \in M^3$

$$-\Delta u + \gamma(u) = -\Delta V$$

Let $\bar{\rho} = \gamma(u)$. We shall exhibit a function $\rho \in K$, such that $\rho \neq \bar{\rho}$ and

$$\inf_{[\rho>0]} \{V - B\rho - j'(\rho)\} = 0$$

Indeed, let $\nu \in M^3$ be the solution of

$$-\Delta\nu + \frac{1}{2}\gamma(\nu) = -\Delta V$$

If we set $\rho = \frac{1}{2}\gamma(\nu)$, it is not difficult to check that ρ satisfies the required properties.

REFERENCES

[1] A. Bamberger. 'Etude de deux équations non linéaires avec une masse de Dirac au second membre'. *Rapport No. 13 du Centre de Mathématiques Appliquées de l'Ecole Polytechnique,* Oct. 1976.

[2] Ph. Benilan, and H. Brezis. Detailed paper on the Thomas-Fermi equation. To appear.

[3] Ph. Benilan, H. Brezis, and M. Crandall. 'A semilinear equation in L^1'. *Ann. Scu. Norm. Sup., Pisa,* 2 (1975), 523-55.

[4] E. Lieb, and B. Simon. 'The Thomas-Fermi theory of atoms, molecules and solids.' *Adv. Math.,* **23** (1977), 22-116.

[5] G. Stampacchia. '*Equations elliptiques du second ordre à coefficients discontinus.*' Presses de l'Université de Montreal (1966).

Chapter 5

Implicit Complementarity Problems and Quasi-Variational Inequalities

I. Capuzzo Dolcetta and U. Mosco

1. INTRODUCTION

The aim of this paper is to point out some connections between *quasi-variational inequalities* (in short, QVI) and what we call *implicit complementarity problems* (in short, ICP); the main feature of these problems is a sort of 'circularity' between solutions and constraints, in the sense that the constraints involved in the problem are not given among the data but depend on the solution itself.

We discuss some theoretical problems concerning existence, regularity and numerical approximation of the solution by referring to some of the examples recently considered in the literature.

Due to the introductory character of the presentation, we will refer for proofs and details to some of the essential literature on the subject without, however, any attempt at completeness.

2. THE STRONG ICP

Let H be a real Hilbert space and a vector lattice under the partial order $\leqslant$ induced by a closed positive cone; for given $f \in H$ and mappings

$$M : H \to H, \qquad A : H \to H,$$

consider the following problem, that will be referred to in what follows as the *strong implicit complementarity problem* (strong ICP):

$$\text{(1)} \qquad \begin{aligned} &\text{find } u \in H \\ &\text{such that} \\ &u \leqslant M(u) \quad A(u) \leqslant f, \quad (A(u) - f, u - M(u)) = 0 \end{aligned}$$

where $(\cdot\,,\,\cdot)$ is the inner product of H.

Problem (1) could be viewed as a generalized complementarity problem for the pair A, $B = I - M$. This, however, will not be our approach to (1), since in most infinite-dimensional problems of this type arising in the applications, such as those related to boundary-value problems for partial differential operators, the roles played by M and A are quite different, as will be made clear by some examples in section 5.

Let us observe, instead, that if one considers for every fixed w in H the standard complementarity problem (in short, CP):

$$\text{(2)} \qquad \begin{array}{l} \text{find } z \in H \\ \text{such that} \\ z \leqslant M(w)\,,\; A(z) \leqslant f, \quad (A(z) - f,\, z - M(w)) = 0 \end{array}$$

then, a (possibly) set-valued mapping

$$w \to z = S(w)$$

in naturally defined.

Due to the fact that (2) can be interpreted (at least when A is the Gateaux differential of some functional J) as the system of Euler inequalities of the constrained variational problem:

$$\text{minimize } \{J(v) - (f, v)\} \quad \text{on} \quad K = \{v \in H : v \leqslant M(w)\}$$

We shall call S the *variational selection* associated with the strong ICP.

It is then immediate to check that a vector u is a solution of (1), if and only if u satisfies the *fixed point equation*:

$$u \in H : u = S(u)$$

3. QVIs AND THE WEAK ICP

As we have already noticed, the simple formulation of the ICP given in section 2 must be modified in order to take into account some boundary-value problems with implicit unilateral constraints arising from applications. To this end consider two spaces H and V such that

$$V \subset H \subset V'$$

where H is as in section 2; V is a closed subspace and a sublattice of H; V' is the dual space of V, equipped with the dual order; $M : H \to H$ is an admissible obstacle operator (that is, for every $w \in H$ there exists $v \in V$ such that $v \leq M(w)$); and $a(\cdot, \cdot)$ is a real functional on $V \times V$ which is linear with respect to the second argument.

In this context we can state the *quasi-variational inequality* (QVI) problem as:

$$
\text{(3)} \qquad \begin{aligned} &\text{find } u \in V \\ &\text{such that} \\ &a(u, v - u) \geq (f, v - u), \quad \forall v \in V \\ &v \leq M(u) \quad u \leq M(u) \end{aligned}
$$

Problems of this type were originally introduced by Bensoussan and Lions [4, 5, 6] in connection with some stochastic impulse control problems. (See also [1] for a different type of application.)

If $a(\cdot, \cdot)$ satisfies suitable continuity properties, then the identity

$$a(u, v) = \langle A(u), v \rangle$$

where $\langle \cdot, \cdot \rangle$ is the duality pairing between V' and V, defines A as an operator acting continuously from V to V'.

In this case one can show, through appropriate choices of the 'test' function v in (3), the equivalence between the QVI (3) and the following *weak implicit complementarity problem*:

$$
\text{(4)} \qquad \begin{aligned} &\text{find } u \in V \\ &\text{such that} \\ &u \leq M(u),\ A(u) \leq f, \quad \langle A(u) - f, u - M(u) \rangle = 0 \end{aligned}
$$

It should be noticed that the symbol $\leq$ appears in (4) with two different meanings: the inequality $u \leq M(u)$ is that of H while in $A(u) \leq f$ it is that of V'.

Let us observe, however, that, if it is known that any solution u of (3) or (4) is 'regular' in the sense that

$$A(u) \in H$$

then QVI, weak ICP and strong ICP are equivalent problems. This corresponds, in this more general case, to the well known relationship existing between complementarity problems and variational inequalities over convex cones [13, 25, 26].

In the same way as in section 2, the QVI and the weak ICP can be treated as a fixed-point problem for the associated variational selection S.

To conclude this section, we want to point out that all the problems considered up to now can be seen as particular cases of general *fixed-points selection problems* as the following *Ky-Fan's implicit inequality*:

$$\text{(5)} \qquad \begin{array}{l} \text{find } u \in V \\ \text{such that} \\ u \in Q(u),\ g(u,v) \leqslant 0,\ \forall\, v \in Q(u) \end{array}$$

where Q is a set-valued mapping defined on a subset C of V and g is some real function on $V \times V$.

The case of QVIs is easily obtained by taking in (5)

$$Q(u) = \{v \in V : v \leqslant M(u)\}$$
$$g(u,v) = \langle A(u), u - v\rangle - \langle f, u - v\rangle$$

Another class of problem that can be put in the form (5) is, for example, those related to the *non-cooperative Arrow-Debreu-Nash equilibria.*

For more details on the Ky-Fan's inequality and other general implicit problems see [27, 28, 21].

4. THE YOUNG-FENCHEL EQUATION AND QVIs IN DUALITY

In this section we want to point out that a 'dual' characterization of the solutions of a QVI can be obtained in the framework of Fenchel's duality theory [29].

To state the result in the simplest form, let us assume more 'regularity' of the obstacle operator, namely that $M(H) \subseteq V$; assume also A to be a 1-1 mapping from V to V'.

Define $A' : V' \to V$ and $M : V' \to V$ by

$$A'(v') = -A^{-1}(-v') \qquad \text{(if } A \text{ is linear then } A' = A^{-1})$$
$$M'(v') = M(-A'(v'))$$

and denote by P (respectively P') the negative cone of V (respectively its polar cone).

If $\delta_{M(u)+P}$ and $\delta_{P'}$, are the indicator functions of the convex cones $M(u) + {}$ $+ P$ and P', then the following problems (i), (ii) are equivalent to the QVI (3), provided u and u' are related by $u' = -A(u)$:

(i) Young-Fenchel equation:

$$u \in V, u' \in V' \quad : \delta_{M(u)+P} + \delta_{P'}(u') = \langle u', u - M(u)\rangle$$

(ii)
$$u' \in P'$$
$$\langle A'(u') + M'(u'), v' - u' \rangle \geq 0, \ \forall v' \in P'$$

(see [14]). It should be noticed that the 'dual' problem (ii) has a quite different nature from the 'primal' QVI: actually in (ii) the constraints *do not depend* on the solution (see [14] for applications of this fact to the numerical analysis of some QVIs).

5. EXISTENCE OF SOLUTIONS

We now specialize the general setting of the previous sections to the case where $H = L^2(\Omega)$ (Ω is a bounded open subset of $\mathbb{R}^N$, with smooth boundary) equipped with the almost everywhere order, $V = H^1(\Omega)$ or $V = H_0^1(\Omega)$, and A is a second-order partial differential operator of the form:

(*)
$$A = -\sum_{i,j=1}^{N} \frac{\partial}{\partial x_i}\left(\alpha_{ij}(x)\frac{\partial}{\partial x_j}\right) + c(x)$$

with bounded measurable coefficients satisfying

$$\alpha_0 \sum_{i=1}^{N} \xi_i^2 \leq \sum_{i,j=1}^{N} \alpha_{ij}(x)\,\xi_i\,\xi_j \leq \alpha_1 \sum_{j=1}^{N} \xi_i^2 \qquad \forall \xi \in \mathbb{R}^N$$

$$c(x) \geq c_0 > 0$$

for some positive constants α_0, α_1, a.e. in Ω.

The purpose is to give a brief account of some existence results for the ICP in a unified, though not fully general, form.

5.1. Constructive methods

The formulation of the ICP as a fixed-point problem (see section 2 and 3) suggests the possibility of finding a solution by means of an iterative scheme such as:

(6)
$$u^0 \in H$$
$$u^n = S(u^{n-1})$$

An obvious case in which the sequence $\{u^n\}$ defined by (6) will converge to the (unique) fixed point of S is when S happens to be a contraction:

this is, in fact, the case when M satisfies a Lipschitz condition with a suitably small constant (see [11] for some examples and applications).

A different situation has been considered by Bensoussan and Lions [4, 5, 6]. They prove that, if M satisfies some order and continuity properties, namely:

$$
(7) \qquad \begin{array}{l} u \leqslant v \Rightarrow M(u) \leqslant M(v) \\ 0 \leqslant v \Rightarrow 0 \leqslant M(v) \\ \left.\begin{array}{l} v^n \searrow v \text{ weakly in } H^1(\Omega) \\ v^n \leqslant M(v^{n-1}) \end{array}\right\} \Rightarrow v \leqslant M(v) \end{array}
$$

then the scheme (6), starting from the solution u^0 of the Neumann problem for A, generates a non-increasing sequence $\{u^n\} \subset H^1(\Omega)$ which is weakly convergent to a fixed point of S or, which is the same, to a solution of (3).

5.2. Non-constructive methods

These methods involve the possibility of proving the existence of solutions to the weak ICP by means of some general fixed-point theorem.

Roughly speaking, we can distinguish between *order methods* and *topological methods* which refer, respectively, to the Birkhoff-Tartar theorem for increasing mappings on order inductive sets (see [26, 35]) and to the Schauder-Tikhonov theorem; it must be noted, however, that in some applications the two approaches are combined (see Examples 2 and 4 below).

The main tool in the order methods are the comparison theorems for variational inequalities, related to the maximum principle property of second-order elliptic operators. On the other hand, the crucial point in the topological methods is to verify the compactness and continuity conditions required by the Schauder-Tikhonov theorem.

We give now some examples of the application of these methods.

Example 1

The QVI of the stochastic impulse control problem [2, 3, 4, 5, 24].

The obstacle operator M is given by

$$
M(v)(x) = 1 + \inf_{\substack{\xi \geqslant 0 \\ x+\xi \in \Omega}} v(x+\xi)
$$

and $a(\cdot, \cdot)$ is the bilinear form associated with the operator A defined in (*). The problem is to find a function u such that:

$$
(8) \qquad \begin{array}{l} u \in H^1(\Omega) \cap L^\infty(\Omega) \\ a(u, v-u) \geqslant (f, v-u), \quad \forall\, v \in H^1(\Omega) \\ v \leqslant M(u) \quad u \leqslant M(u) \end{array}
$$

If $f \in L^{\infty}(\Omega)$, $f \geqslant 0$, then one can show, using the fact that M is increasing with $M(v) \geqslant 0$ for $v \geqslant 0$ and the comparison theorems, that

$$S \text{ is increasing}$$
$$0 \leqslant S(0) \leqslant S(u^0) \leqslant u^0$$

where $u^0 \geqslant 0$ is the solution of the Neumann problem. These facts imply, via the already mentioned Birkhoff-Tartar theorem, the existence of a fixed point for S in the order interval $[0, u^0]$ and hence of a solution of (8), which is unique (see [37]).

It should be noted that the operator M considered in this example satisfies condition (7), so the existence of a solution of (8) can be proved also by the monotonic iteration methods described in section 5.1.

Example 2
The 'dual estimate method' for the QVI of the stochastic impulse control [22].

The existence of regular solution of (8) has been proved by Joly, Mosco, and Troianiello [22] for an operator A with constant coefficients. They show that the convex set

$$C = \{v \in H_0'(\Omega) \cap H^{2,p}(\Omega) : f \geqslant Av \geqslant \min [0, \inf_{\substack{\xi \geqslant 0 \\ x+\xi \in \Omega}} f(x+\xi)]\}$$

is, under suitable assumptions on f, a non-empty bounded subset of $H^{2,p}(\Omega)$ which is invariant with respect to S. The main tool used in proving this fact is the following *dual estimate* for the solution of a variational inequality with obstacle Ψ:

$$f \geqslant Av \geqslant \inf \{f, A\Psi\}$$

which holds in the sense of measures if $\Psi \in H^1(\Omega) \cap C(\overline{\Omega})$ and $A\Psi$ is a measure in $H^{-1}(\Omega)$ (see, for instance, [27]).

Since S is weakly continuous on sequences in C, the existence of a fixed point follows from the Schauder-Tikhonov theorem.

For an extended review of the applications of this methods we refer the reader to [17] an the bibliography therein.

Example 3
Suppose M is a Hammerstein integral operator of the form

$$M(v)(x) = \int_{\Omega} K(x,y) g(y, v(y)) \, dy$$

where g is a Caratheodory function such that

$$|g(x,t)| \leqslant \alpha(x) + \beta |t|, \quad \alpha \in L^2(\Omega) \qquad \beta > 0$$

and K is an L^2 kernel with $\nabla_x K \in L^2(\Omega \times \Omega)$.

Under these conditions M is a continuous operator from $L^2(\Omega)$ into $H^1(\Omega)$. If

$$\beta \left(\int_\Omega \int_\Omega (|K|^2 + |\nabla_x K|^2)\, dx\, dy \right)^{1/2} < \frac{\min \{\alpha_0, c_0\}}{\max \{\alpha_1, \|c\|_{L^\infty}\}}$$

it can be shown that there exists $R > 0$ such that

$$\|S(w)\|_{H^1} \leqslant R \quad \text{for all } w \text{ with } \|w\|_{H^1} \leqslant R \tag{9}$$

Using some results on the continuous dependence of the solution of a variational inequality upon the obstacle (see [33]), one can prove the continuity of the variational selection S of the QVI relative to M as a mapping from $L^2(\Omega)$ into itself. This, together with the *a priori* estimate (9), allows one to apply the Schauder fixed-point theorem to get the existence of a fixed point of S in the ball $B_R \subset H^1(\Omega)$ (see [11] for further details and an other example of the same kind).

It is worthwhile to observe that this approach is independent of the order properties of M and A and thus can be extended to non-monotonic operators of the form

$$A(v) = -\sum_{i=1}^{N} \frac{\partial}{\partial x_i} \left(\alpha(x,v)\, |\nabla v|^{p-2} \frac{\partial v}{\partial x_i} \right) + C(x,v)\, |v|^{p-2}\, v$$

(see [12]). As for non-increasing M, we mention that a QVI for the operator

$$M(v)(x) = K(x) - \int_\Omega v dx$$

has been studied by topological methods in [7, 21].

Example 4

A QVI with implicit obstacle on the boundary (see [21, 26] for details and physical motivation).

The problem is to find a function u such that

$$
(10) \qquad \begin{aligned} & u \in H^1_A(\Omega), \quad u \geqslant h - \langle \varphi , \partial u/\partial n \rangle_{\partial\Omega} \\ & a(u, v-u) \geqslant (f, v-u) \\ & \forall\, v \in H^1(\Omega), \; v \geqslant K - \langle \varphi, \partial u/\partial n \rangle_{\partial\Omega} \end{aligned}
$$

where $H^1_A(\Omega) = \{v \in H^1(\Omega) : A\,v \in L^2(\Omega)\}$, $\partial/\partial n$ is the conormal derivative relative to A which is a bounded linear operator from $H^1_A(\Omega)$ to $H^{-1/2}(\partial\Omega)$, h and φ are given in $H^{1/2}(\partial\Omega)$, f is given in $L^2(\Omega)$.

It is possible to show that, if $\varphi \geqslant 0$ a.e. on $\partial\Omega$, then

$$
D_0 = \{v \in H^1_A(\Omega) : Av = f \text{ in } \Omega, \; \frac{\partial v}{\partial n} \geqslant 0 \text{ a.e. on } \partial\Omega\}
$$

is a non-empty bounded convex subset of $H^1_A(\Omega)$ which is invariant under the selection S associated with (10); moreover, S satisfies the continuity properties needed for the application of a fixed-point argument.

A different approach has been adopted in [12] where the existence of a solution to (10) in proved under the assumption that the norm of φ is sufficiently small (see also [18]).

Remark Many of these examples can also be treated in the case of a system of QVIs; a system of a different type is that arising in the study of Nash equilibria where $M : [L^2(\Omega)]^\ell \to [L^2(\Omega)]^\ell$ is given by

$$
M_i(v) = \beta_i - \sum_{j \neq i} v_j\,, \qquad i,j = i, \ldots, \ell
$$

(see [6, 19, 21] for a similar system).

6. REMARKS ON STABILITY, NUMERICAL APPROXIMATION, AND ALGORITHMS

We give in this section a brief account of some results and problems concerning the stability of solutions of ICP, their approximation, and the algorithms for the numerical solution of the finite-dimensional approximate problems.

6.1. Stability with respect to G-convergence

Let us consider a sequence $(A^n)(u = 1, 2, \ldots, \infty)$ of second-order operators of the form

$$
A^n = -\sum_{i,j=1}^{N} \frac{\partial}{\partial x_i}\left(\alpha^n_{ij}(x)\,\frac{\partial}{\partial x_j}\right) + c^n(x)
$$

whose coefficients satisfy condition (*) of section 5 uniformly in n.

The behaviour in the L^2 topology of the solutions of the corresponding sequence of ICPs with respect to the G-convergence of the sequence (A^n) (see [34, 8]) has been studied by various authors. We mention [10] for the case treated in Example 1, [11] for Example 3, and [6] for the system of Nash equilibria.

6.2. Numerical approximation and algorithms

Consider in relation to Example 1, the iterative methods (see section 5.1):

$$\begin{aligned} &u^n \leqslant M(u^{n-1}) \\ &a(u^n, v-u^n) \geqslant (f, v-u^n), \quad \forall\, v \leqslant M(u^{n-1}) \end{aligned}$$

starting from the solution u^0 of the Neumann problem. As already mentioned $\{u^n\}$ is a monotonically increasing sequence which converges to the solution u of (8); moreover, the following estimate holds (see [20]):

$$\|u^n - u\|_{L^\infty} \leqslant (1-\alpha)^n \, \|u^0 - \bar{u}\|_{L^\infty}$$

where

$$\bar{u} = \frac{1}{c_0} \min(\inf f, 0), \quad \alpha = \frac{c_0}{2\,\|f\|_{L^\infty}}$$

For what concerns the approximation of u^n one can solve the variational inequality

$$\text{(11)} \qquad \begin{aligned} &u_h^n \in V_h\,,\ u_h^n \leqslant M(u_h^{n-1}) \\ &\alpha(u_h^n, v_h - u_h^n) \geqslant (f_h, v_h - u_h^n) \\ &\forall\, v_h \in V_h\,,\ v_h \leqslant M(u_h^{n-1}) \end{aligned}$$

where $V_h (h > 0)$ is, for example, the finite-dimensional subspace of $H^1(\Omega) = V$ spanned by the continuous functions which are piecewise affine on a triangulation of mesh h of Ω. Error estimates are available in this case with respect to the energy norm; in obtaining these estimates an important role is played by *regularity* (see [30]).

Various methods have been proposed for the numerical solution of (11). we mention only those related to relaxation with projection [24] and the complementarity pivot algorithms (see [31] and various contributions in this volume).

A direct approximation of the QVI by the methods described would give rise, one the other hand, to finite-dimensional ICPs of the form

$$(12)\qquad \begin{aligned} &u_h \leqslant M_h(u_h),\ u_h \in \mathbb{R}^{N_h} \\ &A_h u_h \leqslant f_h \\ &(A_h u_h - f_h) \cdot (u_h - M_h(u_h)) = 0 \end{aligned}$$

where $\cdot$ denotes the inner product of $\mathbb{R}^{N_h}$.

If the triangulation of Ω satisfies some conditions (see [31]) then the matrix A_h is of Minkowski type (that is, A has positive principal minors and non-positive off-diagonal elements) and is sparse.

A numerical method for the solution of (12) could consist of selecting between the fixed points u_h of the convex-valued mapping

$$v_h \to Q_h(v_h) = \{w_h \in \mathbb{R}^{N_h} : w_h \leqslant M_h(v_h)\}$$

those which minimize the quadratic function

$$J_h(v_h) = \frac{1}{2} A_h v_h \cdot v_h - f_h \cdot v_h$$

on $Q_h(u_h)$. This leads eventually to a combined use of optimization algorithms and simplicial procedures for the search of fixed points (see [36] for a wide review on this subject).

REFERENCES

[1] C. Baiocchi, and A. Capelo. '*Disequazioni variazionali e quasi variazionali. Applicazioni a problemi di frontiera libera*.' vol. I e II, Pitagora, Bologna (1978).

[2] A. Bensoussan. 'Contrôle impulsionnel et inéquations quasi variationnelles'. *Congrès Unione Mat. Int.*, Vancouver, 1974.

[3] A. Bensoussan. 'Contrôle impulsionnel et temps d'arrêt: inéquations variationnelles et quasi variationnelles d'évolution'. *Lecture Notes at the 'Scuola di Matematica'*. Erice, 1975.

[4] A. Bensoussan, and J. L. Lions. '*Temps d'arrêt optimaux et contôle impulsionnel*.' vol. 3. To appear.

[5] A. Bensoussan, and J. L. Lions. 'Nouvelle formulation des problèmes de contrôle impulsionnel et applications'. *C. R. Acad. Sci., Paris,* **276** (1973).

[6] A. Bensoussan, and J. L. Lions. 'Inéquations quasi variationnelles dépendant d'un paramètre'. *Ann. Scu. Norm. Sup., Pisa,* serie IV, vol. IV (1977).

[7] A. Bensoussan, and J. L. Lions. 'Propriétés des inéquations quasi variationnelles décroissantes'. *Lecture Notes in Economics and Mathematical Systems,* No. 102, Springer-Verlag, Berlin (1974).

[8] A. Bensoussan, J. L. Lions, and G. Papanicolau. '*On some asymptotic problems, homogenization, averaging and applications*.' North-Holland, Amsterdam (1978).

[9] M. Biroli. 'An estimate on convergence for an homogeneization problem for variational inequalities'. *Rend. Sem. Mat. Padova.* To appear.

[10] M. Biroli. 'Sur la *G*-convergence pour des inéquations quasi variationnelles'. *C. R. Acad. Sci., Paris,* **284** (1977).

[11] L. Boccardo, and I. Capuzzo Dolcetta. 'Disequazioni quasi variazionali con funzione d'ostacolo quasi limitata: esistenza di soluzioni e *G*-convergenza'. *Boll. Unione Mat. Ital.* To appear.

[12] L. Boccardo, and I. Capuzzo Dolcetta. 'Esistenza e omogeneizzazione delle soluzioni di alcune disequazioni quasi variazionali non lineari'. To appear.

[13] I. Capuzzo Dolcetta. 'Sistemi di complementarità e disequazioni variazionali'. *Tesi,* Univ. di Roma (1972).

[14] I. Capuzzo Dolcetta, M. Matzeu, and U. Mosco. 'A dual approach to the numerical solution of some quasi variational inequalities'. To appear.

[15] R. Chandrasekaran. '*A special case of the complementarity pivot theory*.' Opsearch (197).

[16] A. Fusciardi, U. Mosco, F. Scarpini, and A. Schiaffino. 'A dual method for the numerical solution of some variational inequalities'. *J. Math. Appl.* **40** (1972).

[17] M. G. Garroni, and G. M. Troianello. 'Some regularity results and a priori estimates for solutions of variational and quasi variational inequalities. A survey'. *Colloquio Metodi recenti di Analisi non lineare e Applicazioni,* Roma (1978).

[18] M. G. Garroni. 'A nonlinear quasi variational inequality with implicit obstacle on the boundary'. To appear.

[19] M. G. Garroni, B. Hanouzet, and J. L. Joly. 'Regularité pour la solution d'un système d'inéquations quasi variationnells'. To appear.

[20] B. Hanouzet, and J. L. Joly. 'Convérgence uniforme des itérés définissant la solution d'une inéquations quasi variationnelle abstraite'. *C. R. Acad. Sci., Paris,* **286** (1978).

[21] J. L. Joly, and U. Mosco. 'A propos de l'existence et de la régularité des solutions de certaines inéquations quasi variationnells'. *J. Funct. Anal..* To appear.

[22] J. L. Joly, U. Mosco, and G. M. Troianiello. 'On the regular solution of a quasi variational inequality connected to a problem of stochastic impulse control'. *J. Math. Anal. & Appl.,* **61** (1977).

[23] J. L. Lions. 'Sur la theorie du contrôle'. *Congrès Unione Mat. Int.,* Vancouver (1974).

[24] J. L. Lions. '*Sur quelques question d'analyse, de mécanique et de contrôle optimal*.' Presses de l'Université de Montreal (1976).

[25] J. J. More. 'The application of variational inequalities to complementarity systems and existence theorems'. *Tech. Rep.,* No 71-110, Cornell Univ., Ithaca (1972).

[26] U. Mosco. 'An introduction to the approximate solution of variational inequalities'. *Centro Internazionale Matematico Estivo (CIME), 1971,* Cremonese, Roma (1973).

[27] U. Mosco. 'Implicit variational problems and quasi variational inequalities'. *Lecture Notes in Mathematics,* vol. 543, Springer-Verlag, Berlin (1976).

[28] U. Mosco. 'Alcuni aspetti della teoria delle disequazioni variazionali e quasi variazionali e delle loro applicazioni'. *Lecture at GNAFA,* Rimini (1978).

[29] U. Mosco. 'Dual variational inequalities'. *J. Math. Anal. & Appl.* **39** (1972).

[30] U. Mosco. 'Error estimates for some variational inequalities'. *Meeting on Mathematical Aspects of the Finite Element Method,* Roma (1975).

[31] U. Mosco, and F. Scarpini. 'Complementarity system and approximation of variational inequalities'. *R.A.I.R.O.*, (DUNOD) 9th year, R-1 (1975).

[32] U. Mosco, and G. Strang. 'One sided approximation and variational inequalities'. *Bull. Am. Math. Soc.*, **80** (1974).

[33] U. Mosco. 'Convergence of convex sets and of solution of variational inequalities'. *Adv. Math.* **3** (1969).

[34] S. Spagnolo. 'Convergence in energy for elliptic operator'. *SYNSPADE* (1975).

[35] L. Tartar. 'Inéquations quasi variationnelles abstraites'. *C. R. Acad. Sci., Paris,* **278** (1974).

[36] M. J. Todd. 'The computation of fixed points and applications'. *Lecture Notes in Economics and Mathematical Systems,* **No.** 124, Springer Verlag, Berlin (1976).

[37] T. Laetsch. 'A uniqueness theorem for elliptic quasi variational inequalities'. *J. Funct. Anal.* **18** (1975).

Chapter 6

A Minimum Time Problem

R. Conti

1. INTRODUCTION

We are given a linear finite-dimensional deterministic control process represented by a family of ordinary autonomous differential equations

$$\mathrm{d}x/\mathrm{d}t - Ax = Bu \qquad (A, B)$$

depending on the parameter u. Here $t \geq 0$ is the time; $x \in \mathbb{R}^n$, the state of a system; $u : t \to u(t)$, $u(t) \in \mathbb{R}^m$, the control on the system. Consequently A and B are, respectively, $n \times n$ and $n \times m$ real matrices.

The set U_Ω of admissible controls is defined by

$$U_\Omega = \{u \in L^\infty_{\mathrm{loc}}(\mathbb{R}, \mathbb{R}^m) : u(t) \in \Omega, \text{ a.e. } t \geq 0\}$$

(a.e. = almost everywhere) where Ω is a compact convex set of $\mathbb{R}^m$ such that

$$0 \in B\Omega \tag{1}$$

The problem is that of driving x along a solution of (A, B) from an initial position $v \in \mathbb{R}^n$ to the origin $0 \in \mathbb{R}^n$ in the shortest possible (i.e., minimum) time by a proper choice of $u \in U_\Omega$.

2. TRANSFER SETS

We start by defining the set $V(t)$ of points from which the transfer to 0 is possible at a time less than or equal to t.

Since the solutions of (A, B) are represented by

$$t \to x(t, v, u) = \mathrm{e}^{tA} \left(v + \int_0^t \mathrm{e}^{-sA} Bu(s)\, \mathrm{d}s \right)$$

$V(t)$ is defined by

$$V(t) = \left\{ - \int_0^t e^{-sA} Bu(s)\, ds \; : \; u \in U_\Omega \right\}$$

Further, the set V of points v from which the transfer to 0 is possible at all by using U_Ω is defined as

$$V = \bigcup_{t \geq 0} V(t)$$

If S denotes the space generated by V, i.e. the set of finite linear combinations of points in V, we obviously have

$$V \subset S \subset \mathbb{R}^n$$

and it is of interest to consider the cases $S = \mathbb{R}^n$, $V = S$, $V = S = \mathbb{R}^n$

It is easily seen that the orthogonal complement $S^\perp$ to S is

$$S^\perp = \{ y \in \mathbb{R}^n : y^* [B \quad AB \quad \ldots \quad A^{n-1}B] = 0 \}$$

Therefore, $S = \mathbb{R}^n$ is equivalent to the Kalman condition

$$(P_1) \qquad \text{rank } [B \quad AB \quad \ldots \quad A^{n-1}B] = n$$

$V(t)$ is a convex set (even if Ω is not) and, because of (1), we have

$$t' < t'' \Rightarrow V(t') \subset V(t'')$$

Therefore, V is also convex.

For $x = (x_1, \ldots, x_n)$, let

$$|x|_1 = \sum_{i=1}^{n} |x_i|$$

Then, it can be shown that, if

$$(2) \qquad \Omega = \{ \omega \in \mathbb{R}^m : |\omega_j| \leq 1 \}$$

we have

$$V = S \Longleftrightarrow (P_2')$$

where

$$(P_2') \qquad \int_0^{+\infty} |B^* e^{-sA^*} y|_1 \, ds = +\infty, \qquad y \in \mathbb{R}^n \backslash S^\perp$$

and

$$V = S = \mathbb{R}^n \iff (\mathrm{P}_2'')$$

where

$$(\mathrm{P}_2'') \qquad \int_0^{+\infty} |B^* e^{-sA^*} y|_1 \, ds = +\infty, \qquad y \in \mathbb{R}^n \setminus \{0\}$$

This is equivalent, in turn, to (P_1) plus the condition that all the real parts of the characteristic roots of A are less than or equal to 0.

3. THE FUNCTION T

For each $v \in V$, we can define the real function $T: v \to T(v)$ by

$$T(v) = \inf \{t \geqslant 0 : v \in V(t)\}$$

In case (2), we have

$$T(0) = 0 < T(v) = T(-v)$$
$$T(v) = \min \{t \geqslant 0 : v \in V(t)\}$$

meaning that the points v, from which x can be transferred into 0 in minimum time, are just the points $v \in V$.

Further,

$$V(t) = \{v \in V : T(v) \leqslant t\}$$

and the boundary of $V(t)$ relative to S is given by

$$\partial_S V(t) = \{v \in V : T(v) = t\}$$

and is called an isochrone.

T is a continuous function on V and

$$T(V) = \mathbb{R}^+ \qquad \text{if } \partial_S V \neq \emptyset$$

Given $v \in \mathbb{R}^n$ there are, in case (2) at least, iterative procedures for recognizing whether $v \in V$ and, if so, for evaluating $T(v)$.

When $T(v)$ is known, one has to solve the problem of determining the corresponding minimum controls, which is not an easy task, even when there is a unique minimum control for each v.

A necessary condition for u to be minimum is represented, in case (2), by the following particular formulation of Pontryagin's maximum principle: for every $v \in V$ there exist $y(v) \in \mathbb{R}^n \backslash S^{\perp}$ such that if u is minimum relative to v, then

$$[B^* e^{-sA^*} y(v)]^* u(s) = |B^* e^{-sA^*} y(v)|_1, \qquad 0 \leqslant s \leqslant T(v).$$

The equation

$$[B^* e^{-sA^*} y]\, u(s) = |B^* e^{-sA^*} y|_1$$

has a unique solution u for every $y \neq 0$ iff

$$(\mathrm{P}_3) \qquad \det (b^j \quad Ab^j \quad \ldots \quad A^{n-1} b^j) \neq 0, \quad j = 1, \ldots, m$$

holds, where b^j is the jth column of B.

Condition (P_3) is the LaSalle normality condition with respect to the cube Ω: it is also equivalent to $V(t)$, being strictly convex.

When (P_3) is satisfied, then the (unique) minimum control is defined by

$$u_j(t) = \operatorname{sgn} \; b^{j*} e^{-tA^*} y(v), \qquad j = 1, \ldots, m$$

where $y(v)$ is any non-zero solution of

$$\int_0^{T(v)} |B^* e^{-sA^*} y(v)|_1 \, ds = - y^*(v)\, v.$$

To determine $y(v)$, there are several iterative methods, among which is the Neustadt-Eaton procedure.

4. REGULARITY OF T

A great deal of research has been done recently, and is still in progress, on the regularity properties of T. This is due to the fact that if the gradient $T_v(v)$ of T exists, then it is a positive multiple of the vector $y(v)$, the normal at v to the isochrone through v.

Further, T can be obtained by solving Bellman's equation, which in case (2) is

$$\max_{\omega \in \Omega} \; T_v(v)\, B\omega = 1 + T_v(v)\, Av$$

and/or Hájek's equation

$$\max_{\omega \in \Omega} T_v(v)\ e^{-T(v)A}\ B\omega = 1$$

Unfortunately, the regularity properties of T are rather scarce.
In fact, all we can say is that T has radial derivatives, finite or not, at each point of V. Sufficient conditions, however, are known for the existence of $T_v(v)$ and in order to ensure that

$$T \in C^{(1)}(V \setminus \{0\})$$

This last property is related to the absence of corner points of the set $B\Omega$, and it does not hold, for instance, in case (2).

Finally, it can be shown that

$$T \in Lip_{loc}(V \setminus \{0\}) \iff 0 \in \text{int}\ B\Omega.$$

5. THE SYNTHESIS PROBLEM

Together with research on the regularity of T, recent papers have appeared on the synthesis problem. Generally speaking, this means defining $f : v \to f(v)$ such that the minimum path from v to 0 coincides with the orbit of the autonomous differential equation

$$dx \setminus dt = Ax + Bf(x)$$

However, here we shall not consider this problem as it is a very difficult one.

REFERENCES

Books:

[1] M. Athans, and P. I. Falb.'*Optimal control*.' McGraw-Hill, New York (1966).

[2] V. G. Boltyanskii.'*Mathematical methods of optimal control*.' Holt-Rinehart-Winston, New York (1971).

[3] R. Conti. '*Problemi di controllo e di controllo ottimale*.' UTET, Torino (1974).

[4] A. Fel'dbaum. '*Principes théoriques des systèmes asservis optimaux*'. MIR, Moscou 1973.

[5] H. Hermes, and J. P. LaSalle. '*Functional analysis and time-optimal control*.' Academic Press, New York (1969)

[6] N. E. Kirin.'*Numerical methods in the theory of optimal control*.' (in Russian.) Izd Leningrad Univ., (1968).

[7] N. N. Krasovskii.'*Theory of the control of motion*.' (in Russian.) Nauka, Moscow (1968).

[8] E. B. Lee, and L. Markus.'*Foundations of optimal control theory*.' Wiley & Sons, New York and Chichester (1967).

[9] L. S. Pontryaghin et al..'*The mathematical theory of optimal processes*.' Interscience, New York (1962).

Papers:

[10] H. A. Antosiewicz. *Arch. Ration. Mech. Anal.*, **12** (1963), 313-24.

[11] T. G. Babunashvili. *Dolk. Akad. Nauk SSSR.*, **155** (1964), 295-8.

[12] R. Bellman, I. Glicksberg, and O. Gross. *Q. Appl. Math.*, **14** (1956), 11-8.

[13] A. A. Belolipetskii. *Zh. Vychisl. Mat. & Mat. Fiz.*, **13** (1973), 1319-23.

[14] P. Brunovsky. *Math. Slovaca*, **28** (1978), 81-100.

[15] J. H. Eaton. *J. Math. Anal. & Appl.*, **5** (1962), 329-44.

[16] R. P. Fedorenko. *Zh. Vychisl. Mat. & Mat. Fiz.*, **7** (1967), 1193-8; **9** (1969) 426-32.

[17] A. Fel'dbaum. *Avtom. & Telemekh.*, **14** (1953), 712-28; **16** (1956), 7-10.

[18] Feng-Hsia-Hsu. *Proc. 14th Can. Math. Congr.*, Univ. of Waterloo, Ont., 1973, *Lecture Notes in Economics*, vol. 106, Springer-Verlag, Berlin (1974).

[19] T. Fujisawa, and Y. Yasuda. *SIAM J. Control*, **5** (1967), 501-12.

[20] R. V. Gamkrelidze. *Izv. Akad. Nauk SSSR, S.M.*, **22** (1958), 449-74.

[21] O. Hájek. *SIAM J. Control*, **9** (1971), 339-50; *Math. Syst. Theor.*, **6** (1973), 289-301; *Funkcialaj Ekvacioj*, **20** (1977), 97-114.

[22] O. Hájek. 'Computation of optimal time', *Preprint No. 286*, TH Darmstadt, Fachbereich Math., June 1976.

[23] H. Halkin. *SIAM J. Control* A, **2** (1965), 199-202.

[24] R. E. Kalman. *Bol. Soc. Mat. Mex.*, (2) **5** (1960), 102-18.

[25] Yu. N. Kiselev. *Diff. Uravnenya,* **7** (1971), 1385-92.

[26] E. Kreindler. *J. Franklin Inst.*, **275** (1963), 314-44.

[27] L. A. Kun, and Yu. F. Pronosin. *Dolk. Akad. Nauk SSSR*, **200** (1971), 1294-7.

[28] J. P. LaSalle. *Contrib. Theor. Nonlinear Oscill.*, **5** (1960), 1-24.

[29] L. W. Neustadt. *J. Math. Anal. & Appl.*, **1** (1960), 484-93.

[30] P. Nistri, *Boll. Unione Mat. Ital.*, (4) **11** (1975), 554-64.

[31] G. J. Olsder. *J. Optimiz. Theor. & Appl.*, **16** (1975), 497-517.
[32] T. Yu. Petrenko. *Diff. Uravnenya,* **9** (1973), 1244-55.
[33] N. N. Petrov. *Prikl. Mat. & Mekh.*, **34** (1970), 820-5.
[34] Yu. F. Pronosin. *Avtom. & Telemekh.*, **12** (1972), 22-32.
[35] A. B. Rabinovich. *Zh. Vychisl. Mat. & Mat. Fiz.*, **6** (1966), 433-45.
[36] E. O. Roxin, and Vera W. de Spinadel. *Rev. Union Mat. Argent.*,**18** (1958), 137-45.
[37] N. Satimov. *Diff. Uravnenya*, **9** (1973), 2176-9.
[38] D. S. Yeung. *J. Optimiz. Theor. & Appl.*, **21** (1977), 71-82.

Chapter 7

Some Recent Developments in Linear Complementarity Theory

R.W. Cottle

1. INTRODUCTION

This paper is a companion to Professor Lemke's survey (Chapter XV) on complementarity. It is a 'mini-survey' of recent developments in the field of linear complementarity. The author admits that the selection is some what limited and shows a bias towards his work and that done in collaboration with two of his students. Moreover, since many of the papers discussed are either in progress or (are to) appear in other publications, the treatment will simply summarize results and comment on their significance.

The results mentioned here are placed under four separate headings: existence, stability, degeneracy, and computational complexity. Even though they are independent developments, each can be related to the important theme of constructive procedures for solving the linear complementarity problem (q, M), that is:

(1) given $q \in \mathbb{R}^n$ and $M \in \mathbb{R}^{n \times n}$ find a solution z of the system

$$q + Mz \geqslant 0, \quad z \geqslant 0, \quad \langle z, q + Mz \rangle = 0$$

The importance of this problems was explained in [7] and is further shown by several of the papers in these proceedings.

2. COMPLEMENTARY SUBMATRICES AND CONES

An alternate formulation of (q, M) is to find a solution w, z of the system

$$Iw - Mz = q \tag{2.1}$$

$$w \geqslant 0 \quad z \geqslant 0 \tag{2.2}$$

(2.3) $$\langle z, w \rangle = 0$$

Since (2.2) and (2.3) imply $z_j w_j = 0$ for $j = 1,2,...,n$, it follows that if (q, M) has any solution at all, there exists at least one matrix C such that

(3) $$C_{\cdot j} \in \{I_{\cdot j}, -M_{\cdot j}\}, \qquad j = 1,2,...,n$$

for which

(4) $$Cv = q, \qquad v \geqslant 0$$

has a solution. Here (and in general) $C_{\cdot j}$ denotes the jth column of C. In (4), one thinks of

$$v_j = w_j \qquad \text{if } C_{\cdot j} = I_{\cdot j}$$

and

$$v_j = z_j \qquad \text{if } C_{\cdot j} = -M_{\cdot j}$$

A matrix satisfying (3) is called a *complementary submatrix* of $[I, -M]$, even though it is only a true submatrix up to a permutation of its columns. If C is a complementary submatrix of $[I, -M]$, the set pos C of all q for which (4) has a solution is called the *complementary cone* spanned (or generated) by C. Each such cone is finitely generated, closed, and convex. The union of all possible complementary cones relative to M is denoted $K(M)$. Thus

$$K(M) = \bigcup_C \text{pos } C$$

Clearly, $K(M)$ is the set of all q for which (q, M) has a solution, as a finite union of finitely generated, closed, convex cones, it is finitely generated and closed, but not necessarily convex. The class of all $M \in \mathbb{R}^{n \times n}$ for which $K(M)$ is convex is denoted $\mathbf{Q}_0$, and the subclass of $\mathbf{Q}_0$ for which $K(M)$ equals $\mathbb{R}^n$ is denoted $\mathbf{Q}$. For the most part, these concepts are due to Murty [19]. Eaves [9] showed that $M \in \mathbf{Q}_0$ if and only if $K(M) = \text{pos } [I, -M]$.

Much of the literature of linear complementarity theory concerns the identification of subclasses of $\mathbf{Q}$ and $\mathbf{Q}_0$. Two of the most important contributions of this sort are the following:

(*a*) The theorem due to Samelson, Thrall, and Wesler [22] showing that (q,M) has a *unique* solution for all $q \in \mathbb{R}^n$ if and only if $M \in \mathbf{P}$ (the class of square matrices with positive principal minors). Thus, $\mathbf{P} \subset \mathbf{Q}$, but the reverse inclusion is false.

(*b*) The theorem of Lemke [18] which shows that the copositive-plus matrices – i.e. those for which $\langle z, Mz \rangle \geqslant 0$ for all $z \geqslant 0$ and $\langle z, Mz \rangle = 0$ for

$z \geq 0$ only if $(M+M^T)z = 0$ – belong to the class $\mathbf{Q}_0$. In the same paper, Lemke showed that the strictly copositive matrices (those for which $0 \neq z \geq 0$ implies $\langle z, Mz \rangle > 0$) are a subclass of $\mathbf{Q}$. It is noteworthy that all positive definite matrices belong to $\mathbf{P}$, and all positive semi-definite matrices are copositive-plus.

3. EXISTENCE: A CHARACTERIZATION OF THE CLASS $\mathbf{P}_0 \cap \mathbf{Q}$

It is natural to consider the generalization of $\mathbf{P}$ to $\mathbf{P}_0$, the class of matrices with *non-negative* principal minors. One can ask the question: what is $\mathbf{P}_0 \cap \mathbf{Q}$? This is the subject of a recent short paper by Aganagic and Cottle [1]. They characterize $\mathbf{P}_0 \cap \mathbf{Q}$ in terms of other interesting classes for which some further notation will be required.

Let $e = e(n) = (1,1,...,1)^T \in \mathbb{R}^n$. Then $M \in \mathbb{R}^{n \times n}$ belongs to $\mathbf{R}$ if and only if for all $t \geq 0$ the only solution of (te, M) is $z = 0$. If $z = 0$ is the only solution of $(0,M)$, then $M \in \mathbf{R}_0$. Obviously, $\mathbf{R} \subset \mathbf{R}_0$. The class $\mathbf{R}$ was introduced by Karamardian [14] who showed $\mathbf{R} \subset \mathbf{Q}$. The class $\mathbf{R}_0$ (denoted differently) was discussed by Garcia [12].

The relationship between these classes is given by the next theorem.

Theorem 1
If $M \in \mathbf{P}_0$, the following statements are equivalent:

(i) $M \in \mathbf{R}_0$

(ii) $M \in \mathbf{R}$

(iii) $M \in \mathbf{Q}$

Proof: A proof of this theorem is given in [1]. The significance of the result is that it gives a finite test for deciding when a $\mathbf{P}_0$ matrix belongs to $\mathbf{Q}$. This is not to say, of course, that such a test is easily executed. Knowing that $M \in \mathbf{P}_0$, one has to show that for all $\alpha \subset \{1,2,...,n\}$, the conditions

$$M_{\alpha\alpha} z_\alpha = 0 \qquad M_{\alpha\alpha} z_\alpha \geq 0, \qquad 0 \neq z_\alpha \geq 0$$

have no solution. If this is so, then $M \in \mathbf{Q}$.

It was formerly thought to be an important and difficult task to state a finite set of conditions by which membership in $\mathbf{Q}$ can be decided. The theoretical importance of this question remains, but its difficulty is merely computational, not conceptual. The key idea behind this observation is due to Professor D. Gale [11]. To decide whether $M \in \mathbf{Q}$ is equivalent to deciding whether $\mathbb{R}^n \sim K(M)$ is empty. If $\mathbb{R}^n \sim K(M)$ is non-empty, then there must exist a point which fails to belong to any of the complementary cones relative to M. Since each of the complementary cones is finitely generated, it follows by Minkowski's theorem [23] saying that each can be described by a finite system of homogeneous linear inequalities. Thus, $M \notin \mathbf{Q}$ if and only if there is a point $q \in \mathbb{R}^n$ which violates at least one inequality from each of these system. If M is a *non-degenerate* matrix (that is, all principal minors of M are non-zero), this could amount to checking n^{2^n} linear inequality system each consisting of 2^n inequalies in n variables. Note that when n has the very modest value 4, the number n^{2^n} is greater than 4 billions. See [2] for further discussion.

4. STABILITY: TOPOLOGICAL PROPERTIES OF Q

In his Ph.D. Thesis, Watson [25] considered the question of whether the class of $\mathbf{Q}$ matrices in open in $\mathbb{R}^{n \times n}$. This led to some surprising results. He showed that the $\mathbf{Q}$ matrices of order 2 do form an open set. For $n = 3$, he gave an example of a $\mathbf{Q}$ matrix which could not be perturbed (no matter how slightly) to another $\mathbf{Q}$ matrix. The example he used was a degenerate matrix, so he narrowed the question to perturbations of non-degenerate $\mathbf{Q}$ matrices. The thesis contained a theorem – later published in [26] – asserting that the class of non-degenerate $\mathbf{Q}$ matrices is open in $\mathbb{R}^{n \times n}$. The proof extended a three-dimensional argument beyond the limits of its validity, and subsequently Kelly and Watson [16, 17] produced a counter-example. This led to their theorem, which follows:

Theorem 2
For $n \geqslant 4$, the set of non-degenerate $\mathbf{Q}$ matrices is neither open nor closed in $\mathbb{R}^{n \times n}$.

Nevertheless, the following can be shown.

Theorem 3
If M is the limit of a convergent sequence of non-degenerate $\mathbf{Q}$ matrices and if M is also non-degenerate, then $M \in \mathbf{Q}$.

Developed in [2], this theorem says that the $\mathbf{Q}$ matrices form a *closed* set in the space of non-degenerate matrices with the relative topology.

5. DEGENERACY: LEAST-INDEX RULES FOR ITS RESOLUTION

If $M \in \mathbb{R}^{n \times n}$, a solution w, z of (2.1) with more that n of its $2n$ components equal to 0 is said to be *degenerate*. (This concept is not related to degeneracy of matrices.) As in linear and quadratic programming, degeneracy is a phenomenon which causes difficulties in the direct (pivoting) methods frequently used for the solution of linear complementarity problems. It can (but need not) lead to circling: the infinite repetition of a sequence of bases.

The traditional approach to handling degeneracy has been through the perturbation of q or through lexicographic ordering techiques [8, 9]. Recently, though, Bland [3] published a double least-index approach for resolving the degeneracy problem in the simplex method of linear programming. According to Bland's rule, the entering variable is chosen as the candidate with the smallest index (i.e. subscript). The leaving variable is also chosen as the candidate with the smallest index. With these pivot-selection rules, the simplex method is finite.

There is a simplicial method for quadratic programming due to Keller [15] which extends earlier convex quadratic programming methods of Dantzig [8], and van de Panne and Whinston [24]. It, too, requires special handling of the degeneracy problem. But, it is possible to adopt Bland's double least-index rule for use in Keller's algorithm. Indeed, Chang and Cottle [5] have proved the following theorem.

Theorem 4
Keller's quadratic programming algorithm with Bland's least-index pivot-selection rule is finite.

This result is mentioned here because the format of Keller's algorithm is really that of a linear complementarity problem which happens to possess special structure.

For the linear complementarity problem (q, M) with $M \in \mathbf{P}$, Murty [20] introduced and proved the finiteness of a principal pivoting algorithm according to which the pivot location is chosen as the candidate with the lowest index. Subsequently, Chang [4] has invented natural least-index rules for the principal pivoting method of Cottle and Dantzig [7] as well as for Lemke's complementary pivoting method [18]. Chang's result goes as follows.

Theorem 5
If $M \in \mathbf{P}$ or if M is positive semi-definite, then with the least-index rule, the principal pivoting method of Cottle and Dantzig and Lemke's complementary pivoting method are finite.

The hypotheses on M stated in Theorem 5 are all to which the Cottle-Dantzig method applies, but this is not so for Lemke's method. However, it has been found that the theorem is not true for all the types of matrix to which Lemke's method is applicable in the absence of degeneracy. Hence, a more powerful method of resolving degeneracy is being sought.

6. COMPUTATIONAL COMPLEXITY: WORST-CASE BEHAVIOUR OF COMPLEMENTARITY ALGORITHMS

Murty [21] and Fathi [10] have produced problems (q, M), i.e. data $q \in \mathbb{R}^n$ and $M \in \mathbb{R}^{n \times n} \cap \mathbf{P}$ on which the complementarity pivoting methods of Lemke [18] and Murty [20] require 2^n and $2^n - 1$ pivot steps, respectively. While this worst-case behaviour is not typical of these methods, the Murty-Fathi examples call attention to the fact that simple-looking complementarity problem can be nasty. It is not possible to bound the number of steps required by either of the aforementioned algorithm by a polynomial in n.

When $M \in \mathbf{P}$, Lemke's method is equivalent to a parametric principal pivoting method for (q, M). This variant differs from the original by just one pivot step. For brevity, Lemke's method in this parametric form and Murty's method are denoted CPM-I and CPM-II, respectively.

Murty uses the $n \times n$ matrix $M(n) = (m_{ij})$ where

$$m_{ij} = \begin{cases} 0 & \text{if } i < j \\ 1 & \text{if } i = j \\ 2 & \text{if } i > j \end{cases}$$

This matrix is lower traingular and belongs to $\mathbf{P}$. Defining $q(n) = (q_i)$ by

$$q_i = -\sum_{j=n+1-i}^{n} 2^j; \qquad i = 1,2,\ldots,n$$

Murty shows that CPM-I uses $2^n - 1$ pivot steps to solve $(q(n), M(n))$; he also shows that CPM-II requires $2^n - 1$ pivot steps to solve $(-e(n), M(n))$.

Fathi's examples are based on Murty's. He defines $F(n) = M(n)M(n)^T$. The matrix $F(n)$ is a symmetric $\mathbf{P}$ matrix with positive elements. Fathi obtains Murty's results with CPM-I applied to $(q(n), F(n))$ and CPM-II applied to $(-e(n), F(n))$. In fact, Fathi shows that q can be chosen from a set of positive measure.

It is interesting to consider the sequence of complementary bases (non-singular complementary submatrices) that arise as CPM-I and CPM-II are used to solve the special problems mentioned above.

For any complementary basis C, one can assign a binary n-vector $g(C)$ as follows:

$$g_j(C) = \begin{cases} 1 & \text{if } C_{\cdot j} = -M_{\cdot j} \\ 0 & \text{if } C_{\cdot j} = I_{\cdot j} \end{cases}$$

Notice that when $M \in \mathbf{P}$, this is completely unambiguous (since $I_{\cdot j} \neq -M_{\cdot j}$) and the function g maps the complementary bases one-to-one onto the set of all binary n-vectors. This association leads to an observation, the details of which are worked out in [6].

Theorem 6
Under the bijection g, the sequences of complementary bases generated by

(i) CPM-I applied to $(q(n), M(n))$ and $(q(n), F(n))$ and

(ii) CPM-II applied to $(-e(n), M(n))$ and $(-e(n), F(n))$

correspond to the binary Gray code representations [13] of the integers $0,1,\ldots,2^n - 1$ in that order.

The sequence of binary n-vectors by which the integers $0,1,\ldots,2^n - 1$ are represented in binary Gray code is a particular Hamiltonian path on the unit n-cube. Quite possibly other problems may give rise to different Hamiltonian paths.

REFERENCES

[1] M. Aganagic, and R.W. Cottle. 'A note on Q-matrices'. *Technical Report SOL 78-5,* Systems Optimization Laboratory, Department of Operations Research, Stanford University, 1978 *Math. Program.* To appear.

[2] M. Aganagic and R.W. Cottle. 'Further properties of Q-matrices'. Forthcoming.

[3] R.G. Bland. 'New finite pivoting rules for the simplex method'. *Math. Oper. Res.,* 2, (1977) 103-7.

[4] Y.Y. Chang. *Ph.D. Thesis,* Department of Operations Reserch, Stanford University. Forthcoming.

[5] Y.Y. Chang, and R.W. Cottle. 'Least-index resolution of degeneracy in quadratic programming'. *Technical Report 78-3,* Department of Operations Reserch, Stanford University, 1978.

[6] R.W. Cottle. 'Observations on a class of nasty linear complementarity problems' Forthcoming.

[7] R.W. Cottle, and G.B. Dantzig. 'Complementary pivot theory of mathematical programming'. *Linear Algebra & Appl.,* **1** (1968) 103-25.

[8] G.B. Dantzig *'Linear Programming and Extensions'.* Princeton University Press, Princeton, NJ (1963).

[9] B.C. Eaves. 'The linear complementarity problem'. *Manage. Sci.,* **17**, (1971) 612-34.

[10] Y. Fathi 'Computational complexity of LCP's associated with positive definite symmetric matrices'. *Technical Report 78-2,* Department of Industrial and Operations Engineering, University of Michigan, 1978.

[11] D. Gale. Private communication, 1978.

[12] C.B. Garcia 'Some classes of matrices in linear complementarity theory'. *Math. Program.,* **5**, (1978) 299-310.

[13] M. Gardner. 'Mathematical games: The curious properties of the Gray code and how it can be used to solve puzzles'. *Sci. Am.* (August 1972), 106-9.

[14] S. Karamardian. 'The complementarity problem'. *Math. Program.,* **2**, 107-29.

[15] E.L. Keller. 'The general quadratic optimization problem'. *Math. Program.,* **5** (1973) 311-37.

[16] L.M. Kelly, and L.T. Watson. 'Erratum: Some pertubation theorems for Q-matrices' *SIAM J. Appl. Math.,* **34** (1978), 320-1.

[17] L.M. Kelly, and L.T. Watson. 'Q-matrices and spherical geometry'. To appear.

[18] C.E. Lemke. 'Bimatrix equilibrium points and mathematical programming'. *Manage. Sci.,* **11** (1965), 681-9.

[19] K.G. Murty. 'On the number of solutions to the complementarity problem and spanning properties of complementary cones'. *Linear Algebra & Appl.,* **5** (1978), 65-108.

[20] K.G. Murty. 'Note on a Bard-type scheme for solving the complementarity problem . *Opsearch,* **11** (1974) 123-30.

[21] K.G. Murty. 'Computational complexity of complementary pivot methods'. *Math. Program. Stud.,* **7** (1978), 61-73.

[22] H. Samelson, R.M. Thrall, and O. Wesler. 'A partition theorem for Euclidean n-space'. *Proc. Am. Math. Soc.* **9** (1958) 805-7.

[23] J. Stoer, C. Witzgall. *'Convexity and Optimization in Finite Dimensions'.* Springer-Verlag, Berlin, Heidelberg, New York (1970).

[24] C. van de Panne, and A. Whinston. 'The symmetric formulation of the simplex method for quadratic programming'. *Econometrica,* **37** (1969), 507-27.

[25] L.T. Watson. *Ph.D. Thesis,* University of Michigan (1974).

[26] L.T. Watson. 'Some pertubation theorem for Q-matrices'. *SIAM J. Appl. Math.,* **31** (1976) 379-84.

Chapter 8

Some Stochastic Strategies in Optimization Problems

M. Cugiani

1. INTRODUCTION

Let us first consider the following problem: find x^* such that

$$\text{(P1)} \qquad f(x^*) = \min_{x \in K} f(x)$$

where K is a compact set and f is continuous on K. If $f(x)$ is strictly convex, then there is only one local minimum to which many algorithms can be shown to converge.

For general non-convex functions (e.g. multimodal functions), the usual optimization techniques converge to any of the local extrema and are therefore likely to fail to find the global minimum. The global optimization problem has been gaining increasing attention in the last few years and many algorithms have been proposed in the literature [8, 9].

Before outlining the main numerical approaches so far suggested, we consider carefully the statement of the problem.

Firstly, one has to remark that (P1) is not well posed, in the classical sense: the continuous dependence of x on the data of the problem cannot in general be ensured, for example let

$$f(x) = \cos \pi x - \delta x, \qquad x \in [-2, 2]$$

If δ is a small positive number, $x^* \simeq 1$; if δ is negative, $x^* \simeq -1$. In what follows we shall mainly be concerned with the problem: find

$$\text{(P2)} \qquad f^* = \min_{x \in K} f(x)$$

Provided that $f(x)$ is continuous on K, this problem is well posed as shown by the relation $|f^* - g^*| \leqslant \|f - g\|_\infty$, where $\|\cdot\|_\infty$ denotes the supremum of $|\cdot|$ on K.

Quantitative assumptions are required to ensure an approximation to f^*, of determined accuracy. Let, for example, $f(x)$ be valued at points x_i, $i = 1, \ldots, n$ such that

$$\min_{i=1,\ldots,n} f(x_i) - f^* < \varepsilon$$

It is not difficult to construct a function $g(x)$ such that

$$g(x_i) = f(x_i)$$

but

$$\min_{i=1,\ldots,m} g(x_i) - g^* > \varepsilon$$

In the following section, *a priori* 'quantitative' conditions on the objective function will be considered: they will be shown to be instrumental in the design of algorithms, which can ensure in a finite number of steps an approximation to f^* of predetermined accuracy. For general problems, we can look only for a probabilistic estimate of f^*: stochastic strategies for obtaining it will be discussed in sections 2.2 and 2.3.

2. DETERMINISTIC STRATEGIES

2.1. The use of a priori conditions

Let $f(x) \in C[a, b]$ and $W(f, \delta)$ be its modulus of continuity

$$W(f, \delta) = \sup_{\substack{|x-y| \leqslant \delta \\ x, y \in [a, b]}} |f(x) - f(y)|$$

A function $W(\delta)$ is assumed to exist such that:

$$\lim_{\delta \to 0^+} W(\delta) = 0 \tag{1}$$

and

$$W(f, \delta) \leqslant W(\delta) \qquad \delta \geqslant 0$$

For any predetermined $\varepsilon > 0$ we can choose $\delta > 0$ such that $W\delta \leqslant \varepsilon$. Next f

is evaluated at the points $x_i \in [a, b], j = 1, \ldots, n$ such that

$$\max_{x \in [a, b]} \min_{j=1,\ldots,n} |x - x_j| \leqslant \delta \tag{2}$$

A partition of $[a, b]$, satisfying this relation, can be accomplished, for instance, taking[1]

$$n = \left\lceil \frac{b-a}{2\delta} \right\rceil$$

and

$$x_j = a + (2j - 1)\delta \qquad j = 1, \ldots, n$$

If

$$\hat{f}_n = \min_{j=1,\ldots,n} f(x_j)$$

we have

$$\hat{f}_n - f^* \leqslant f(\bar{x}_n) - f^*$$

where

$$|x^* - \bar{x}_n| = \min_{j=1,\ldots,n} |x^* - x_j|$$

for some $x^* \in X^*$, $X^* \equiv \{x \in [a, b] : f(x) = f^*\}$.

By (1) and (2) we may derive:

$$0 \leqslant \hat{f}_n - f^* \leqslant W(f, \delta) \leqslant \varepsilon$$

If $f \in C^1([a, b])$ and $|f'(x)| \leqslant M$ we may assume $W(\delta) = M\delta$. The number of function evaluations required to approximate f^* within the prescribed accuracy is about

$$\frac{b-a}{2\varepsilon} M$$

If $|f^{(r)}(x)| \leqslant M$, for $r > 1$, it may be shown [7] that the maximum number of

(1) $\lceil * \rceil$ means the least integer not less than $*$; $\lfloor * \rfloor$ the greatest integer not greater than $*$.

function evaluations required to ensure an accuracy ε is about

$$(M/2\varepsilon)^{1/r} \ (b-a)$$

In the following we shall be concerned with the wider class of functions for which a Lipschitz condition holds with constant ℓ for some distance ρ, over a compact $K \subset \mathbb{R}^N$, i.e.

$$f \in L_{\ell,\rho}(K) \equiv \{f: K \to \mathbb{R}, \ |f(x)-f(y)| \leq \ell\rho(x,y) \ \text{for} \ x,y \in K\}$$

where $\rho(x,y)$ satisfies the following conditions for any $x, y, z \in K$:

(i) $\rho(x,y) = \rho(y,x)$ (ii) $\rho(x,y) \geq 0$ for $x \neq y$, $\rho(x,x)=0$

(iii) $\rho(x,y) \leq \rho(x,z) + \rho(y,z)$, (iv) $\rho \in C(K \times K)$

Condition (iv) implies that if $f \in L_{\ell,\rho}(K)$ then f is continuous on K.

For this class of functions a theoretical framework will be developed, centered around the notion of 'strategy' and its guaranteed accuracy, in which the performance of deterministic methods will be evaluated.

For any n, a *strategy* is any vector-valued function with domain $L_{\ell,\rho}(K)$ and range[(2)] K^n. If the strategy is constant in $L_{\ell,\rho}(K)$, that is, it maps $L_{\ell,\rho}(K)$ into a single point of K^n, then the strategy is termed *passive*. Passive strategies mapping $L_{\ell,\rho}(K)$ into $(x_1, x_2, \ldots, x_n)$, $x_i \in K$, will be indicated in the following by S_n; also the set of points $(x_1, x_2, \ldots, x_n)$ will be called a *passive strategy*, or briefly a *strategy*, and denoted $S_n \equiv (x_1, x_2, \ldots, x_n)$. When the strategy is allowed to depend on $f \in L_{\ell,\rho}(K)$ so that its values $(\hat{x}_1, \hat{x}_2, \ldots, \hat{x}_n)$, $\hat{x}_j \in K$ are

$$\begin{aligned} \hat{x}_2 &= \chi(\hat{x}_1, f(\hat{x}_1)) \\ &\ \vdots \\ \hat{x}_n &= \chi(\hat{x}_1, \ldots, \hat{x}_{n-1}, f(\hat{x}_1), \ldots, f(\hat{x}_{n-1})) \end{aligned}$$

where χ is some selection rule, then the strategy is termed *sequential* and indicated by $\hat{S}_n$. We remark that passive strategies can be viewed as particular sequential ones. The set of all sequential strategies ranging in K^n will be indicated by $\hat{K}_n$.

(2) K^n means the Cartesian product $K \times K \times \ldots \times K$ (n times).

Let $S_n \equiv (x_1, x_2, \ldots, x_n)$ be a passive strategy. We measure its effectiveness for the function f by the accuracy

$$A(f, S_n) = \min_{i=1,\ldots,n} f(x_i) - f^*$$

A more significant index associated with S_n is its performance with respect to the whole class of functions under consideration:

$$A(S_n) = \sup_{f \in L_{\ell,\rho}(K)} A(f, S_n)$$

which is called the *guaranteed accuracy* of S_n in $L_{\ell,\rho}(K)$.

We may introduce the following definitions of optimality for strategies.

Definition 1 S_n^* is said to be A-optimal (A is for 'respect to accuracy') when

$$A(S_n^*) = \inf_{S_n} A(S_n)$$

Definition 2 Given $\delta > 0$, if there exists n^* and a strategy S_{n^*} such that $A(S_{n^*}) \leqslant \delta$, while no strategy S_n exists, for $n > n^*$, such that $A(S_n) \leqslant \delta$, then S_{n^*} is said to be n-optimal (n is for 'respect to the number of points').

By the following lemma, can be proved a general existence theorem which is given next.

Lemma 1 For any strategy $S_n = (x_1, x_2, \ldots, x_n)$

$$A(S_n) = \sup_{a \in K} \min_{i=1,\ldots,n} \ell\rho(x_i, a) \tag{3}$$

Theorem 1

Let K be a compact set in $\mathbb{R}^N$. Then

(a) for any n an A-optimal strategy exists;

(b) for any δ an n-optimal strategy exists.

Proof: For the proofs of Lemma 1 and Theorem 1, see [4, 19].

From the definitions and the existence of both A-optimal and n-optimal strategies some interesting properties easily follow. We list here three of them.

1. The sequence $A(S_n^*)$ is not increasing. Indeed, if we consider the strategy $S_n^* = (x_1^*, x_2^*, \ldots, x_n^*)$ and the strategy $\overline{S}_{n+1} = (x_1^*, x_2^*, \ldots, x_n^*, x_n), x_n \in K$ then

$$A(S_n^*) \geqslant A(\overline{S}_{n+1}) \geqslant A(S_{n+1}^*)$$

2. $A(S_n^*) \to 0$ as $n \to \infty$. Indeed, for any $\delta > 0$, by the existence of n-optimal strategies we can find n^* and S_{n^*} such that $A(S_{n^*}) \leqslant \delta$; thus $A(S_{n^*}^*) \leqslant \delta$ and

by property 1 we have $A(S_n^*) \leqslant \delta$ for any $n \geqslant n^*$

3. n-optimal strategies can be constructed from A-optimal ones. Indeed, let $\delta > 0$ and, for any n, let S_n^* be an A-optimal strategy and n^* be such that

$$n^* = \min \{n : A(S_n^*) \leqslant \delta\}$$

The number n^* exists by property 2 and $A(S_{n^*}^*) \leqslant \delta$; as $A(S_n^*) > \delta$ for any $n < n^*$, we have also $A(S_n) > \delta$ for any $n < n^*$ and for any strategy S_n; this implies that $S_{n^*}^*$ is n-optimal.

Let K be a compact set in $\mathbb{R}^N$. We recall that a covering of K is a family of sets, D_α, $\alpha \in A$, such that

$$\bigcup_{\alpha \in A} D_\alpha \supset K$$

In what follows, we shall consider coverings $C_{\delta, n} = \{D^\delta(x_i)\}$, $\quad i = 1, \ldots, n$, where

$$D^\delta(x_i) = \{x : \rho(x_i, x) \leqslant \delta\}, \qquad x_i \in K$$

We introduce the following definition.

Definition 3 δ is the *radius* of $C_{\delta,n}$ and x_i, $i = 1, \ldots, n$, are its *centres*. Two optimality criteria can be stated for these coverings as follows.

Definition 4 A covering $C_{\delta^*,n}$ of K is said to be r-optimal (r is for 'respect to radius') if no covering $C_{\delta,n}$ of K exists with $\delta < \delta^*$.

Definition 5 A covering $C_{\delta,n}$ of K is said to be n-optimal if no covering $C_{\delta,n}$ of K exists with $n < n^*$.

As ρ is bounded on K, for each strategy $S_n = (x_1, x_2, \ldots, x_n)$ a set Δ of positive numbers exists such that, for $\delta \in \Delta$, it is

$$\bigcup_{i=1}^{n} D^\delta(x_i) \supset K$$

Let $\bar{\delta} = \inf \Delta$: we call $C_{\bar{\delta}, n}$ the *covering characteristic* for S_n.

Thus relation (3) says that, for any strategy S_n, $A(S_n)$ is just ℓ times the radius of the covering characteristic for S_n. On the other hand, to any covering we may link the strategy whose points are the centres of the covering.

The following theorem can be proved [4].

Theorem 2

Let S_n be an A-optimal strategy: its characteristic covering $C_{\delta^*, n}$ is r-optimal with $\delta^* = A(S_n)/\ell$.

We remark that, by theorem 2, optimal strategies do not depend on ℓ, but only on ρ and K.

2.2. Construction of optimal strategies

First we consider the A-optimality: let $K = [0, 1]^N$ and

$$\rho(x, y) = \max_{i=1,\ldots,N} |x^i - y^i|, \qquad x = (x^1, \ldots, x^N), \qquad y = (y^1, \ldots, y^N)$$

The construction of an A-optimal strategy is connected, by theorem 2, with the construction of an r-optimal covering of K with n centres. In [19] the r-optimal covering of K with n hypercubes

$$D(x_i, r) = \{x : \rho(x, x_i) \leqslant r\}$$

is given for any n; the result can be stated as follows.

Theorem 3
(a) If $n=\upsilon^N$ for some natural number υ then the strategy $S_n^* = (x_1^*, x_2^*, \ldots, x_n^*)$, where x_i^*, $i = 1, \ldots, n$, are the centres of the hypercubes with edges of length $1/\upsilon$ partitioning K, is A-optimal.
(b) If $\upsilon^N < n < (\upsilon + 1)^N$, then the strategy $S_n^* = (x_1^*, x_2^*, \ldots, x_n^*)$, where x_i^*, $i = 1, \ldots \upsilon^N$, are the same as in (a) and x_i^*; $i = \upsilon^N + 1, \ldots, n$ are selected arbitrarily, is A-optimal.

In both case the guaranteed accuracy is

$$A(S_n^*) = \frac{\ell}{2[\sqrt[N]{n}]}$$

The method in which the function to be optimized is evaluated at the points of the strategy S_n^* for $n = \upsilon^N$ is usually termed the 'grid search' method.

With the above assumptions about K and ρ, it is easy to derive, by property 3 at the end of section 2.1 n-optimal strategies. For any $\delta > 0$, let $n^* = \min\{n : A(S_n^*) \leqslant S\}$ and it turns out that

$$n^* = \lceil 1/2\delta \rceil^N$$

Thus the strategy $S_{n^*}^*$ is n-optimal.

For a general domain, the problem of finding an r-optimal covering of K, and hence an A-optimal strategy, is not easy to solve; it leads to the problem of optimizing the function of $n \times N$ variables

$$\max_{a \in K} \quad \min_{i=1,\ldots,n} \quad \rho(x_i, a)$$

In [4] a technique is introduced by which n-optimal coverings C_{δ, n^*} of K, and therefore n-optimal strategies with guaranteed accuracy not larger than δ/ℓ, can be calculated through successive approximations, each of them implying the solution of an integer linear programming problem.

2.3. The effectiveness of sequential strategies against passive ones

The concept of accuracy and guaranteed accuracy, as well as the optimality criteria introduced above, can be easily extended to sequential strategies:

$$A(f, \hat{S}_n) = \min_{i=1,\ldots,n} f(\hat{x}_i) - f^*, \qquad \hat{S}_n(f) = (\hat{x}_1, \hat{x}_2, \ldots, \hat{x}_n)$$

is the accuracy given by $\hat{S}_n$ for the particular function f;

$$A(\hat{S}_n) = \sup_{f \in L_{\ell,\rho}(K)} A(f, \hat{S}_n)$$

is the accuracy guaranteed by $\hat{S}_n$ in $L_{\ell,\rho}(K)$.

A strategy $\hat{S}_n$ is said to be an A-optimal sequential strategy when

$$A(\hat{S}_n) = \inf_{\hat{S}_n \in \hat{K}_n} A(\hat{S}_n)$$

Given $\delta > 0$, if there exists an n^* and a sequential strategy $\hat{S}_n^*$ such that $A(\hat{S}_{n^*}) \leqslant \delta$, while no strategy $\hat{S}_n$ exists for $n < n^*$ such that $A(\hat{S}_n) \leqslant \delta$, then $\hat{S}_{n^*}$ is said to be n-optimal.

One could think that sequential strategies improve the performance of passive ones, in terms of guaranteed accuracy: this is not the case, as shown by Theorems 4 and 5. Theorem 4 states [19] that no sequential strategy can guarantee a better accuracy than an A-optimal passive one with the same number of points; theorem 5 states [4] that no sequential strategy can guarantee a predetermined accuracy $\delta > 0$ with less points than those required by an n-optimal passive strategy.

Theorem 4
Let

$$m_1 = \inf_{\hat{S}_n \in \hat{K}_n} A(S_n) \qquad \text{and} \qquad m_2 = \inf_{S_n \in K^n} A(S_n)$$

then $m_1 = m_2$.

Theorem 5
Let

$n_1 = \min\{n$ for which a passive strategy S_n exists such that $A(S_n) \leq \delta\}$

and

$n_2 = \min\{n$ for which a sequential strategy $\hat{S}_n$ exists such that $A(\hat{S}_n) \leq \delta\}$

then $n_1 = n_2$.

The above results are concerned only with the guaranteed accuracy in $L_{\ell,\rho}(K)$: thus for particular functions in $L_{\ell,\rho}(K)$, sequential strategies may perform better than passive ones: for this reason, some research has been devoted to developing sequential algorithms [4, 18].

3. PROBABILISTIC METHODS

The probabilistic approach to global optimization relies on a random scanning of the search domain (usually drawn from a uniform distribution), whose behaviour can be summarized by the following result.

Let μ be the Lebesgue measure on a compact $K \subset \mathbb{R}^n$ and $A \subset K$ measurable such that

$$\frac{\mu(A)}{\mu(K)} = \alpha > 0$$

Given N points x_i drawn from a uniform distribution on K, the probability $P(A, N)$ that at least one of those points falls in A is given by

$$P(A, N) = 1 - (1 - \alpha)^N \tag{4}$$

Moreover, it can be shown that

$$\text{prob}\ \{\min_{i=1,\ldots,N} \text{dist}\ (A, x_i) = 0 \text{ for sufficiently large } N\} = 1$$

where

$$\text{dist}\ (A, x_i) = \inf_{x \in A} \|x - x_i\|$$

Even if it can be shown that a uniform sampling process eventually reaches a neighbourhood of the global minimum, later on indicated as D^*, of any positive measure, this result does not imply that any finite sample will ensure a good approximation.

This can be gained, only within some predetermined probability level, by (4), by which one could compute how many function evaluations are required to reach, in probability, D^*.

Actual algorithms should usually gain a better performance by an effective management of the information gathered during the sampling process and the use of local routines.

Probabilistic algorithms for global optimization usually follow the general pattern outlined below.

Phase 1 Draw N_1 points from a uniform distribution in K and evaluate $f(x)$ at these points.

Phase 2 Select the N_2 'most promising' points and from any of these start a local routine obtaining a least value f_c.

Phase 3 Test if f_c is 'probably' the global minimum. If so, stop the algorithm, otherwise go to *Phase 1*.

Many algorithms of different complexity and effectiveness can be fitted in the above general framework depending on the strategies chosen at *Phase 2* and *Phase 3*.

In the simplest algorithm, that point yielding the least value is picked up in *Phase 2* and the test in *Phase 3* is satisfied if no improvement on f_c is observed by evaluating $f(x)$ at N_3 new randon points.

If f_c is not 'validated' as the global minimum, the new sampled point yielding the least value is considered in *Phase 2*.

This test relies upon (4). as N_3 increases, the probability of not improving f_c decreases, unless f_c is the value of the global minimum.

Now we outline two strategies, which can be merged in a highly effective global optimization algorithm, to use respectively in *Phase 2* and *Phase 3*.

4 THE CLUSTERING APPROACH

A good feature of the above algorithm is that only one local search is performed in the region of attraction of a local minimum: unfortunately, at the cost of wasting most information about $f(x)$ contained in the sample.

A good compromise between these two conflicting goals can be reached structuring *Phase 2* of the general scheme as follows:

(1) Few iterations of a local routine are performed from any sampled point; this process transforms the uniformly distributed initial sample into a number of separate groups of 'terminal' points, come together in the process

of converging to the same minimum.
(2) A statistical procedure known as 'clustering' is used to determine clusters of terminal points, calculating first the average distance between these points; clusters are then determined as groups of points with less than the average distance from one another (actually, the distance from the so-called 'seed point' in each cluster in crucial).
(3) A certain fraction of points from each cluster is discarded and step 1 is repeated performing local optimization from the remaining points. The process stops when a prescribed number of clusterings has been performed or when a predetermined number of function evaluations is reached. After another termination criterion, the process is continued until two consecutive clusterings give the same number of clusters, then the best point is considered from each cluster and a local routine is allowed to converge.

In an ideal situation one cluster would correspond to each local minimum. thus all points but one of any cluster could be discarded. But even when all local minima are eventually located, this ideal set-up may not established until some late iterations.

At the beginning there are usually changes from one clustering to next; some clusters split, others merge, individual points may pass from one cluster to another.

In general, the user of the global optimization method is faced with the problem of setting the right balance between many conflicting factors: the number of initial points and the number of iterations (i.e. clusterings) the rate of reduction of points in a cluster, the number of steps performed by the local optimization routine between subsequent clusterings. For details about the actual implementation of the clustering approach see [20, 12, 11].

5. THE ψ FUNCTION APPROACH

As far as *Phase 3* is concerned, we describe here in some detail an interesting method which gives, by sequential samples of the search domain, an approximation in probability to the value of the global minimum: this approximation can be used to control whether the best local optimum obtained at *Phase 2* is 'probably' the global one.

For further reading about this approach see [1, 2, 4].

5.1. A measure theoretic analysis of $\psi(\xi)$

Let f be a Lebesgue measurable function, defined on a compact set $K \subset \mathbb{R}^n$ mapping K into $\mathbb{R}^1$. In this section we are concerned with the regularity properties of the function

$$\psi(\xi) = \frac{m(E(\xi))}{m(K)}, \qquad \xi \in \mathbb{R}^1$$

where $m(\cdot)$ is the Lebesgue measure on K and

$$E(\xi) = \{x \in K : f(x) \leqslant \xi\}$$

The function $\psi(\xi)$ is completely defined, as $E(\xi)$ (account taken of the measurability of f) belongs to the σ - algebra of Lebesgue sets in K.

The compactness of K implies that $\psi(\xi)$ is finite for every ξ.

If $\xi < f^*$, the essential infimum of f in K, then $\psi(\xi) = 0$; $m((f^{-1}(f)) = 0$ implies $\psi(\xi) = 0$ iff $\xi \leqslant f^*$. The properties of $\psi(\xi)$; which are relevant to the next sections, are given below; for the proofs see, for example, W. Rudin, *Real and Complex Analysis,* McGraw-Hill, New York, 1966, Chapter 8:

(i) $\psi(\xi)$ is non-decreasing in $\mathbb{R}^1$;
(ii) $\psi(\xi)$ is a.e. (almost everywhere) differentiable;
(iii) $\psi(\xi)$ is continuous in $\mathbb{R}^1$, provided that no set $H \subset K$ exists, such that $m(H) > 0$ and $f(x) =$ constant, $x \in H$.

5.2. The stochastic evaluation of $\psi(\xi)$

Now we focus our attention on the estimation of f^* such that

$$\psi(f^*) = 0, \quad \psi(\xi) > 0 \quad \text{for } \xi > f^*$$

The function $\psi(\xi)$; apart from some trivial cases, cannot be evaluated analytically in order to approximate it, we set up a stochastic framework.

Let z be a random n-dimensional vector, uniformy distributed in K. Let $m(\cdot)$ be a set function proportional to the Lebesgue measure such that the normalization condition $m(K) = 1$ holds.

If $P(\xi) = \text{prob } \{f(z) \leqslant \xi\}$ is the probability of hitting the region $E(\xi)$, the uniformity of the distribution in K of the trial points z implies that:

$$\psi(\xi) = m(E(\xi)) = P(\xi)$$

In order to approximate $\psi(\xi)$ we carry out a sample S_q evaluating $f(x)$ at the random trial points z_j, $j = 1,\ldots,q$.

Let

$$S_\ell = \min \{f(z_j), \quad j = 1, \ldots q\}$$

and

$$S_u = \max \{f(z_j), \quad j = 1, \ldots, q\}$$

and

$$\xi \in [s_\ell, s_u]$$

If p is the number of trial points in S hitting $E(\xi)$ we approximate $\psi(\xi)$ by $\psi(\xi) = p/q$.

The random variable $\psi(\xi)$ is an unbiased estimator of $\psi(\xi)$, it follows a binomial distribution with parameter $\psi(\xi)$. By the law of large numbers we get:

$$\text{prob} \; \{ \lim_{q \to \infty} \; (p/q = \psi(\xi)\} = 1$$

The error $|(p/q) - \psi(\xi)|$ is of the order $1/\sqrt{q}$: its expected value is 0 and its standard deviation is $\sqrt{\psi(\xi)(1 - \psi(\xi))/q}$.

In this stochastic framework $\psi(\xi)$ turns out to be an unknown regression function, and we are looking for its root ξ^* such that

$$\psi(\xi^*) = 0 \qquad \text{and} \qquad \psi(\xi) > 0, \qquad \xi > \xi^*$$

Several stochastic approximation techniques have been developed for finding roots or minima of unknown regression functions (see, for example, [3, 6]), but here the problem is fairly difficult for the following reasons.

The root we are looking for is located outside the actually sampled interval (stochastic extrapolation).

The 'dispersion' of the sampled data (the ratio of the standard deviation to the mean), of the order $1/\sqrt{q\psi(\xi)}$, diverges for a fixed sample size, as ξ approaches ξ^*.

The usual stochastic approximation procedures cannot be shown to converge. In order to overcome this difficulty we derive, on the basis of the sampled data, an explicit approximation of $\psi(\xi)$, whose root can be evaluated analytically. Therefore, we look for an approximation of $\psi(\xi)$ in the form of a linear combination of some coordinate functions $\Phi_\ell(\xi)$, $\ell = 1,\ldots,k$. In order to find the coefficients of the combination, we minimize the error function [3]:

$$D(\lambda) = \int_{f^*}^{\|f\|_\infty} \left(\psi(\xi) - \sum_{\ell=1}^{k} \lambda_\ell \, \Phi_\ell(\xi) \right)^2 d\xi$$

As $\psi(\xi)$, $\|f\|_\infty$ and f^* are unknown, $D(\lambda)$ itself cannot be minimized; Therefore, it is replaced by the empirical error (risk) function:

$$J(\lambda) = \sum_{i=1}^{r} \left(\widetilde{\psi}_i - \sum_{\ell=1}^{k} \lambda_\ell \, \Phi_\ell(\xi_i) \right)^2$$

where r, $\widetilde{\psi}_i$ and ξ_i are defined in a sequential sampling process in the way outlined below.

Let $S^i_{q_i} = \{f(z^i_{j_i}), j_i = 1, \ldots, q_i\}$, $i = 1, \ldots, r$ be the ith sample performed, where $z^i_{j_i}$, $i = 1, \ldots, r$, $j_i = 1, \ldots, q_i$, are independent random points uniformly distributed in K. We define

$$Z^r_l = \min_{i=1,\ldots,r} \; \min_{j_i=1,\ldots,q_i} \; \{f(z^i_{j_i})\}$$

$$Z^r_u = \max_{i=1,\ldots,r} \; \max_{j_i=1,\ldots,q_i} \; \{f(z^i_{j_i})\}$$

and consider a random number ξ_i uniformly distributed in the interval $[Z^r_l - \epsilon, Z^r_u + \delta]$, $\epsilon, \delta > 0$; as $r \to \infty$ the following lemma ensures that $[Z^r_l - \epsilon, Z^r_u + \delta] \supset [f^*, \|f\|_\infty]$ with probability 1, for a sufficiently large r.

Lemma 2 Let $m_i = \min\{f(x_1), \ldots, f(x_i)\}$ and $M_i = \max\{f(x_1), f(x_2), \ldots, f(x_i)\}$ where $x_j, j = 1, \ldots, i$, are independent random points uniformly distributed in K. Then, as $i \to \infty$, $m_i \to f^*$ and $M_i \to \|f\|_\infty$ with probability 1.

We are now in the condition of defining an unbiased estimator of ψ at the points

$$\xi_i : \widetilde{\psi}_i = \widetilde{\psi}(\xi_i) = p_i / q_i$$

where p_i is the number of trial points of $S^i_{q_i}$ hitting the region $E(\xi_i)$. As the samples $S^i_{q_i}$, $i = 1, \ldots, r$, of $f(x)$ are independent we have thus defined an in-

(3) $\|f\|_\infty$ indicates $\sup_{x \in K} f(x)$.

dependent sample $(\xi_i, \widetilde{\psi}(\xi_i))$, $i = 1, \ldots, r$, of $\xi, \psi(\xi)$.
Let

$$A_K(\xi) = \sum_{\ell=1}^{k} \lambda_l^* \, \Phi_l(\xi), \lambda^* \quad \text{such that} \quad J(\lambda^*) \leqslant J(\lambda)$$

the value β_k such that $A_k(\beta_k) = 0$ and $A_k(\xi) > 0$ for $\xi > \beta_k$ may be considered as an approximation to f^*. We remark that a root of $A_k(\xi)$ does not necessarily exist: thus more generally β_k may be defined by the condition

$$A_k(\beta_k) \leqslant \epsilon_k \quad \text{and} \quad A_k(\xi) > \epsilon_k, \xi > \beta_k$$

where $\epsilon_k \to 0$ as $k \to \infty$.

Clearly our approximation model should be such that new sampled information about $\psi(\xi)$ improves the approximation to f^*: as $r \to \infty$ the sequence β_k should converge to f^*, when $k \to \infty$.

This asymptotic property is assured if β_k is defined in the general way outlined above and if $A_k(\xi)$ converges uniformly to the regression function $\psi(\xi)$ when $r, k \to \infty$.

The theorems of Karshiladze-Lozinskii state that $A_k(\xi)$ defined by a least-squares approximation, with a polynomial or trigonometric basis, may fail to converge uniformly to $\psi(\xi)$. Moreover, $\psi(\xi)$ may display a non-polynomial behaviour in the neighbourhood of the minima of $f(x)$.

For example, let $f(x) \in C^2(K)$, and $\overline{x}$ be any interior minimum. In a neighbourhood of $\overline{x}$ we have a quadratic behaviour of $f(x)$: hence, if $\xi = f(x)$ and $|x - \overline{x}| \sim h$ we get

$$\xi = f(\overline{x}) + g(h^2) \quad \text{with} \quad 0 < k_1 < \frac{|g(h^2)|}{h^2} < k_2 .$$

Therefore $\psi(\xi) = m(E(\xi)) = m(E(f(\overline{x})) + 0(h^n)$ and

$$\psi(\xi) = m(f(\overline{x})) + 0(\xi - f(\overline{x}))^{n/2} .$$

Therefore, for n odd, $\psi(\xi)$ displays a non-polynomial behaviour in a right neighbourhood of $\xi = f(\overline{x})$ and hence also of the global minimum f^*.

In order to obtain an effective approximation of $\psi(\xi)$ for finite k and r and the uniform convergence of the sequence $A_k(\xi)$ as $k \to \infty, r \to \infty$ to $\psi(\xi)$, we use a system of averaging spline functions.

For every N, $A_N(\xi)$ is expressed in terms of spline functions $S_N(\xi)$ of order $(2m - 1)$ with N equidistributed knots $\zeta_j, j = 1, \ldots, N$, in the approximation interval $[\zeta_0, \zeta_{N+1}]$ with $\zeta_0 = z_u^r - \epsilon$, $\zeta_{N+1} = z_l^r + \delta$; at these knots we require the continuity of $s_N(\xi)$ and its first $(2m - 2)$ derivatives.

The spline approximation $s_N(\xi)$ depends upon $(2m + N)$ parameters

$\lambda_1, \lambda_2, ..., \lambda_{2m+N}$ which are chosen in order to minimize the empirical error (risk) function

$$J(s_N) = \sum_{i=1}^{r} (\widetilde{\psi}_i - s_N(\xi_i))^2$$

Let $s_N^r(\xi)$ be the least-squares spline approximation with N equidistributed knots, such that

$$J(s_N^r) \leqslant J(s_N)$$

for given r and N. The feasibility of the spline approach in order to approximate $\psi(\xi)$ relies upon the following result [13]: let $s_{N(r)}$ be a spline function in which the number of knots $N(r)$, equidistributed in some interval $[a, b]$, increases as $r \to \infty$. Let $f(x)$ be a continuous function on $[a, b]$ and (x_i, f_i) an independent sample of $f(x)$, such that f_i is an unbiased estimator of $f(x_i)$: the sequence $s_{N(r)}^r$ such that $J(s_{N(r)}^r) \leqslant J(s_{N(r)})$ converges uniformly in $[a, b]$ with probability 1 to $f(x)$, provided that the following condition holds:

$$\lim_{r \to \infty} \frac{N^2(r) \log r}{r} = 0$$

We are now able to apply this result to the approximation of $\psi(\xi)$:
(1) $\psi(\xi)$ is continuous in wide conditions;
(2) $(\xi_i, \widetilde{\psi}_i)$ is an independent sample in view of the independence of $S_{qi}^i, \widetilde{\psi}_i$.
(3) As the interval $[a, b]$ we can consider $[z_1^r - \varepsilon, z_u^r + \delta]$ which contains, as shown in the above lemma, the interval $[f^*, \|f\|_\infty]$ for a sufficiently large r, with probability 1.

6. GLOBAL OPTIMIZATION VIA STOCHASTIC MODELS

The performance of 'space covering' methods, whose theoretical framework has been outlined in section 2, is to be evaluated in terms of 'guaranteed accuracy'. i.e. keeping account of the worst possible case. Therefore, it turns out that n function evaluations imply a bound on the error $f_n^* - f^*$ of the magnitude of $1/2\ \ell\ [\sqrt[N]{n}]$ where ℓ is the Lipschitz constant and N is the number of variables.

The algorithms outlined in section 3, even if they can result in good numerical performance in general problems, are still missing a comprehensive theoretical framework, by which the error could be bounded, at least in a probabilistic sense.

Now we are going to outline a third class of methods which are beginning to receive some attention, based on suitable stochastic models [14, 15]. The objective function is to be considered as a realization of a stochastic process $f(x, \omega), x \in \mathbb{R}^n$, where ω belongs to some probability space (Ω, β, p).

In this case we are looking for algorithms which minimize the expected error or maximize the probability that the error is kept within prefixed bounds. Here we will only outline some basic ideas, which in any case enable the design of actual algorithms for 1-dimensional problems. The computational evidence is still limited but the first results are encouraging.

As stochastic model, we shall consider the classical Wiener process.

6.1. The basic structure of the model

Let $x(t)$, $t \in T = [0, \bar{t}\,]$ and $x(t) \in \mathbb{R}^1$, be a Wiener process for which:

(i) $x(0) = 0$

(ii) $x(t) - x(u) \sim N(0, \sigma^2 \,|t-u|)$

Let $k(t,s) = \mathrm{E}\,(x(t) \cdot x(s))$ be the covariance kernel of the process: it turns out $k(t,s) = \sigma^2 \min(t,s)$.

Given n observation $x_1, x_2, \ldots, x_n$ of the process at the points $t_1, t_2, \ldots, t_n$ $(t_i < t_{i+1})$ the joint distributions of the x_i is normal with zero mean and covariance matrix $K = (k_{ij})$ with

$$k_{ij} = k(t_i, t_j)$$

Now we compute the *a posteriori* probability distribution of $x(t)$ conditioned by $x(t_j) = x_j$, $j = 1,2,\ldots,n$.

Let

$$K^+ = \begin{pmatrix} K_{11} & K_{12} \\ K_{21} & K_{22} \end{pmatrix}$$

where $K_{11} = \sigma^2 t$, $K_{22} = K$, $K_{12} = (k(t,t_1), k(t,t_2), \ldots, k(t,t_n))$, $K_{21} = K_{12}^T$.

Let

$$\mathrm{V} = (K^+)^{-1} = \begin{pmatrix} V_{11} & V_{12} \\ V_{21} & V_{22} \end{pmatrix}$$

It turns out that the density function of $x(t)$ conditioned by z is given by:

$$p\left(x(t)\,|\,z\right) = \frac{|V_{11}|^{1/2}}{\sqrt{2\pi}} \exp\left\{-\frac{1}{2}\left(x(t) + V_{11}^{-1} V_{12} z^T\right)^2 V_{11}\right\}$$

where $z = (x_1, x_2, \ldots, x_n)$, $V_{11}^{-1} = K_{11} - K_{12} K_{22}^{-1} K_{21}$ and $V_{11}^{-1} V_{12} = - K_{12} K_{22}^{-1}$. Therefore

$$\mathrm{E}(x(t)\,|\,z) = K_{12} K_{22}^{-1} z$$

and

$$\mathrm{var}\,(x(t)\,|\,z) = K_{11} - K_{12}^{-1} K_{21}$$

Cumbersome but straightforward computations lead to the final result:

$$\mathrm{E}(x(t)\,|\,z) = x_i \frac{t_{i+1} - t}{t_{i+1} - t_i} + x_{i+1} \frac{t - t_i}{t_{i+1} - t}$$

$$\mathrm{var}\,(x(t)\,|\,z) = \sigma^2 \frac{(t - t_i)(t_{i+1} - t)}{t_{i+1} - t_i}, \qquad (t_i < t < t_{i+1})$$

We remark that the assumption $x(0) = 0$ does not influence, by the Markov property, the behaviour of the process for $t \geqslant t_1$.

6.2. Optimal strategies for probabilistic models

In the framework of the Wiener model, the strategies (or, equivalently, algorithms) come to be considered as statistical decision functions.

Definition A strategy is a vector $d = (d_0, d_1, \ldots, d_{N-1})$ where d_i $(i = 0, 1, \ldots, N-1)$ are measurable functions (d_0 is the initial decision corresponding to a point $t_1 \in T$) mapping $(T \times \mathbb{R})^i$ into T

$$t_2 = d_1(z_1), \ldots, t_N = d_{N-1}(z_{N-1}).$$

where

$$z_i = \{t_1, x(t_1); t_2, x(t_2), \ldots, t_i, x(t_i)\}$$

A first optimality criterion can be derived considering the risk function

$$R_1(d) = \mathrm{E}\,\{x(t_N) - \inf_{t \in T} x(t)\}$$

d^* is called an E-optimal strategy (E is for optimality with respect to the expected value of the objective function) when

$$R_1(d^*) = \inf_{d \in \Lambda} R_1(d)$$

where Λ is the set of all measurable vector functions.

The problem of finding

$$\min_{d \in \Lambda} R_1(d)$$

can be reduced to that of finding

$$\min_{d \in \Lambda} \mathsf{E}\{x(t_N)\} = \min_{d \in \Lambda} \mathsf{E}\{\mathsf{E}\{\ldots \mathsf{E}\{\mathsf{E}\{x(t_N)|z_{N-1}\}|z_{N-2}\}\ldots|z_1\}\}$$

Therefore, d^* can be expressed reducing the minimization of $\mathsf{E}\{x(t_N)\}$ to the following chain of recursive minimizations

$$\begin{aligned}
u_N(z_{N-1}) &= \min_{t_N} \mathsf{E}\{x(t_N) \mid z_{N-1}\} \\
u_{N-1}(z_{N-2}) &= \min_{t_{N-1}} \mathsf{E}\{u_N(z_{N-1}) \mid z_{N-2}\} \\
&\vdots \\
u_2(z_1) &= \min_{t_2} \mathsf{E}\{u_3(z_2) \mid z_1\} \\
u_1 &= \min_{t_1} \mathsf{E}\{u_2(z_1)\}
\end{aligned}$$

The actual computation of d^* by the above relations can be too costly: a simplified version for computing an approximation to d^* has been proposed by Mockus.

Another optimality criterion (P-optimality) can also be considered.

Let

$$A_{n,\epsilon} = \{x : x(t_N) < \inf_{t \in T} x(t) + \epsilon\}$$

$$R_2(d) = \mathsf{E}\{X_{A_{n,\epsilon}}(x)\} = \text{meas } A_{n,\epsilon} = P\{x \in A_{n,\epsilon}\}$$

where $X_{A_{n,\epsilon}}$ is the characteristic function of the set $A_{n,\epsilon}$.

The decision function d^* is considered P-optimal when

$$R_2(d^*) = \sup_{d \in \Lambda} R_2(d)$$

P-optimal strategies can still, in principle, be computed by a chain of recursive maximization

$$
\begin{aligned}
u_N(z_{N-1}) &= \max_{t_N} \; \mathrm{E}\,\{X_{A_{n,\epsilon}}(x) \mid z_{N-1}\} \\
u_{N-1}(z_{N-2}) &= \max_{t_{N-1}} \; \mathrm{E}\,\{u_N(z_{N-1}) \mid z_{N-2}\} \\
&\;\;\vdots \\
u_2(z_1) &= \max_{t_2} \; \mathrm{E}\{u_3(z_2) \mid z_1\} \\
u_1 &= \max_{t_1} \; \mathrm{E}\,\{u_2(z_1)\}
\end{aligned}
$$

P-optimal means optimal 'in probability': indeed the first term of the above recursion of maximization implies, by the meaning of $X_{A_{n,\varepsilon}}$

$$\text{prob}\;\{x(t_N) < x^* + \varepsilon \mid z_{N-1}\} = \max_{t \in T}\;\text{prob}\;\{x(t) < x^* + \varepsilon \mid z_{N-1}\}$$

The probability distribution required to compute

$$\mathrm{E}\;\{X_{A_{n,\varepsilon}}(x) < x^* + \varepsilon \mid z_{N-1}\}$$

is difficult to obtain for the Wiener model.

A computational simplification could be set up considering, in place of the first relation, the following:

$$\tilde{u}_N(z_{N-1}) = \max_{t_N}\;\text{prob}\;\{x(t_N) < x^*_{N-1} - \varepsilon \mid z_{N-1}\}$$

where

$$x^*_{N-1} = \min(x_1, x_2, \ldots, x_{N-1})$$

$\tilde{u}_N(z_{N-1})$ is easily computed if $x(t)$ is a Wiener process over T, by the formulae for the conditional density given in the first part of this section.

6.3. Probabilistic bounds of the error

The assumption that $x(t)$ is a Wiener process allows the introduction of rather simple formulae for probabilistic bounds of the error $x^*_N - x^*$.

This is accomplished by means of the following theorem, whose proof is in a forthcoming paper by Archetti and Betrò. This theorem generalizes, in a suitable sense, a result given in [11].

Theorem 6
Let $x(t), t \in [a, \delta]$ be a Wiener process such that: $x(a) = x_a$ and consider the distribution

$$F(y) = P_a \left\{ \min_{a \leqslant t \leqslant b} x(t) \leqslant y \mid x(\delta) = x_b \right\}$$

It turns out that

$$F(y) = \begin{cases} 1 & y \geqslant \min(x_a, x_b) \\ \exp\left[-2 \left(\dfrac{(x_a - y)(x_b - y)}{\sigma^2 (b - a)} \right) \right] & y < \min(x_a, x_b) \end{cases}$$

Now we are in conditions of computing a probabilistic bound of the error $x_N^* - x^*$.

Let us denote with τ_j the points t_i ordered so that $\tau_j < \tau_{j+1}$ (we suppose for sake of simplicity that the points a and b are included in the set of the t_i). Setting

$$\Delta_j = [\tau_j, \tau_{j+1}]$$

we have

$$\text{prob} \left\{ \min_{a \leqslant t \leqslant b} x(t) < x_N^* - \varepsilon \mid z_N \right\} = 1 - \prod_{j=1}^{N-1} (1 - P_j)$$

where

$$P_j = \text{prob} \left\{ \min_{t \in \Delta_j} x(t) < x_N^* - \varepsilon \mid z_N \right\}$$

REFERENCES

[1] F. Archetti. 'A sampling technique for global optimization'. In [8]. (1975).

[2] F. Archetti, and B. Betró. 'Recursive stochastic evaluation of the level sets measure optimization problem'. *Paper A-21,* Dept of Operations Research, Univ. of Pisa, Italy, 1975.

[3] F. Archetti, and B Betrò. 'Convex programming via stochastic regularization'. *Paper A-17,* Dept of Operations Research, Univ. of Pisa, Italy, 1975.

[4] F. Archetti, and B. Betrò. 'A priori analysis of deterministic strategies for global optimization problems'. In [9] . (1978).

[5] F. Archetti, and F. Frontini. 'A global optimization method and its applications to techonological problems'. In [9]. (1977).

[6] B. Betrò, and L. De Biase. 'A Newton-like method for stochastic optimization'. In [9]. (1977).

[7] R.P. Brent. *'Algorithms for Minimization without Derivatives'.* Prentice-Hall, Englewood Cliffs, N.J. (1973).

[8] L.C.W. Dixon, and G.P. Szegö (eds). *'Towards Global Optimization'.* vol I, North-Holland, Amsterdam (1975).

[9] L.C.W. Dixon, and G.P. Szegö (eds) *'Towards Global Optimization'.* vol II, North-Holland, Amsterdam (1978).

[10] Yu.G. Evtushenko 'Numerical methods for finding global extrema'. *USSR Comput. Math. & Math. Phy.,* **11** (1971), 6.

[11] I.I. Gikhman, and A.V. Skorokhod *'Introduction to Random Processes Theory'.* Nauka, Moscow (1965).

[12] J. Gomulka. 'Numerical Experience with Torn's clustering algorithm and two implementations of Branin's method'. In [9]. (1977).

[13] A.I. Mikhal'skii. 'The method of averaged splines in the problem of approximating dependences on the basic of empirical data'. *Autom. & Remote Control,* **35**, n. 3 (1974) 1.

[14] J. Mockus. 'On Bayesian methods of optimization'. In [9] (1978).

[15] J. Mockus, V. Tiesis, and A. Zilinskas. 'The application of Bayesian methods for seeking the extremum'. In [9] (1978).

[16] W.L. Price. 'A controlled random search procedure for global optimization'. In [9], (1977).

[17] B.O. Shubert. 'A sequential method seeking the global maximum of a function'. *SIAM J. Numer. Anal.,* **9** (1972) 3.

[18] R.G. Strongin. 'On the convergence of an algorithm for finding a global extremum'. *Eng. Cybern.* **4** (1973) 549-55.

[19] A.G. Sukharev. 'Optimal strategies of the search for an extremum'. *USSR Comput. Math. & Math. Phys.,* **11** (1971) 4.

[20] A. Torn. 'A search clustering approach to the global optimization problem'. In [9]. (1977).

Chapter 9

Pricing Underemployed Capacity in a Linear Economic Model

G.B. Dantzig and P.L. Jackson

1. INTRODUCTION

This paper presents a method of associating a new set of dual variables with the optimal primal solution of a linear program so that the shadow price associated with an item in excess supply need not be zero. The general motivation for the method is the observation that, in real economies, resource prices must be positive if the owners are to have sufficient incentive to supply their resources to the market. A solution to the system in which a resource in excess supply had non-zero price clearly represents a dis-equilibrium situation which would give rise to competition among the owners that would presumably drive the price down toward zero. However, for technical or institutional reasons this competition may not exist, at least in the short run. Barriers to competition can take the form of market entry costs, resource differentiation, conversion costs, and the existence of large bargaining units, all of which effects may not be captured or even represented in the linear economic model. In the absence of competition there may be a range of prices which 'work' and we propose to allow a small amount (actually, an infinitesimally small amount) of substitution among the resources and capacities that are actually employed and to select a price system which reflects the marginal substitution possibilities within this amount.

2. THE METHOD

We are given the primary optimization problem: maximize Z subject to $X \geqslant 0$, $Y \geqslant 0$, $s \geqslant 0$ and

		DUAL VARIABLES
(1)	$a\theta + BX + IY = K$	$\sigma \geqslant 0$
(2)	$-gY + s = 0$	$\tau \geqslant 0$
(3)	$\theta = Z(\max)$	

The objective is to maximize a vector output from the system, the bill-of-goods vector $a\theta$. The vector X may represent both production activities and certain exogenous consumption activities. Similarly, K may represent both resource/capacities (upper bounds) and required outputs (the negative of lower bounds) of the system. The vector Y measures excess supply of the resources commodities of the system and the scalar variable s is a weighted total of this excess. Equations (1) and (3) form a very general representation of a single-period linear economic system.

Equation (2) defines a variable s as a composite measure of the slack capacity of the system. In general, the zero components of g correspond to (end-use) commodities and the positive components correspond to resources. We will refer to s as a measure of the availability of 'generalized capacity'.

Let the solution to this problem be denoted θ^0, X^0, Y^0, and s^0 with dual variables σ^0, and τ^0. In general, this solution will satisfy:

$$s^0 \geqslant 0 \tag{4}$$

$$\tau^0 = 0 \tag{5}$$

$$\sigma^0 a = 1 \tag{6}$$

$$\sigma^0 K = \theta^0 \tag{7}$$

Furthermore, the dual variables partition the matrices B and I, and the corresponding vectors X and Y, into two groups, indexed by subscripts α and β according as the columns of B and I price out positive or zero, respectively:

$$\sigma^0 B_\alpha > 0 \quad \text{and} \quad \sigma^0 I_\alpha > 0 \tag{8}$$

$$\sigma^0 B_\beta = 0 \quad \text{and} \quad \sigma^0 I_\beta = 0 \tag{9}$$

Note that $(X_\alpha^0, Y_\alpha^0) = 0$.

We require that the system be efficient with regard to the availability of generalized capacity, by which we mean that it is not possible to attain $\max(Z) = \theta^0$ with $s < s^0$. Such a solution can be obtained by a secondary optimization: maximize W subject to $\theta \geqslant 0$, $X_\beta \geqslant 0$, $Y_\beta \geqslant 0$, and

		DUAL VARIABLES
(10)	$a\theta + B_\beta X_\beta + I_\beta Y_\beta = K$	$\sigma \geqslant 0$
(11)	$-g_\beta Y_\beta + s = 0$	$\tau \geqslant 0$
(12)	$s = W(\max)$	

Note that the original optimal solution θ^0, X_{β^0}, Y_{β^0}, and s^0 also satisfies (10) and (11) so that this system is feasible. Moreover, multiplying (10) by σ^0 yields:

$$\sigma^0 a\theta + \sigma^0 B_\beta X_\beta + \sigma^0 I_\beta Y_\beta = \sigma^0 K$$

which upon substituting (6), (7), and (9) yields $\theta = \theta^0$ so that the solution obtained also optimizes the original problem.

Let the solution to this secondary problem be denoted θ^1, X^1_β, Y^1_β, s^1 with dual variables σ^1, and τ^1. This solution will satisfy:

(13) $$\tau^1 = 1$$

(14) $$\sigma^1 a \geqslant 0$$

(15) $$\sigma^1 B_\beta \geqslant 0$$

(16) $$\sigma^1 I_\beta \geqslant g_\beta$$

(17) $$\sigma^1 K = s^1$$

We choose as basis the latter, which is optimal for both the primary and secondary optimizations. Note that the previously obtained prices (σ^0, τ^0) hold in the original problem for this basis so that no revision of these prices is necessary.

We next assume that in the short run, for technical or institutional reasons, the system is 'sticky' with respect to generalized capacity. By this, we mean that if the data a, B or K should change by small amounts during the period for which the model is defined the availability of generalized capacity, as measured by s, would be unable (in the short run) to adjust for these changes. So, given s^1, the real system is 'better' represented by the problem: maximize Z subject to $X_\alpha \geqslant 0$, $X_\beta \geqslant 0$, $Y_\alpha \geqslant 0$, $Y_\beta \geqslant 0$, $t \geqslant 0$, and

		DUAL VARIABLES
(18)	$a\theta + B_\alpha X_\alpha + B_\beta X_\beta + I_\alpha Y_\alpha + I_\beta Y_\beta = K$	σ
(19)	$-g_\alpha Y_\alpha - g_\beta Y_\beta + t = -s^1$	τ
(20)	$\theta = Z(\max)$	

where we have set $t = s - s^1$. Because of our choice of s^1, max (Z) yields $t = 0$ and (19) is now an active constraint. With $t = 0$ this constraint may be written:

$$g(Y - Y^0) = 0$$

This relationship implies that even though a resource may be in technical excess supply, an increase in its use can only be accompanied by a decrease in the use of one or more other resources, where the substitution occurs according to the vector g. Effectively, what this constraint has done is remove all the slack capacity from the system but permit substitution among the capacities actually in use (1).

(1) No substitution actually takes place in this problem since the original primal solution, with $Y = Y^0$, is still optimal. Our device for generating a new system of prices consists of forcing an infinitesimally small amount of substitution to take place.

We do not propose to solve this problem directly because there can be numerical difficulties during computation due to round-off of s^1.

By construction, the original basis (with t taking the place of s in the basis) is optimal for both the primary and secondary optimizations and so will be optimal for this modified problem. Since at an optimum $t = 0$, the basic solution is degenerate and the optimal prices are not necessarily unique. As we will demonstrate the set of the optimal values for the dual variables can be easily parametrized by a single parameter, λ. We will resolve the price ambiguity by proposing a stability condition which, in turn, will be seen to imply a particular value of the parameter λ, easy to compute.

To begin, we claim that the optimal dual variables for problem will be of the form:

$$\sigma^2 = \sigma^0 + \lambda\sigma^1 \tag{21}$$

$$\tau^2 = \lambda \tag{22}$$

where λ is chosen from an interval $[0, L]$ for some positive real number L. This can be seen by combining (6)-(9) and (13)-(17):

$$\sigma^2 a = \sigma^2 a + \lambda\sigma^1 a \geqslant 0 \tag{23}$$

$$\sigma^2 B_\beta = \sigma^0 B_\beta + \lambda\sigma^1 B_\beta \geqslant 0 \tag{24}$$

$$\sigma^2 I_\beta - \tau^2 g_\beta = \sigma^0 I_\beta + \lambda(\sigma^1 I_\beta - g_\beta) \geqslant 0 \tag{25}$$

$$\sigma^2 K - \tau^2 s^1 = \sigma^0 K + \lambda(\sigma^1 K - s^1) = \theta^0 \tag{26}$$

which hold for all values of $\lambda \geqslant 0$. In addition, for optimality we require

$$\sigma^2 B_\alpha = \sigma^0 B_\alpha + \lambda\sigma^1 B_\alpha \geqslant 0 \tag{27}$$

$$\sigma^2 I_\alpha - \tau^2 g_\alpha = \sigma^0 I_\alpha + \lambda(\sigma^1 I_\alpha - g_\alpha) \geqslant 0 \tag{28}$$

which hold for at least $\lambda = 0$. Since $\sigma^0 B_\alpha$ and $\sigma^0 I_\alpha$ are both strictly positive there is some $L > 0$ for which the relations (27) and (28) hold for all $\lambda \in [0, L]$. Relations (23), (24), (25), (27), and (28) represent dual feasibility and equation (26) shows that be solution satisfies strong duality. Thus, (σ^2, τ^2) is optimal for all $\lambda \in [0, L]$. The proof that there is a maximal $L < +\infty$ rests on a further assumption, to the made shortly. Let λ^2 equal the maximal such L, assuming it exists. From (27) and (28) it can be seen that:

$$\begin{aligned} \lambda^2 &\equiv \max (L) \\ &= \min \left[\min_{\substack{i \in \alpha \\ \sigma^1 B_i < 0}} \frac{\sigma^0 B_i}{-\sigma^1 B_i}, \quad \min_{\substack{i \in \alpha \\ \sigma^1 I_i - g_i < 0}} \frac{\sigma^0 I_i}{-(\sigma^1 I_i - g_i)} \right] \end{aligned} \tag{29}$$

The numerator in each expression within the brackets of (29) is the per-unit amount that the objective function for the primary optimization (θ) is reduced by the introduction of a non-basic activity corresponding to $i \epsilon \alpha$. The denominator is the per-unit amount that the objective function for the secondary optimization (s) is increased by the introduction of the activity. To emphasize this trade-off, which λ^2 optimizes, write:

$$\lambda^2 \equiv \frac{\Delta\theta}{-\Delta s} \tag{30}$$

To motivate the next step, we propose a stability condition. Imagine that the generalized resource is owned by some monopolistic agent in the economy. Under the original basis with $\tau^2 = 0$, this agent receives nothing. However, since t is basic and zero in the solution, these prices may not be unique. By reducing the amount of the generalized capacity which the agent supplies to the economy by some small amount ϵ, the agent can effect a change in prices resulting in $\tau > 0$. Under the new price system, the agent's return is then maximized by supplying as much of the resource as he now has available (that is, by letting ϵ tend to zero). If there is a price system which is optimal for all ϵ in an interval $[0, \epsilon_1]$ for some sufficiently small ϵ_1 we will define this to be the stable system of prices we seek at $\epsilon = 0$.

The next step, then, is to remove an infinitesimal amount ϵ ($\epsilon > 0$) of the generalized capacity from the system [2]. Rewritten, the problem becomes: maximize Z subject to $X_\alpha \geqslant 0$, $X_\beta \geqslant 0$, $Y_\alpha \geqslant 0$, $Y_\beta \geqslant 0$, $t \geqslant 0$, and

DUAL VARIABLES

$$a\theta + B_\alpha X_\alpha + B_\beta X_\beta + I_\alpha Y_\alpha + I_\beta Y_\beta = K \qquad \sigma \tag{31}$$

$$- g_\alpha Y_\alpha - g_\beta Y_\beta + t = -(s^1 + \epsilon) \qquad \tau \tag{32}$$

$$\theta = Z(\max) \tag{33}$$

Assume that this system is feasible for sufficiently small ϵ. Since s^1 is maximal for $\theta = \theta^1$ any solution to (31)-(32) must have $\theta < \theta^1$ and t non-basic. Again, we do not propose to solve this system directly because of numerical problems due to redundancy and round-off errors in representing s^1 exactly and in choosing 'small' values of ϵ.

Defining σ^2 and τ^2 as before, we note that there is some $L > 0$ such that (σ^2, τ^2) is a dual feasible solution for all $\lambda \epsilon [0, L]$. That is, relations (27) and (28) hold for all $\lambda \epsilon [0, L]$ and the relations (23)-(25) hold for all $\lambda \geqslant 0$. With the new right-hand side, relation (26) now becomes:

$$\begin{aligned} \sigma^2 K - \tau^2(s^1 + \epsilon) &= \sigma^0 K + \lambda(\sigma^1 K - s^1) - \lambda\epsilon \\ &= \theta^0 - \lambda\epsilon \end{aligned} \tag{34}$$

(2) An alternative approach would be to 'introduce into' rather than 'remove from' the system. However, no change in prices would result since t is already in the basis.

If there is no upper bound for L then (σ^2, τ^2) is dual feasible for all $\lambda \geqslant 0$ and equation (34) shows that the dual program is unbounded. Thus, the assumption that (31)-(32) has a feasible solution, and hence a bounded dual, implies that λ^2 as defined in (29) does exist and that L has a finite upper bound.

Proposition
If, using the original basis, all variables other than t are strictly positive and if the system (31)-(32) is feasible for small ϵ, then there exists a sufficiently small $\epsilon_1 > 0$ such that the dual variables given by $\sigma^2 = \sigma^0 + \lambda^2 \sigma^1$ and $\tau^2 = \lambda^2$, for λ^2 as defined in (29), will be optimal for all ϵ in the interval $[0, \epsilon_1]$.

Proof: Consider the original basis applied to the system (31)-(32). Since t was basic, the solution value of t is given by the inner product of the 't row' of the basis inverse and the right-hand side. It is easily verified that the 't row' of the basis inverse must be $(\sigma^1 ; \tau^1)$ since this basis is optimal for the secondary optimization. Hence the solution value of t is given by:

$$\begin{aligned} t &= \sigma^1 K - \tau^1 (s^1 + \epsilon) \\ &= \sigma^1 K - s^1 - \epsilon \\ &= -\epsilon \end{aligned}$$

using the substitutions (13) and (17). Given the non-negativity constraint on t this basis is not feasible for any positive ϵ.

Let X_i, or possibly Y_i, be the non-basic variable which minimizes the expression for λ^2 (29) and consider the effect of increasing this variable. For convenience, let us assume this variable is X_i. Denote the column of coefficients in the problem associated with this variable by P. Note that in general the vector P is either of the form $[B_i', 0]'$ or $[I_i', g_i]'$. Premultiplying this column by the basis inverse yields the representation of P in terms of the basis. It may also be interpreted as the column of 'substitution factors' [1]. The substitution factor for the basic variable θ will be the inner product of the 'θ row' of the basis inverse and the vector P. Since this basis is optimal for the primary optimization, it is easily verified that (σ^0, τ^0) is the 'θ row' of the basis inverse. It follows that the substitution factor for θ, $(\sigma^0, \tau^0)P$, is $\Delta\theta$ as defined in (29) and (30):

(35) $$\theta = \theta^0 - \Delta\theta \; X_i$$

Similarly, the substitution factor for t, $(\sigma^1, \tau^1)P$, is Δs as defined in (29) and (30):

(36) $$t = -\epsilon - \Delta s \; X_i$$

Consequently, we can maintain feasibility ($t \geq 0$) by setting $t = 0$ and $X_i = \epsilon/(-\Delta s) \geq 0$.

Assume first that all basic variables other than t are strictly positive so that there is a range $[0, \overline{X}_i]$ over which X_i can be increased without forcing other basic variables negative. Let $\epsilon_1 = \Delta s\ \overline{X}_i$. For any $\epsilon \in [0, \epsilon_1]$ we can maintain primal feasibility by pivoting X_i into the basis at level $\epsilon/(-\Delta s)$ and dropping t from the basis.

It is easily verified that the dual variables corresponding to this new basis must be (σ^2, τ^2) with $\tau^2 = \lambda = \lambda^2$. We have already shown that this solution is dual feasible. From (30), (34), and (35) the right-hand side prices out:

$$
(37) \qquad \begin{aligned} \sigma^2 K - \tau^2 (s^1 + \epsilon) &= \theta^0 - \lambda^2 \epsilon \\ &= \theta^0 - \epsilon \Delta\theta/(-\Delta s) \\ &= \theta^0 - \Delta\theta \cdot X_i \\ &= \theta \end{aligned}
$$

demonstrating that strong duality holds. The new basis must be optimal for all $\epsilon \in [0, \epsilon_1]$. Q.E.D.

In general, however, we cannot expect all the primal basic variables to be positive, so the possibility remains that this new basis will be optimal for $\epsilon = 0$ but not optimal for any positive ϵ. Since we do not propose actually to perform the optimization of (31), (32), and (33) for some pre-selected ϵ, we will content ourselves with defining the new prices to be the same as those obtained in the non-degenerate case – namely (σ^2, τ^2) with $\tau^2 = \lambda^2$. In either case, since $\tau^2 > 0$, any agent owning generalized capacity has less incentive to withhold the resource than under the original price system ($\tau = 0$).

Combining relations (25) and (28) reveals that we have achieved our objective:

$$
(38) \qquad \begin{aligned} &\tau^2 > 0 \\ &\text{and} \\ &\sigma^2 I_i \geq \tau^2 g_i > 0 \qquad \text{for all } i \text{ such that } g_i > 0 \end{aligned}
$$

Under the original price system we had $\sigma^0 I_\beta = 0$ which meant that items in excess supply received a zero price. Under the new price system ($\tau^2 > 0$) we have that a resource (in general, any item i for which $g_i > 0$) will receive a positive price regardless of technical excess supply. Interpreting τ^2 as the price per unit of generalized capacity and g_i as the physical conversion factor for specific resource i, (38) states that the price of a resource must not be less than its value as generalized capacity. Furthermore, we may interpret $\tau^2 s^1$ as a transfer payment to the owners of slack capacity, suggesting that our device will provide prices more compatible with models using an 'institutional arrangements' approach [2].

3. SUMMARY

In summary, equilibrium prices of a linear economic model can be unstable. Small changes in capacities or resources can induce wide variations in prices. As an alternative, we have looked at an economy where an absence of competition prevents changes in slack capacity from optimal (equilibrium) levels except for some potential substitution among capacities in use. We proposed new prices obtained through a device of forcing an infinitesimally small amount of substitution to take place among the capacities in use. These new prices are given by (21) and (22) with $\lambda = \lambda^2$ as given by (29). These new prices are stable in the sense that they are invariant to smal changes in available resources and capacities.

REFERENCES

[1] G.B. Dantzig. '*Linear Programming and Extensions*.' Princeton University Press, Princeton N.J. (1963), p. 268.
[2] G.B. Dantzig. 'An instituzionalized divvy economy'. J. Econ. Theor. **11** no. 3 (1975).

Chapter 10

The Dam Problem with Variable Permeability

A. Friedman

1. THE STATIONARY PROBLEM

Let $Q = \{(x, y): 0 < x < a, 0 < y < H\}$. Let $k(t)$ be a positive-valued function, h a positive number less than H, and define

$$g(0, y) = \int_y^H k(t)(H - t)\,dt \qquad \text{if } 0 < y < H$$

$$g(a, y) = \begin{cases} \displaystyle\int_y^h k(t)(h - t)\,dt & \text{if } 0 < y < h \\ 0 & \text{if } h < y < H \end{cases}$$

$$g(x, 0) \text{ is linear in } x \qquad 0 < x < a$$

$$g(x, H) = 0 \qquad \text{if } 0 < x < a$$

Let

$$K = \{w \in H^1(Q): w = g \text{ on } \partial Q, w \geqslant 0 \text{ in } Q\}$$

Consider the variational inequality (V.I.):

$$\text{(1)} \qquad \begin{aligned} & w \in K \\ & \iint_Q \frac{1}{k} \nabla w \cdot \nabla (v - w)\,dx\,dy \geqslant - \iint_Q (v - w)\,dx\,dy \qquad \forall\, v \in K \end{aligned}$$

This V.I. has a unique solution, and

$$p = -w_y / k$$

represents the pressure in the dam Q when a reservoir of water at level $y = H$ is at the wall $x = 0$ and another reservoir of water at level $y = h$ is at the wall $x = a$; $k(t)$ is the permeability (see [1, 2]).

We are interested in the situation when $k(y)$ is piecewise constant and $k(y) \to \infty$ in some y intervals. This problem was studied by Caffarelli and Friedman [3]. We quote here two results for the case

$$k(y) = \begin{cases} 1 & \text{in } Q_1 \\ k & \text{in } Q_2 \\ 1 & \text{in } Q_3 \end{cases} \qquad \begin{matrix} y = H \\ y = y_1 \\ y = y_2 \end{matrix} \qquad \begin{array}{|c|} \hline Q_1 \\ \hline Q_2 \\ \hline Q_3 \\ \hline \end{array} \; - y = h$$

Set $w_i = w\big|_{Q_i}$. Denote by ζ^2 the solution of the modified dam problem:

$$(2) \qquad \begin{array}{ll} \zeta^2 \in W^{2,p}(Q_2) & \text{for any } p < \infty \\ \Delta\zeta^2 = 1 & \text{if } \zeta^2 > 0 \\ \zeta^2 \geqslant 0 & \text{in } Q_2 \\ (\Delta\zeta^2 - 1)\,\zeta^2 = 0 & \text{a.e. (almost everywhere) in } Q_2 \\ \zeta^2(0, y) = \frac{1}{2}\,[(H - y)^2 - (H - y_1)^2] & \text{if } y_2 < y < y_1 \\ \zeta^2(x, y_2) & \text{is linear in } x \\ \zeta^2(a, y) = 0 & \text{if } y_2 < y < y_1 \\ \zeta^2(x, y_1) = 0 & \text{if } 0 < x < a \end{array}$$

Let S_1 denote the line segment $y = y_1$, $0 < x < a$, and denote by ζ^1 the solution of the V.I. with Dirichlet-Neumann conditions defined by the inequality

$$(3a) \qquad \iint_{Q_1} \nabla\zeta^1 \cdot \nabla(v - \zeta^1)\,dx\,dy \geqslant -\iint_{Q_1} (v - \zeta^1)\,dx\,dy - \int_{S_1} \zeta^2_y\,(v - \zeta^1)\,dx$$
$$\text{for any } v \geqslant 0 \qquad v \in H^1(Q_1)$$

and by

$$(3b) \qquad \begin{array}{ll} \zeta^1 \geqslant 0 \quad \zeta^1 \in H^1(Q_1) & \\ \zeta^1(0, y) = \frac{1}{2}(H - y)^2 & \text{if } y_1 < y < H \\ \zeta^1(0, y) = 0 & \text{if } y_1 < y < H \\ \zeta^1(x, H) = 0 & \text{if } 0 < x < a \end{array}$$

Theorem 1

As $k \to \infty$,

$$(4) \qquad \begin{array}{lll} w^1 \to \zeta^1 & \text{uniformly in } Q_1 & \\ \dfrac{w^2}{k} \to \zeta^2 & \text{uniformly in } Q_2 & \\ \dfrac{w^3}{k} \to N\,\dfrac{a - x}{2a} & \text{uniformly in } Q_3 & (N = (H - y_2)^2 - (H - y_1)^2) \end{array}$$

Consider next the case when y_2 is fixed but $y_1 \to y_2$, i.e. $\varepsilon \equiv y_1 - y_2 \to 0$. It is not difficult to show that if $k\varepsilon \to 0$ then $w \to \zeta$ where ζ is the solution of the dam problem with $k \equiv 1$. Thus, the middle layer is asymptotically negligible. In order to consider the case when $k\varepsilon \to \alpha$, $0 < \alpha < \infty$, define a function $\tilde{g}$ on ∂Q by

$$\tilde{g}(0, y) = \begin{cases} \frac{1}{2}(H - y)^2 & \text{if } y_2 < y < H \\ \frac{1}{2}(H - y)^2 + (H - y_2)\alpha & \text{if } 0 < y < y_2 \end{cases}$$

$$\tilde{g}(a, y) = \tfrac{1}{2}(h - y)^2 \qquad \text{if } 0 < y < h$$

$$\tilde{g}(x, 0) \text{ is linear in } x \qquad 0 < x < a$$

$$\tilde{g} = 0 \qquad \text{elsewhere on } \partial Q$$

We introduce the closed convex set

$$K = \{\xi = (\xi^1, \xi^2) : \xi^i \in H^1(G_i), \xi \geq 0, \xi = \tilde{g} \text{ on } \partial Q\}$$

and the bilinear form

$$a(\xi, \nu) = \iint_Q \nabla \xi \cdot \nabla \nu + \frac{1}{\alpha} \iint_S (\xi^1 - \xi^2)(\nu^1 - \nu^2)$$

for $\xi = (\xi^1, \xi^2)$, $\nu = (\nu^1, \nu^2)$ with ξ^i, ν^i in $H^1(G_i)$; here

$$S = \{(x, y_2) : 0 < x < a\}$$
$$G_2 = Q_2 \qquad G_1 = Q \backslash \overline{Q_2}$$

and for definitness we take $0 < h < y_2$.

Consider the variational inequality

$$\min_{\xi \in K} \left(a(\xi, \xi) + \iint_Q \xi \right) = a(\zeta, \zeta) + \iint_Q \zeta \qquad \zeta \in K \tag{5}$$

This can be written in the form

$$a(\zeta, \xi - \zeta) + \iint_Q (\xi - \zeta) \geq 0 \qquad \text{for any } \xi \in K, \zeta \in K \tag{6}$$

Lemma 1 There exists a unique solution $\zeta = (\zeta^1, \zeta^2)$ of (6) and

$$\zeta^i \in W^{2,p}(G_i) \qquad \text{for any } p < \infty \tag{7}$$

$$\zeta_y^2 = \zeta_y^1 \qquad \text{on } S \cap \{\zeta^1 > 0\} \cap \{\zeta^2 > 0\} \tag{8}$$

(9) $$\zeta^2 - \zeta^1 + \alpha\zeta_y^2 = 0 \quad \text{on } S \cap \{\zeta^1 > 0\} \cap \{\zeta^2 > 0\}$$

The conditions (8), (9) are called *transmission conditions.*
We can now state the following theorem.

Theorem 2
If $k\varepsilon \to \alpha$, $0 < \alpha < \infty$ then

(10) $$w^1 \to \zeta^1 \quad \text{uniformly in } Q \cap \{y > y_2 + \delta\}$$

(11) $$w^2 \to \zeta^2 \quad \text{uniformly in } Q \cap \{y < y_2 - \delta\}$$

for any $\delta > 0$, where $\zeta = (\zeta^1, \zeta^2)$ is the solution of the variational inequality (6); ζ satisfies (7)-(8).
In case $k\varepsilon \to \infty$,

$$\frac{w^3}{k\varepsilon} \to \frac{a - x}{a} \, (H - y_2)$$

and $w^1 \to \hat{\zeta}^1$, where $\hat{\zeta}^1$ solves a V.I. with Neumann condition in S.

2. NON-STEADY FLOW

The results described in this section will appear in a joint paper with Jensen [4].
Let $k(x)$ be a C^1 function in x, $0 < x < \infty$, satisfying

$$k(x) > 0 \qquad (0 < x < \infty)$$

Let $g(x)$ be a C^2 function for $0 < x < b$, satisfying

$$g(x) > x \qquad \text{if } 0 < x < b$$
$$g(b) = b$$

where b is a fixed positive number. Finally, let $\ell(t)$ be a C^1 function for $0 < t < \infty$, satisfying

$$\ell(t) > -1 \qquad (0 \leq t < \infty)$$

(12) $$-\ell(0) = g'(0)$$

$$k(0) \int_0^t \ell(s)\,ds > \tfrac{1}{2}b^2 - b - \int_0^b g(x)\,dx \qquad \text{for all } t > 0$$

Consider the free-boundary problem: find functions $u(x, t)$, $s(t)$ satisfying:

(13) $\quad u_t - (k u_x)_x = 0 \quad$ if $0 < x < s(t), t > 0$

(14) $\quad u = s \quad$ if $x = s(t), t > 0$

(15) $\quad k u_x = -\dot{s} \quad$ if $x = s(t), t > 0$

(16) $\quad u(x, 0) = g(x) \quad$ if $0 < x < b$

(17) $\quad u_x(0, t) = -\ell(t) \quad$ if $t > 0$

For any $T > 0$, let

$$\Omega_T = \{(x, t): 0 < x < s(t), 0 < t < T\}$$

By a solution (13)-(17) we always mean a classical solution, that is, $s(t)$ is continuously differentiable for $t \geqslant 0$, and, for any $T > 0, u$ and u_x are continuous in $\overline{\Omega}_T$ and (13)-(17) are satisfied.

Under some conditions on k, g, ℓ, one can prove the existence of a unique solution.

The solution describes the flow of liquid in a porous vertical pipe, with pressure $u(x, t) - x$; we refer to this problem as the hydraulic problem.

Consider now the three cases:

(18) $$k(x) = \begin{cases} 1 & \text{if } 0 \leqslant x \leqslant 1 \\ k & \text{if } x > 1 \end{cases}$$

(19) $$k(x) = \begin{cases} k & \text{if } 0 \leqslant x \leqslant 1 \\ 1 & \text{if } x > 1 \end{cases}$$

(20) $$k(x) = \begin{cases} 1 & \text{if } 0 \leqslant x \leqslant 1 \quad \text{if } x \geqslant \beta \\ k & \text{if } 1 < x < \beta \quad \text{for } \beta > 1 \end{cases}$$

Here k is a fixed positive number larger than 1. We refer to these cases as the cases $(1, k)$, $(k, 1)$ and $(1, k, 1)$, respectively.

In case $(1, k)$ we assume that

$$g'(x) = 0 \quad \text{if } 1 < x < b, b > 1$$
$$g'(x) < 1 \quad \text{if } 0 < x < 1$$

In case $(k, 1)$ we require that

$$g'(x) = 0 \quad \text{if } 0 < x < 1$$
$$g'(x) < 1 \quad \text{if } 1 < x < b, b > 1$$
$$\ell(t) \equiv 0$$

In case $(1, k, 1)$ we require that

$$g'(x) = 0 \qquad \text{if } 1 < x < \beta$$
$$g'(x) < 1 \qquad \text{if } 0 < x < 1 \text{ or if } \beta < x < b,\ \beta < b$$

Then by approximating the function $k(x)$ (in each of the cases (18)-(20)) by smooth functions $k_n(x)$ one can show that for any value of the parameter k there exists a unique solution $(u, s) = (u^k, s^k)$ of (13)-(17). The solution satisfies a transversality condition at the points when $k(x)$ has a jump.

We now describe the behaviour of the solution as $k \to \infty$.

Consider the problem:

$$(21)\qquad \begin{aligned} &w_t - w_{xx} = 0 && \text{if } 0 < x < 1, T > 0 \\ &w_x(0, t) = -\ell(t) && \text{if } t > 0 \\ &w(x, 0) = g(x) && \text{if } 0 < x < 1 \\ &w_x + ww_t = 0 && \text{if } x = 1, t > 0 \end{aligned}$$

The solution w is taken in the sense that it is continuous up to the boundary, w_x is continuous for $0 \leqslant x \leqslant 1, t > 0$, and w_t is continuous for $0 < x \leqslant 1, t > 0$.

One can show that there exists a unique solution of (21) for all $t < \tilde{t}$, where

$$\tilde{t} = \sup\ \{t\colon w(1, s) > 0 \text{ if } 0 \leqslant s \leqslant t\}$$

Let

$$\sigma_1 = \sup\ \{t\colon w(1, s) \geqslant 1 \text{ for } 0 \leqslant s \leqslant t\}$$

Then one can show that the free-boundary problem

$$(22)\qquad \begin{aligned} &z_t - z_{xx} = 0 && \text{if } 0 < x < 1, t > \sigma_1 \\ &z_x(0, t) = -\ell(t) && \text{if } t > \sigma_1 \\ &z(x, \sigma_1) = w(x, \sigma_1) && \text{if } 0 < x < 1 \\ &z = s && \text{if } x = s(t), t > \sigma_1 \\ &z_x = -\dot{s} && \text{if } x = s(t), t > \sigma_1 \end{aligned}$$

has a unique solution with $s(t) < 1$ for some interval $\sigma_1 < t < \sigma_1 + \delta \quad (\delta > 0)$. Let τ_1 be the last time that $s(t)$ remains less than or equal to 1 for all $\sigma_1 < t \leqslant \tau_1$.

For $t > \tau_1$ one considers the problem (21) with $w(x, 0)$ replaced by $z(x, \tau_1)$. Then one proceeds to solve for w until the time σ_2 when $w(1, t)$ becomes smaller than 1. Next one defines z for $t > \sigma_2$ as in (22) with $z(x, \sigma_2)$ given by $w(x, \sigma_2)$ etc.

Denote the function thus defined by $\tilde{u}(x, t)$ and its free boundary by $\tilde{s}(t)$. One can show that the time σ_i, τ_i cannot converge to a finite positive number.

Theorem 3
The solution $u=u^k$, $s=s^k$ for the problem $(1, k)$ satisfies:

(23) $u^k(k, t) \to \tilde{u}(x, t)$ uniformly in compact sets of $0 \leq x < \tilde{s}(t)$, $t \geq 0$

(24) $s^k(t) \to s(t)$ uniformly in t, $0 \leq t \leq T$, for any $T > 0$.

Consider next the case $(k, 1)$. We have the same assertions (for $1 < x < \tilde{s}(t)$) (23), (24) but now $\tilde{u} = w$, $\tilde{s} = s$ where w, s form the solution of the free-boundary problem:

$$
(25) \qquad
\begin{aligned}
& w_t - w_{xx} = 0 && \text{if } 1 < x < s(t),\ t > 0 \\
& w(x, 0) = g(x) && \text{if } 1 < x < b \\
& w_x = w_t && \text{if } x = 1,\ t > 0 \\
& w = s(t) && \text{if } x = s(t),\ t > 0 \\
& w_x = -\dot{s}(t) && \text{if } x = s(t),\ t > 0
\end{aligned}
$$

This is the hydraulic problem with $k(x) \equiv 1$, except that the condition $w_x = -\ell(t)$ on the fixed left endpoint is replaced by the condition $w_x = w_t$ at the left endpoint. (One can show that (25) has a unique solution for all $t > 0$.)

Finally, in the case $(1, k, 1)$ $(k \to \infty)$, the limit of (u^k, s^k) satisfies a free-boundary problem obtained by superposing the above two special cases.

REFERENCES

[1] C. Baiocchi, V. Comincioli, E. Magenes, and G. A. Pozzi. 'Free boundary problems in the theory of fluid flow through porous media: Existence and uniqueness theorems'. *Ann. Mat. Pura & Appl.*, **96** (1973), 1-82.

[2] V. Benci. 'On a filtration problem through a porous medium'. *Ann. Mat. Pura & Appl.*, **100** (1974), 191-209.

[3] L. A. Caffarelli, and A. Friedman. 'Asymptotic estimates for the dam problem with several layers'. *Indiana Univ. Math. J.* **27** (1978), 551-80.

[4] A. Friedman, and R. Jensen. 'A non-steady flow permeability'. *J. Differ. Equations.* To appear.

Chapter 11

Stochastic Control with Partial Observations

A. Friedman

1. INTRODUCTION

The results contained in the present chapter are based on two joint papers with Anderson [1, 2].

A Brownian motion $\xi(t)$ is developing in time with cost $f(\xi(t))$ per unit time. It is assumed that $\xi(t) = (x(t), y(t))$ where $x(t)$ and $y(t)$ are independent Brownian motions. The component $x(t)$ is being continuously observed, whereas the position of $y(t)$ can be discovered only by making observations at random times σ_n with incurred cost $\beta(\xi(\sigma_n))$. Thus, σ_n is a stopping time with respect to the σ-field generated by $x(t)$, $t \geqslant 0$, and the random variables $y(\sigma_1), \ldots, y(\sigma_{n-1})$. A set A is given, and at the time σ_n the following policy is executed: (*a*) continue with the process until the next inspection if $y(\sigma_n) \epsilon A$, or (*b*) stop and shut-off the process with cost $\gamma(\xi(\sigma_n))$ if $y(\sigma_n) \notin A$. The problem considered here is that of finding an optimal sequence of inspections $\{\sigma_n\}$.

At the end we shall briefly consider strategies other than (*a*), or (*b*).

2. THE PROBLEM

Let A be a bounded Borel measurable set in the Euclidean k-dimensional space $\mathbb{R}^k$ and let α be a positive real number. Let $\beta(x, y)$, $\gamma(x, y)$ be bounded continuous functions of (x, y), $x \epsilon \mathbb{R}^m$, $y \epsilon \mathbb{R}^k$ and let $f(x, y)$ be a Borel measurable function satisfying

$$(1) \quad \int_0^\infty e^{-\alpha t} dt \int_{\mathbb{R}^m} \int_{\mathbb{R}^k} g_m(t, x - x_0)\, g_k(t, y - y_0)\, |f(x, y)|\, dx\, dy < M < \infty$$

where M is a constant independent of (x_0, y_0) and

(2) $$g_j(t,z) = \frac{1}{(2\pi t)^{j/2}} \exp\left(-\frac{|z|^2}{2t}\right) \qquad t > 0 \qquad z \in \mathbb{R}^j$$

If, for instance,

(3) $$|f(x,y)| \leq C(1 + |x| + |y|)^{\mu} \qquad (C > 0, \mu > 0)$$

then f satisfies (1).

Let $\Omega = C([0, \infty); \mathbb{R}^m \times \mathbb{R}^k)$ and (x, y): $[0, \infty) \times \Omega \to \mathbb{R}^m \times \mathbb{R}^k$ be defined by

$$(x(t, \omega), y(t, \omega)) = (\omega_1(t), \omega_2(t))$$

where $\omega_1(t)$ is the block of first m components of $\omega(t)$ and $\omega_2(t)$ is the block of last k components of $\omega(t)$. Let

$$\widetilde{\mathcal{M}}_t = \sigma(x(s), y(s), 0 \leq s \leq t)$$

and let $(p^{x,y})(x, y) \in \mathbb{R}^m \times \mathbb{R}^k$ be the probability measure on $(\Omega, (\widetilde{\mathrm{M}}_t)_{t \geq 0})$ such that for each $(x_0, y_0) \in \mathbb{R}^m \times \mathbb{R}^k$ and for any Borel sets $E \subset \mathbb{R}^m$ and $F \subset \mathbb{R}^k$

$$P^{x_0, y_0}(x(0) = x_0, \ y(0) = y_0) = 1$$

$$P^{x_0, y_0}(x(t) \in E, \ y(t) \in F) = \int_E \mathrm{d}x \int_F \mathrm{d}y \ g_m(t, x - x_0)\, g_k(t, y - y_0)$$

and for $t \geq s \geq 0$,

$$P^{x_0, y_0}(x(t) \in E, \ y(t) \in F \,|\, \widetilde{\mathcal{M}}_s) = p^{x(s), y(s)}(x(t-s) \in E, \ y(t-s) \in F)$$

Let $\widetilde{\mathcal{F}}_t = \sigma(x(s), 0 \leq s \leq t)$.

Definition 1. $\tau = (\tau_1, \tau_2, \ldots)$ is an inspection sequence if τ_1 is an $\widetilde{\mathcal{F}}_t$ stopping time, τ_{n+1} is an $\widetilde{\mathcal{F}}_t^n$ stopping time where $\widetilde{\mathcal{F}}_t^n = (\widetilde{\mathcal{F}}_t, y(\tau_1), \ldots, y(\tau_n))$, and if

$$0 < \tau_1 < \tau_2 < \ldots < \tau_n \ldots, \qquad \tau_n \uparrow \infty \text{ as } n \uparrow \infty$$

For any inspection sequence τ we define a cost

$$J(x, y; \tau) = E^{x,y}\left[\int_0^{\tau_1} e^{-\alpha u} f(x(u), y(u))\, \mathrm{d}u\right.$$

(4) $$+ e^{-\alpha\tau_1}[\beta(x(\tau_1), y(\tau_1)) + I_{y(\tau_1) \in A^c}\, \gamma(x(\tau_1), y(\tau_1))]$$

$$+ \sum_{n=1}^{\infty} \sum_{\ell=1}^{n} I_{y(\tau_\ell) \in A}\left(\int_{\tau_n}^{\tau_{n+1}} e^{-\alpha u} f(x(u), y(u))\, \mathrm{d}u\right.$$

$$+e^{-\alpha\tau_{n+1}}[\beta(x(\tau_{n+1}), y(\tau_{n+1})) + I_{y(\tau_{n+1})\in A^c}\,\gamma(x(\tau_{n+1}), y(\tau_{n+1}))]\Big)\Big]$$

where A^c is the complement of A in $\mathbb{R}^k$. The problem we wish to consider is that of studying

$$V(x, y) = \inf_{\tau} J(x, y; \tau) \tag{5}$$

and finding $\bar{\tau}$ such that

$$V(x, y) = J(x, y; \bar{\tau}) \tag{6}$$

3. IMBEDDING THE PROBLEM AND DERIVING THE QUASI-VARIATIONAL INEQUALITY

Define for $(x, y) \in \mathbb{R}^m \times \mathbb{R}^k$ and $s \geqslant 0$

$$P_s^{x,y} = \int_{\mathbb{R}^k} dz\, g_k(s, z-y)\, P^{x,z}$$

i.e. $P_s^{x,y}$ is the probability measure on Ω when the x-process starts with the initial measure δ_x and the y-process starts with the initial measure $N(s, y)$. It can be easily verified that, if $0 \leqslant x \leqslant$ t,

$$\begin{aligned} P_s^{x,y}(x(t)\in A,\ y(t)\in B|\widetilde{\mathscr{M}}_v) &= P_0^{x(v),y(v)}(x(t-v)\in A,\ y(t-v)\in B) \\ &= P^{x(v),y(v)}(x(t-v)\in A, y(t-v)\in B) \end{aligned} \tag{7}$$

For technical reasons, we wish to work with the right-continuous σ-fields

$$\mathscr{F}_t = \bigcap_{\varepsilon>0} \widetilde{\mathscr{F}}_{t+\varepsilon}, \mathscr{M}_t = \bigcap_{\varepsilon>0} \widetilde{\mathscr{M}}_{t+\varepsilon}$$

instead of with $\widetilde{\mathscr{F}}_t, \widetilde{\mathscr{M}}_t$.

If $\tau: \Omega\to[0, \infty)$, define $\phi_\tau: \Omega\to\Omega$ by $(\phi_\tau\omega)(t) = \omega(t+\tau(\omega))$. If $(g_t)_{t\geqslant 0}$ is a right-continuous family of σ-fields, then (see [7] p. 74):

$$\begin{aligned} &\text{if } \sigma \text{ and } \tau \text{ are } g_t \text{ stopping times then} \\ &\tau + \sigma(\phi_\tau) \text{ also is a } g_t \text{ stopping time} \end{aligned} \tag{8}$$

Definition 2 $\tau = (\tau_1, \tau_2, \ldots)$ is a Markov inspection sequence if τ_1 is an $\mathscr{F}_t$ stopping time,

$$(9)\ \tau_{n+1} = \begin{cases} \tau_n + \zeta_n(\phi_{\tau_n}, y(\tau_1), y(\tau_2), \ldots, y(\tau_n)) & \text{if } \prod_{\ell=1}^{n} I_A(y(\tau_\ell)) = 1 \\ \infty & \text{if } \prod_{\ell=1}^{n} I_A(y(\tau_\ell)) = 0 \end{cases}$$

where $\tau_n : \Omega \times \overset{n}{\underset{1}{\oplus}} \mathbb{R}^k \to [0, \infty), \zeta_n(\cdot, y_1, \ldots, y_n)$ is an $\mathcal{F}_t$ stopping time for fixed $y_1, \ldots, y_n$ and $\zeta_n(\omega, \cdot, \ldots, \cdot)$ is Borel measurable in $y_1, \ldots, y_n$ for fixed ω; we further require that

$$(10) \qquad 0 < \tau_1 < \tau_2 < \ldots < \tau_n < \ldots, \qquad \tau_n \uparrow \infty \text{ if } n \uparrow \infty$$

Remark One can show that, for any n, τ_{n+1} is a stopping time with respect to the σ-fields

$$\mathcal{F}_t^n = \sigma(\mathcal{F}_t, y(\tau_1), \ldots, y(\tau_n))$$

In particular, τ_{n+1} is an $\mathcal{M}_t$ stopping time.

We now define a cost function

$$(11) \qquad J(s, x, y; \tau) = E_s^{x,y} \Bigg[\int_0^{\tau_1} e^{-\alpha u} f(x(u), y(u))\, du$$

$$+ e^{-\alpha\tau_1} [\beta(x(\tau_1), y(\tau_1)) + I_{A^c}(y(\tau_1))\, \gamma(x(\tau_1), y(\gamma_1))]$$

$$+ \sum_{n=1}^{\infty} \prod_{\ell=1}^{n} I_A(y(\tau_\ell)) \Bigg(\int_{\tau_n}^{\tau_{n+1}} e^{-\alpha u} f(x(u), y(u))\, du$$

$$+ e^{-\alpha\tau_{n+1}} [\beta(x(\tau_{n+1}), y(\tau_{n+1})) + I_{A^c}(y(\tau_{n+1}))\, \gamma(x(\tau_{n+1}), y(\tau_{n+1}))] \Bigg) \Bigg]$$

Note that when $s = 0$ this definition coincides with (4). Note also that Definition 2 of a Markov inspection sequence (which we shall adopt from now on) is slightly more restrictive than Definition 1 of an inspection sequence.

Set

$$(12) \qquad V(s, x, y) = \inf_{\tau \in \mathcal{A}} J(s, x, y; \tau)$$

where $\mathcal{A}$ is the set of all Markov inspection sequences.

For any function $h(x, y)$, define

$$(13) \qquad T_{y,s} h(x, y) = \int_{\mathbb{R}^k} g_k(s, y - z)\, h(x, z)\, dz$$

and let

$$(14) \qquad \tilde{f}(s, x, y) = T_{y,s} f(x, y)$$

$$(15) \qquad G_h(s, x, y) = T_{y,s} [\beta(x, y) + I_A(y) h(x, y) + I_{A^c}(y)\, \gamma(x, y)]$$

We now proceed with the heuristic arguments of dynamic programming to derive a quasi-variational inequality (in short, QVI) for $V(s, x, y)$.

For each (s, x, y) we have two alternatives:

(i) make an inspection; or

(ii) continue for a while without inspection.

In the second case, if we continue for a small time δ and thereafter proceed optimally, we obtain the inequality

$$V(s,x,y) \leqslant E_s^{x,y} \Big(\int_0^\delta e^{-\alpha u} f(x(u), y(u))\, du$$

$$+ e^{-\alpha\delta} [\beta(x(\delta), y(\delta)) + I_{A^c}(y(\delta))\, \gamma(x(\delta), y(\delta))$$

$$+ I_A(y(\delta))\, V(0, x(\delta), y(\delta))] \Big)$$

Applying Ito's formula, dividing by δ and letting $\delta \to 0$, we obtain

$$\frac{\partial V}{\partial s} + \frac{1}{2} \Delta_x V - \alpha V + \tilde{f} \geqslant 0 \tag{16}$$

where Δ_x is the Laplacian

$$\sum_{j=1}^{m} \frac{\partial^2}{\partial x_j^2}$$

If we make the first choice above, we get

$$V(s,x,y) \leqslant E_s^{x,y} [\beta(x,y) + I_A(y)\, V(0,x,y) + I_{A^c}(y)\, \gamma(x,y)]$$

i.e.

$$V(s,x,y) \leqslant G_{V(0,\,.\,,\,.)}(s,x,y) \tag{17}$$

Finally, since the optimal choice is either (i) or (ii),

$$\left(\frac{\partial V}{\partial s} + \frac{1}{2} \Delta_x V - \alpha V + f \right) \; (V - G_{V(0,\,.\,,\,.)}) = 0 \tag{18}$$

Theorem

Suppose (i) there exists a continuous function $U(s, x, y)$ in (s, x, y) a $[0,\infty) \times \mathbb{R}^m \times \mathbb{R}^k$ with bounded derivatives U_s, U_{x_i}, $U_{x_i x_j}$ in every strip $0 < s < T$ and satisfying the QVI (16)-(18) a.e. (almost everywhere) (ii) the hitting time of the set

$$\{(s,x,y) \colon U(s,x,y) = G_{U(0,\,.\,,\,.)}(s,x,y)\}$$

by the path $((t, x(t), y), t > 0)$ is a uniformly positive function $t = t(x,y)$ for $x(0) = x \in \mathbb{R}^m$, $y(0) = y \in A$.

Then $V(t,x,y) = U(t,x,y)$ and the optimal inspections are given by $\tau_{n+1} = \tau_n + t^*(x(\tau_n), y(\tau_n))$ if $y(\tau_n) \in A$.

We cannot establish in general the existence of a solution U having derivatives U_t, U_{x_i}, $U_{x_i x_j}$. We therefore proceed in a different manner.

(i) First we derive for V the equation

$$V(s,x,y) = \inf_{\sigma} E_s^{x,y} \Big(\int_0^{\sigma} e^{-\alpha u} f(x(u), y(u))\, du + \tag{19}$$
$$+ e^{-\alpha\sigma} [\beta(x(\sigma), y(\sigma)) + I_{A^c}(y(\sigma))\, \gamma(x(\sigma), y(\sigma)) + I_A(y(\sigma))\, V(0, x(\sigma), y(\sigma))] \Big)$$

and we show that V is the unique bounded continuous solution of (19).

(ii) Next we show that the solution of (19) can be approximated by solutions to a partial differential equation which can be viewed as 'penalized problem' approximating the QVI (16)-(18).

In carrying out (i), and (ii) we use known techiques for variational inequalities (such as in [5, 8]).

From (i), and (ii) it follows that V is a 'generalized' solution of the QVI (16)-(18); furthermore, the optimal policy of inspections is precisely that described at the end of the theorem quoted above.

The boundary of the set $\{V < G_V\}$ is called the free boundary. The study of the location, shape, and regularity of this boundary is an open problem. In case the $x(t)$ process does not appear in the model and $y(t)$ is one-dimensional, the free boundary is a curve $t = \varphi(y)$, and some properties of φ are known. But even in this case, it is not known, for instance, whether φ is a continuous function.

The methods of [1, 2] apply to more general cost functions. For instance, to the model where at the time of inspection τ_i we are allowed the following options:

(1) replace $y(\tau_i)$ by moving it into any desired location on the ray $y(\tau_i) - \eta$, with cost $\gamma(x(\tau_i), y(\tau_i) - \eta)$ where $\eta = (\eta_1, \ldots, \eta_k)$, $\eta_i \geqslant 0$;

(2) do not change the position of $y(\tau_i)$.

The corresponding QVI is again (16)-(18) except that in the definition of G_V the initial conditions (on $s = 0$) are changed into

$$G_V(x,y,0) = \beta(x,y) + \min \Big\{ \min_{\eta \geqslant 0} [\gamma(x,y-\eta) + V(x,y-\eta,0],\ V(x,y,0) \Big\}$$

The corresponding QVI can be solved by the method of [5] and that of Mainaldi [6].

A problem of this type (but with delay) was also studied by Robin [9].

Some earlier models in connection with surveillance theory were introduced in [3, 4, 10].

REFERENCES

[1] R. F. Anderson, and A. Friedman. 'Optimal inspections in stochastic control with costly observations'. *Math. Oper. Res.*, **2** (1977), 155-90.
[2] R. F. Anderson, and A. Friedman. 'Optimal inspections in a stochastic control problem with costly observations. II'. *Math. Oper. Res.* **3** (1978), 67-81.
[3] G. R. Mantelman, and I. R. Savage. 'Surveillance problems: Wiener processes'. *Nav. Res. Logist. Q.*, **12** (1965), 35-55.
[4] J. A. Bather. 'A control chart model and a generalized stopping problem for Brownian motion'. *Math. Oper. Res.*, **1** (1976), 209-24.
[5] A. Bensoussan, and J. L. Lions. 'Temps d'arrêt et contrôle impulsionnnel: Inéquations variattionnelles et quasi variationnelles d'évolution'. *Univ. Paris IX, Cah. Math. Décision*, no. 7523 (1975).
[6] J. L. Mainaldi. *Thesis,* IRIA, Domaine de Voluceau, Rocquencourt, France. To appear.
[7] P. A. Meyer. *'Probability and Potentials'.* Blaisdell Publishing Co., Waltham, Mass. (1966).
[8] M. Robin. 'Controle impulsionnel de processus de Markov'. *Thesis,* IRIA, Domaine de Voluceau, Rocquencourt, France (1977).
[9] M. Robin. 'Optimal maintenance and inspection for Markov deterioration models.' *Conf. on Optimization*, Würzburg, Germany, 1977. To appear.
[10] J. R. Savage. 'Surveillance problems'. *Nav. Res. Logist. Q.*, **9** (1962), 189-209.

Chapter 12

Theorems of Alternative, Quadratic Programs and Complementarity Problems

F. Giannessi

1. INTRODUCTION

A theorem of alternative for non-linear systems is considered under general assumptions both for the unknown and for the involved functions. Some known results are shown to be straightforward consequences of it. Connections with duality and applications both to extremum and to vector extremum problems are discussed.

In the case of a finite-dimensional unknown, particular attention is devoted both to quadratic programming and to quadratic complementarity problems. Some difficulties of convexification of the former problems are exhibited by such an analysis. Moreover, it will be shown how to reduce the latter problems to a finite sequence of quadratic programs.

2. THEOREMS OF ALTERNATIVE FOR NON-LINEAR SYSTEMS

Consider a set [1] $X \subseteq \mathbb{R}^n$, and real-valued functions $f : X \to \mathbb{R}^\ell$, $g : X \to \mathbb{R}^m$; let $x = (x_1, \ldots, x_n)$ denote an element of X and define the set

$$H = \{(u,v) \in \mathbb{R}^{\ell+m} : u > 0;\ v \geqslant 0\}$$

Denote by $G : \mathbb{R}^{\ell+m} \to \mathbb{R}$ a real-valued function, such that $G(u, v) > 0$, iff $(u, v) \in H$.

(1) Throughout the paper X is a subset of the real Euclidean space; however, some results hold under weaker assumptions; int $*$ and cl $*$ will denote interior of $*$ and closure of $*$, respectively; $\langle *, ** \rangle$ will denote the inner product between $*$ and $**$.

Theorem 1

Let X, f and g be given. The system

$$(1) \qquad f(x) > 0, \quad g(x) \geqslant 0, \quad x \in X$$

is impossible, iff the inequalities

$$(2) \qquad G(f(x), g(x)) \leqslant 0, \quad \forall\, x \in X$$

hold.

Proof: If (1) is possible, i.e. if $\exists\, \overline{x} \in X$ such that $\overline{u} \triangleq f(\overline{x}) > 0$ and $\overline{v} \triangleq g(\overline{x}) \geqslant 0$, then $(\overline{u}, \overline{v}) \in H$ and hence $G(f(\overline{x}), g(\overline{x})) = G(\overline{u}, \overline{v}) > 0$, so that (2) is false. If (1) is impossible, i.e. if $(u, v) \triangleq (f(x), g(x)) \notin H, \forall\, x \in X$, then $G(f(x), g(x)) = G(u, v) \leqslant 0, \forall\, x \in X$, so that (2) is true. This completes the proof.

Note that the former part of the proof shows *weak alternative,* i.e. the intersection between the set of solutions to (1) and the set $\{x \in X : G(f(x), g(x)) \leqslant 0\}$ is empty; the latter shows *strong alternative*, i.e. the union of the above sets equals X. Then we can say that the alternative holds, iff both weak and strong ones hold.

Theorem 1 would seem just what we need to state alternative. Unfortunately, G is impracticable for applications, so that its definition must be weakened. A first case is obtained when G is replaced by a function $\Gamma : \mathbb{R}^{\ell+m} \to \mathbb{R}$, such that $\Gamma(u, v) > 0$ iff $(u, v) \in \operatorname{int} H$; alternative is still guaranteed if in (2) the strict inequality is imposed on H-int H. The replacement of G by Γ leads to a slight improvement: Γ has an open positivity level set. This brings us less far from practically meaningful theorems of alternative. However, further relaxation is necessary. To this aim consider the set

$$K = \{(u, v) \in \mathbb{R}^{\ell+m} : u = f(x)\,;\; v = g(x)\,;\; x \in X\}$$

A function $S : \mathbb{R}^{\ell+m} \to \mathbb{R}$ will be called a *separation function* (in short, S-function), iff

$$S(u, v) \leqslant 0 \qquad \forall (u, v) \in K - H \cap K$$

and

$$S(u, v) > 0 \qquad \forall (u, v) \in H$$

If H is replaced by a set containing it or by any its subset, S will be called a *weak* (or *strong*) *separation function,* respectively. A weak (strong) S-function guarantees weak (strong) alternative. Of course, the separation property depends on K and hence on f, g. G is an S-function and Γ a strong S-function independently of f, g; every relaxation will lead to more practicable functions, but will lose such an independence. To smooth away this difficulty it is natural to try to characterize a set of f, g and a corresponding class of functions (defined, for instance, by a parameter), such that, whatever f and g may be in the set, a (weak, strong) S- function can be found in the class. This is partially achieved by the next theorem and completely by Theorem 3.

Theorem 2
System (1) is impossible, iff there exist functions $\lambda : X \to \mathbb{R}_+^\ell$, $\mu : X \to \mathbb{R}_+^m$, with $(\lambda(x), \mu(x)) \neq 0$, $\forall x \in X$, and such that

$$(3) \qquad \langle \lambda(x), f(x) \rangle + \langle \mu(x), g(x) \rangle \leq 0, \quad \forall x \in X$$

where the inequality must be strictly satisfied $\forall x \in X$ such that $\lambda(x) = 0$.

Proof: If (1) is possible, i.e. if $\exists\, \bar{x} \in X$ such that $f(\bar{x}) > 0$ and $g(\bar{x}) \geq 0$, then $\langle \lambda(\bar{x}), f(\bar{x}) \rangle$ is positive or zero (and correspondingly the left-hand side of (3) is positive or non-negative) according to whether $\lambda(\bar{x}) \neq 0$ or $\lambda(\bar{x}) = 0$, respectively, so that (3) is false. If (1) is impossible, i.e. if $(u, v) \triangleq (f(x), g(x)) \notin H$, $\forall x \in X$, then there exists a closed half-space containing (u, v), while H is contained in the complement. It follows that $\exists\, \lambda \in \mathbb{R}_+^\ell$ and $\mu \in \mathbb{R}_+^m$ with $(\lambda, \mu) \neq 0$, such that $\langle \lambda, u \rangle + \langle \mu, v \rangle \leq 0$ or $\langle \mu, v \rangle < 0$ according to whether $\lambda \neq 0$ or $\lambda = 0$, respectively. Because of the arbitrariness of x, it follows that (3) is true. This completes the proof.

When f, g are concave and X is convex, λ and μ may be assumed to be constant, as is shown by the following theorem.

Theorem 3
Let f and g be concave and X convex. System (1) is impossible, iff there exist $\lambda \in \mathbb{R}_+^\ell$ and $\mu \in \mathbb{R}_+^m$, with $(\lambda, \mu) \neq 0$, such that

$$(4) \qquad \langle \lambda, f(x) \rangle + \langle \mu, g(x) \rangle \leq 0, \qquad \forall x \in X$$

where the inequality must be strictly verified, if $\lambda = 0$.

Proof: Set

$$E(x) = \{(u, v) \in \mathbb{R}^{\ell + m} : u \leq f(x); v \leq g(x)\}$$

$$E = \bigcup_{x \in X} E(x)$$

The concavity of f, g and the convexity of X imply the convexity of E. By noting that E is the intersection of supporting half-spaces of K of the kind $\langle\lambda, u\rangle + \langle\mu, v\rangle \leqslant k$, with $(\lambda,\mu) \geqslant 0$ and $(\lambda,\mu) \neq 0$, a well known theorem of separation for convex sets enables one to achieve (4). This completes the proof.

Remark 1 The crucial point of the above proof consists in showing the convexity of E; any other assumption which guarantees such a convexity guarantees (4) too. This happens in the following.

Corollary 1 Let X, f, and g be given. Assume that there exists a convex $Y \subseteq X$, such that the restrictions of f and g to Y are concave; and that, for every $x \in X$, there exists at least one $y \in Y$, such that $f(x) \leqslant f(y)$ and $g(x) \leqslant g(y)$. Then, system (1) is impossible, iff there exist $\lambda \in \mathbb{R}^{\ell}_{+}$ and $\mu \in \mathbb{R}^{m}_{+}$, with $(\lambda,\mu) \neq 0$, such that condition (4) holds.

Proof: When f, g, and X are replaced respectively by the restrictions of f, g to Y and Y, then obviously Theorem 3 can be applied; in such a case let E_Y denote the set which replaces E. To complete the proof it is enough to note that $E \subseteq E_Y$.

Remark 2 Of course, the convexity of the above set E is only a sufficient condition for (4) to hold. This is shown by the following corollary of Theorem 3, where, for the sake of simplicity, we consider the case $\ell = 1$.

Corollary 2 Assume that there exist scalars $a_i > 0$; b_i, $i = 1,\dots, m$; a_0; and $\hat{x} \in X$, such that

(i) $$f(x) \leqslant a_0 - \sum_{i=1}^{m} a_i\,[g_i(x) - b_i]^2 \;, \quad \forall\, x \in X$$

(ii) $$g_i(\hat{x}) = (1 - \delta_i)\, b_i \;, \quad i = 1, \dots, m$$

(iii) $$f(\hat{x}) = a_0 - \sum_{i=1}^{m} a_i\, \delta_i\, b_i^2$$

where $\delta_i = 0$ if $b_i > 0$ and $\delta_i = 1$ if $b_i \leqslant 0$. Then, system (1) is impossible, iff there exist $\lambda \in \mathbb{R}_+$ and $\mu \in \mathbb{R}^m_+$, with $(\lambda,\mu) \neq 0$, such that condition (4) holds.

Proof: Set

$$\mathcal{E} = \{(u,v) \in \mathbb{R}^{1+m} : u \leqslant a_0 - \sum_{i=1}^{m} a_i(v_i - b_i)^2\}$$

It will be shown that $H \cap K = \emptyset \iff H \cap \mathcal{E} = \emptyset$; then Theorem 3 can be applied, with the convex set $\mathcal{E}$ in place of K, to get the thesis. (i)$\Rightarrow K \subseteq \mathcal{E}$ so that $H \cap \mathcal{E} = \emptyset \Rightarrow H \cap K = \emptyset$. If the vice versa is not true, then $\exists\ \tilde{u} > 0$, $\tilde{v} \geqslant 0$ such that

$$\tilde{u} \leqslant a_0 - \sum_{i=1}^{m} a_i (\tilde{v}_i - b_i)^2$$

and hence

$$a_0 - \sum_{i=1}^{m} a_i \delta_i b_i^2 > 0$$

Account taken of (ii) and (iii), this inequality implies $f(\hat{x}) > 0$, $g(\hat{x}) \geqslant 0$, $\hat{x} \in X$, so that $H \cap K \neq \emptyset$; the absurd completes the proof.

Note that assumption (i) of Corollary 2 may be written as $Q(f(x), g(x)) \leqslant 0$, where Q is a particular positive-semidefinite quadratic form whose rank is m. Results more general than Corollary 2 may be achieved by adopting a more general function of the kind of Q. Note also that, when $\ell > 1$, an extension of Corollary 2 may be obtained by making, for every element of f, assumptions like (i) and (iii).

Remark 3 Let $T: \mathbb{R} \to \mathbb{R}$ be a real-valued function, such that $T(z) \gtreqless 0$ according to $z \gtreqless 0$, respectively. Obviously, (1) is equivalent to

$$T(f_i(x)) > 0,\ i = 1, \ldots, \ell;\ T(g_i(x)) \geqslant 0,\ i = 1, \ldots, m;\ x \in X$$

A particular case is $T(z) = 1 - \exp(-\alpha z)$, where α is a positive scalar; this represents an *exponential transformation* of the left-hand sides of (1). When f and g are not concave, but their transformations are, then Theorem 3 can be applied (to the transformed system); see Example 1. If the transformations are concave only on the subset Y defined in Corollary 1, this can be applied.

Remark 4 In system (1) some inequalities can be aggregated, i.e. replaced by only one. For the sake of simplicity and without loss of generality, consider the case of two inequalities. To this end set

$$H_2 = \{(u_1, v_1) \in \mathbb{R}^2 : u_1 > 0;\ v_1 \geqslant 0\}$$

$$f(x) = (f_1(x),\ \bar{f}(x))$$

$$g(x) = (g_1(x),\ \bar{g}(x))$$

and denote by $G_1 : \mathbb{R}^2 \to \mathbb{R}$ a real-valued function, such that $G_1(u_1, v_1) > 0$

iff $(u_1, v_1) \in H_2$, so that G_1 is an S-function (of two variables) like G, i.e. independent of f, g. Then, it is easy to show that (1) is impossible iff the system

$$(5)\qquad \begin{aligned} &G_1(f_1(x), g_1(x)) > 0 \\ &\bar{f}(x) > 0, \quad \bar{g}(x) \geq 0, \quad x \in X \end{aligned}$$

is impossible. Hence, (1) may be replaced by (5), which has $(\ell + m - 1)$ inequalities.

Remark 5 All the known theorems of alternative for linear systems can be quickly deduced from Theorem 3. For instance, consider the following non-homogeneous Farkas theorem.

Corollary 3 Let A be an $m \times n$ matrix; b an $m \times 1$ vector; c a $1 \times n$ vector; and γ a scalar. The system

$$(6)\qquad \langle c, x\rangle > \gamma, \quad Ax \leq b$$

is impossible iff at least one of the following systems is possible:

$$(7a)\qquad tA = c, \quad \langle t, b\rangle \leq \gamma, \quad t \geq 0$$

or

$$(7b)\qquad tA = 0, \quad \langle t, b\rangle < 0, \quad t \geq 0$$

where t is a $1 \times m$ vector.

Proof: Set $X = \mathbb{R}^n$, $\ell = 1$, $f(x) = \langle c, x\rangle - \gamma$, $g(x) = b - Ax$, and apply Theorem 3. If we set $t = \mu$ when $\lambda = 0$, and $t = (1/\lambda)\mu$ when $\lambda > 0$, (4) becomes either

$$\langle tA - c, x\rangle \geq \langle t, b\rangle - \gamma, \qquad t \geq 0, \qquad \forall x \in \mathbb{R}^n$$

if $\lambda > 0$; or

$$\langle tA, x\rangle > \langle t, b\rangle, \qquad t \geq 0, \qquad \forall x \in \mathbb{R}^n$$

if $\lambda = 0$; (7) now follows.

Remark 6 Another particular case is met when $X = \mathbb{R}^n$, f and g are concave and such that

$$f(\alpha x) = \alpha f(x), \quad g(\alpha x) = \alpha g(x)$$

$$\forall \alpha \in \mathbb{R} \quad \forall x \in \mathbb{R}^n$$

$$\left.\begin{array}{l} f(x') \geqslant 0 \\ f(x'') \geqslant 0 \\ \alpha, \beta \in \mathbb{R}_+ \end{array}\right\} \quad \Rightarrow \quad f(\alpha x' + \beta x'') \geqslant 0$$

$$\left.\begin{array}{l} g(x') \geqslant 0 \\ g(x'') \geqslant 0 \\ \alpha, \beta \in \mathbb{R}_+ \end{array}\right\} \quad \Rightarrow \quad g(\alpha x' + \beta x'') \geqslant 0$$

In such a case, the sets

$$C = \{x \in \mathbb{R}^n : f(x) \geqslant 0\}$$
$$D = \{x \in \mathbb{R}^n : g(x) \geqslant 0\}$$

are convex cones. System (1) is possible, iff

(8) $$(\text{int } C) \cap D \neq \emptyset$$

The cones

$$C^* = \{y \in \mathbb{R}^n : \langle y, x\rangle \leqslant 0, \quad \forall\, x \in C\}$$

$$D^* = \{y \in \mathbb{R}^n : \langle y, x\rangle \leqslant 0, \quad \forall\, x \in D\}$$

are the polars of C and D, respectively. Under the preceding assumption on f and g, Theorem 3 can be applied. It is not difficult to show that (4) is equivalent to

(9) $$\begin{array}{l} \lambda \geqslant 0, \quad \mu \geqslant 0, \quad \lambda \neq 0 \\ \langle \lambda, f(x)\rangle + \langle \mu, g(x)\rangle = 0, \quad \forall\, x \in \mathbb{R}^n \end{array}$$

In fact, if the left-hand side of equation (9) is negative at x, a contradiction is obtained by evaluating it at $-x$; the same conclusion is met if $\lambda = 0$. It is easy to show that (9) is possible iff

(10) $$(-C^*) \cap D^* \neq \{0\}$$

In fact, we have $C \subseteq C' \triangleq \{x \in \mathbb{R}^n : \langle \lambda, f(x)\rangle \geqslant 0\}\ \forall\, \lambda \in \mathbb{R}^{\ell}_{+}$; $D \subseteq D' \triangleq \{x \in \mathbb{R}^n : \langle \mu, g(x)\rangle \geqslant 0\}\ \forall\, \mu \in \mathbb{R}^m$; and that (9) implies

$$D' = \{x \in \mathbb{R}^n : -\langle \lambda, f(x)\rangle \geqslant 0\} = \sim \text{int } C'$$

($\sim$ denotes the complement). This fact enables us to conclude, if we note that the thesis is trivial at $n=2$ and that we can consider equivalently the fact on the set of all the hyperplanes of $\mathbb{R}^n$ containing the origin. Now, because of Theorem. 3, we have that (8) is true iff (10) is false. This result has been recently stated by Lehmann and Oettli [21]; when $g(x)\equiv 0$ and f is linear, such a result is the well known Gordan (1873) theorem of alternative.

In a quite similar way, and always considering a real Euclidean space and ordinary convexity, it is possible to show that Theorem 2.1 of [2] is a straightforward consequence of Theorem 3.

Remark 7 (1) is not the only system for which a theorem of alternative can be stated. Let $\mathbb{Z}$ and $\mathbb{Q}$ denote the sets of integers and rationals; and $\lfloor * \rfloor$ or $\lceil * \rceil$ denote the vectors whose elements are the greatest integers less than or the smallest integers greater than the corresponding elements of vector $*$, respectively. In (1) set $X = \mathbb{R}^n$ and replace '>0' and '$\geqslant 0$' with '$\notin \mathbb{Z}^\ell$' and '$\in \mathbb{Z}^m$', respectively; in (2) replace '$\leqslant 0$' with '$\in \mathbb{Z}$'. Moreover, define here $H=\{(u,v): u \notin \mathbb{Z}^\ell;\ v \in \mathbb{Z}^m\}$ and $G: \mathbb{R}^{\ell+m} \to \mathbb{R}$, such that $G(u,v) \in \mathbb{Z}$ iff $(u,v) \notin H$. Theorem 1 holds here too. Of course, G is again impracticable for applications, so that we may be obliged to weaken its definition. In the particular case $\ell=1$ a function which fulfills the above definition is

$$G(u,v) = u + \langle \mu, \lfloor v \rfloor \rangle + \delta(\lceil u \rceil - u), \text{ with } \mu \in \mathbb{Z}^m$$

where $\delta = 0$ if $v \in \mathbb{Z}^m$ and $\delta = 1$ if $v \notin \mathbb{Z}^m$. For instance, we may set $\delta = \lceil \sigma/(1+\sigma) \rceil$, where $\sigma = v_1 - \lfloor v_1 \rfloor + \ldots + v_m - \lfloor v_m \rfloor$. Note that the above function G is an S-function. Just to have an idea of this completely open area of study consider the following particular case: set $f(x) = \langle c^0, x \rangle$, $g_i(x) = \langle c^i, x \rangle$, $i=1,\ldots,m$, where c^i is an n-vector. It is easy to show that (2) becomes: $\exists\, \mu \in \mathbb{Z}^m$ such that

$$\langle c^0 + \sum_{i=1}^{m} \mu_i c^i, x \rangle \in \mathbb{Z}, \qquad \forall x \text{ such that } g(x) \in \mathbb{Z}^m$$

$$\lceil \langle c^0, x \rangle \rceil + \sum_{i=1}^{m} \mu_i \lfloor \langle c^i, x \rangle \rfloor \in \mathbb{Z}, \qquad \forall x \text{ such that } g(x) \notin \mathbb{Z}^m$$

The second part of this condition is redundant, if $g(x) \in \mathbb{Z}^m$ is possible.

Thus, it is not easy to state if this condition is true or not (i.e. to state alternative), as it requires a partitioning of the set of x's. In the special case when the elements of all vectors and matrices belong to $\mathbb{Q}$, the condition $g(x) \in \mathbb{Z}^m$ is always possible and hence the above condition collapses to: $\exists \mu \in \mathbb{Z}^m$, such that

$$c^0 + \mu_1 c^1 + ... + \mu_m c^m = 0$$

In fact, in the contrary case, if $\bar{x}$ is such that $g(\bar{x}) \in \mathbb{Z}^m$, by a well known property of diophantine systems, $\langle c^0 + \sum_{i=1}^{m} \mu_i c^i, \bar{x} \rangle \in \mathbb{Z}$ is contradicted. This result is quoted in [7] as integer theorem of Farkas, and is implicitly already contained in a paper of Kronecker (1899) [20]. At last note that, if we replace G with the function $\Gamma(u, v) = u + \langle \mu, v \rangle$, alternative is no longer ensured; however *weak alternative* is still guaranteed, i.e. Γ is a weak S-function. In fact, if (1) is possible, i.e. if $\exists\, \bar{x} \in X$ such that $f(\bar{x}) \notin \mathbb{Z}$ and $g(\bar{x}) \in \mathbb{Z}^m$, then $f(\bar{x}) + \langle \mu, g(\bar{x}) \rangle \notin \mathbb{Z}, \forall\, \mu \in \mathbb{Z}^m$, and hence (2) is false.

3. APPLICATIONS TO EXTREMUM PROBLEMS AND TO VARIATIONAL INEQUALITIES

Consider the real-valued function $\varphi : X \to \mathbb{R}$, and the following extremum problem [2]

(11) $$\min\ \varphi(x), \qquad x \in R \triangleq \{x \in X : g(x) \geqslant 0\}$$

Theorems 1 and 2 may be used to state a necessary and sufficient optimality condition for problem (11). To this end it is enough to set $\ell = 1$ and $f(x) = \varphi(\hat{x}) - \varphi(x)$. Then, Theorem 1 (or 2) may be read this way: $\hat{x} \in R$ *is an optimal solution of* (11) *iff* (2) (or (3)) *holds.* In the same way, from Theorem 3 we deduce that, *when X is convex and $-\varphi$, g concave, $\hat{x} \in R$ is an optimal solution of* (11) *iff* (4) *holds*; in the particular case of φ, g affine, the last statement collapses to a well known fact [1, 4, 9].

These propositions may be used not only to state whether a given $\hat{x}$ is or is not an optimal solution of (11), but also to decompose (11) into a sequence of 'easier' subproblems [1, 3, 4, 9, 11, 12, 13]. Later it will be shown how such a decomposition may be obtained for quadratic programs and for complementarity problems. Now let us discuss some general aspects of the application of a theorem of alternative to an extremum problem. To

(2) Of course, when necessary, min and max must be replaced by inf and sup, respectively.

this end, consider some special weak S-functions. Set

$$\Gamma_1(u, v; \lambda, \mu, \alpha) = \lambda u + \sum_{i=1}^{m} \mu_i[1-\exp(-\alpha_i v_i)]$$

$$\Gamma_2(u, v; \lambda, \mu) = \lambda u + \langle \mu, v \rangle$$

where the scalar λ and vectors $\mu = (\mu_1, \ldots, \mu_m)$, $\alpha = (\alpha_1, \ldots, \alpha_m)$ are non-negative parameters. Note that, if $(u, v) = (0, 0)$ or $(u, v) \notin \text{cl } H$ then the inequality $\Gamma_1 \leq 0$ can be fulfilled with $\lambda = 1$ and $\mu, \alpha > 0$; if $(u, v) \neq (0, 0)$ and $(u, v) \in \text{cl } H - H$ then $\Gamma_1 \leq 0$ implies either $\mu_i = 0$ or $\alpha_i = 0$ when $v_i \neq 0$; if $(u > 0, v = 0) \notin K$ is a limit point of K then the inequality $\Gamma_1 \leq 0$ requires $\lambda = 0$ and is fulfilled as strict inequality: $\Gamma_1 < 0$. It follows that (2) *holds, iff there exist non-negative* λ, μ, α *such that*

$$X^+ \cup X^0 \cup \text{lev}_{\leq 0} f = X \tag{12}$$

where

$$X^+ \triangleq \{x \in X : \Gamma_1(f, g; 1, \mu, \alpha) \leq 0\}$$

$$X^0 \triangleq \{x \in X : \Gamma_1(f, g; 0, \mu, \alpha) < 0\}$$

The esponential transformation $\bar{g}_i(x) \triangleq 1 - \exp[-\alpha_i g_i(x)]$ of the constraining function g_i modifies the set E of Remark 1 in the sense that the i-th element of vector v is reduced to a value less than 1, or remains unchanged, or (arbitrarily) decreases, according to it is $\gtreqless 0$, respectively, Hence, if some suitable assumption has been made which ensures that no element $(u, v) \neq 0$ of K belongs to $\text{cl } H - H$ and no $(u > 0, v = 0) \notin K$ is a limit point of K, then (12) becomes

$$\Gamma_2(f(x), \bar{g}(x); \lambda, \mu) \leq 0, \ \forall x \in X$$

In fact, when α has been fixed, $\Gamma_2(f, \bar{g}; \lambda, \mu) = \Gamma_1(f, g; \lambda, \mu, \alpha)$. This means that in the non-convex case, if the above mentioned assumption has been fulfilled, a linear weak S-function can be adopted to state optimality after a suitable exponential transformation of the constraining functions has been made. It is useful to state when it is necessary to transform all the g_i or simply a part of them in order to state optimality by means of Γ_2. *The above inequality becomes a necessary and sufficient condition for* (2) *to hold, if* X *is convex and* $-\varphi$, g *concave.*

Remark 8 System (1) may be equivalently written as

$$f(x) > 0$$

$$g_i(x) - y_i^2 = 0, \qquad i = 1, \ldots, m \tag{13}$$

$$(x, y) \in X \times \mathbb{R}^m$$

where $y = (y_1, \ldots, y_m)$. Thus, if H is now replaced by

$$\{(u, v) \in \mathbb{R}^{1+m}: u > 0; \ v = 0\}$$

then a weak S-function is the following

$$\Gamma_3(u, v; r) = u - r\langle v, v\rangle \tag{14}$$

where r is a positive scalar. Of course, the function (14) is 'simpler' than Γ_1. As a consequence of the preceding trick, it would seem that, in the non-convex case, the development of a theory for constraining inequalities should be useless. It has been already noted [27] that this is not true. In fact, if the preceding trick is adopted, on the feasible region, which now is

$$R = \{(x, y) \in \mathbb{R}^{n+m} : g_i(x) - y_i^2 = 0, \ i = 1, \ldots, m\}$$

we have

$$\Gamma_3(f(x), g_1(x) - y_1^2, \ldots, g_m(x) - y_m^2; \ r) = f(x) = \varphi(\hat{x}) - \varphi(x)$$

whatever r may be. Hence, we cannot expect to have a concave or convex (weak) S-function if φ is neither convex nor concave, and so on; moreover, as φ does not depend on y, we have to minimize, under equality constraints, an objective function, which is never strictly convex or concave.

Remark 9 In the preceding remark, instead of (14), consider the function

$$\Gamma_4(u, v; \lambda, r) = u + \langle\lambda, v\rangle - r\langle v, v\rangle \tag{15}$$

where $\lambda = (\lambda_1, \ldots, \lambda_m)$ is non-negative and r again is a positive scalar. All that has been noted for (14) holds for (15) too. The only difference is that, in the optimality condition, r may be obliged to become large in the case (14), so that (2) becomes a numerically ill-posed problem, while in the case (15) this may be avoided. (14) corresponds to the classical penalty function approach and (15) to the augmented Lagrangian one [27, 29, 31]. Finally, note that the transformation described in Remark 3 contains a penalty approach both for convex and non-convex cases.

Before going into some details of the use of theorems of alternative for studying the optimality of (11), it is useful to consider a connection with duality which has recently been stated in an abstract form [24].

Remark 10 If

$$g(x) = \begin{pmatrix} \theta(x) \\ -\theta(x) \end{pmatrix}$$

i.e. $g(x) \geqslant 0$ is the inequality formulation of a system of equalities $\theta(x) = 0$, then $x \in R$ implies

$$\Gamma_1(f(x), g(x); \lambda, \mu, \alpha) = \lambda f(x), \qquad \forall \lambda, \mu, \alpha$$

so that the disadvantages of the approach (13) pointed out in Remark 8 apply to the inequalities approach too. This happens not only when Γ_1 is adopted, but in every case where the adopted (weak) S-function is linear with respect to u. Thus, to be able to handle equalities we need a (weak) S-function, which is non-linear with respect to u; otherwise, as pointed out in section 2 of [31] from a duality viewpoint, we get a convex or concave (weak, strong) S-function by chance.

Remark 11 According to Remark 3, if f (or any of g_i) is not concave, and if the concavity of $1 - \exp[-\alpha f(x)]$ may be shown, then Theorem 3 can be applied even if (11) is not convex (see Example 1). Moreover, note that in this case the set E of Remark 1 (which is not convex) becomes convex under the exponential transformation. Hence, conditions which guarantee the convexity, under exponential transformation, of E (or at least of the set E_Y considered in the proof of Corollary 1) might be extremely useful. Note also that the introduction of the weak S-function Γ_1 may be interpreted like a two-step approach: the functions

$$f(x) = \varphi(\hat{x}) - \varphi(x); \qquad g_i(x), \qquad i = 1, \ldots, m$$

are transformed exponentially; then, a weak S-function Γ_2 is applied. Finally, note that another useful transformation consists in replacing φ with $\Phi(\varphi)$ where $\Phi(*)$ is any increasing function of $*$; hence in (1) f is replaced with $\Phi(\varphi(\hat{x})) - \Phi(\varphi(x))$.

For the sake of simplicity in the sequel of this section, it is assumed that some (regularity) condition is fulfilled which ensures that $(u, v) \notin \text{cl } H - H - \{0\}$ and moreover that no $(u > 0, v = 0) \notin K$ is a limit point of K, so that in (2) G may be equivalently replaced by a weak S-function. When this is not true, suitable changes must be made to the following deductions. With this in mind, in (2) replace G by Γ_1; moreover, denote by Y the domain of parameters of Γ_1 and define

$$\psi(\mu, \alpha) = \min_{x \in X} \left\{ \varphi(x) - \sum_{i=1}^{m} \mu_i [1 - \exp(-\alpha_i g_i(x))] \right\}$$

The function between braces will be called an *auxiliary function*. The above-mentioned equivalence between theorems of alternative and duality thorems implies that (11) and the problem

(16) $$\max \psi(\mu, \alpha), \qquad (\mu, \alpha) \in Y$$

are *dual* in the sense that

(17*a*) $$\{x \in R; (\mu, \alpha) \in Y\} \Rightarrow \{\psi(\mu, \alpha) \leqslant \varphi(x)\}$$

(17*b*) $$\{x \in R; (\mu, \alpha) \in Y; \psi(\mu, \alpha) = \varphi(x)\} \Rightarrow \left\{\begin{array}{l} x \text{ and } (\mu, \alpha) \text{ are optimal for} \\ \text{(11) and (16) respectively} \end{array}\right\}$$

Note that, if the adopted function is not an S-function, but merely a weak (strong) S-function, so that it guarantees only a weak (strong) alternative, then only a *weak (strong) duality* is guaranteed, i.e. only (17*a*) (or (17*b*)) is satisfied. For instance, if (11) is not convex and Γ_2 is adopted, then only weak duality is guaranteed (see Example 2). In the convex case, by adopting Γ_2, also strong duality (and hence duality) is guaranteed. The same happens when the condition of Corollary 1 (or Corollary 2) is fulfilled.

Note also that, in the convex case, (16) becomes a well known part of a saddle point condition:

(18) $$\max_{\mu \geqslant 0} \min_{x \in X} [\varphi(x) - \langle \mu, g(x) \rangle]$$

Problem (16) is interesting mainly in the non-convex case (see Example 3), where there is a degree of freedom in choosing a weak S-function. On the other hand, we must pay attention to the kind of auxiliary function which will appear. Of course, it is to be desired to get a convex auxiliary function, like in the convex case. When this is not achievable, it may be useful to be able to guarantee at least the strict concavity. In section 5 this is discussed for quadratic programs.

Remark 12 In [10] it is shown that a non-linear extremum problem with binary unknowns is equivalent to a concave extremum problem with real unknowns. Hence, the preceding duality scheme, with Γ_1 as a weak S-function, may be adopted, via the above equivalence, for a binary program.

Remark 13 Even if throughout the paper X is a subset of the real Euclidean space, much more general situations may be reduced to the alternative between systems. For instance, with obvious notation and under suitable assumptions, let us touch slightly the following classical problem:

$$\min \varphi(x) \quad \text{s.t.} \quad x \in X \quad g(t, x) \geqslant 0$$

where $x = x(t) = (x_1(t), \ldots, x_n(t))$ is a vector of unknown functions defined on T;

$$\varphi(x) = \int_T F(t, x(t), \dot{x}(t))\,dt$$

and X a given set of functions $x(t)$. The system

$$f(x) \triangleq \varphi(\bar{x}) - \varphi(x) > 0, \qquad g(t, x) \geqslant 0, \qquad x \in X$$

(which is impossible iff $\varphi(\bar{x})$ is a strong minimum of the preceding problem) may be put in alternative with the other

$$\Gamma_5(x; \lambda, \mu) \triangleq \lambda[\varphi(\bar{x}) - \varphi(x)] + \int_T \langle \mu(t, x), g(t, x) \rangle\, dt \leqslant 0, \qquad \forall x \in X$$

where the inequality must be strongly satisfied when $\lambda = 0$. The weak alternative may be shown as in Theorem 2, the strong one is obtained by slight changes.

Now consider a real-valued function $F: X \to \mathbb{R}^n$, and the problem which consists in finding an $\bar{x} \in R$, such that

$$\langle F(\bar{x}), x - \bar{x} \rangle \geqslant 0 \qquad \forall x \in R \tag{19}$$

where R is defined in (11); (19) is a general finite-dimensional setting for a *variational inequality*. Note that $\bar{x} \in R$ is a solution of (19), iff system

$$f(x) \triangleq \langle F(\bar{x}), \bar{x} - x \rangle > 0 \qquad g(x) \geqslant 0 \qquad x \in X \tag{20}$$

is impossible. We can apply Theorem 1 to (20) and obtain that $\bar{x} \in R$ is a solution of (19), iff

$$G(\langle F(\bar{x}), \bar{x} - x \rangle,\ g(x)) \leqslant 0 \qquad \forall x \in X \tag{21}$$

When g is concave and X convex, Theorem 3 can be applied to (20) to obtain that $\bar{x} \in R$ is a solution of (19), iff there exist $\lambda \in \mathbb{R}_+$ and $\mu \in \mathbb{R}^m_+$, with $(\lambda, \mu) \neq 0$, such that

$$\lambda \langle F(\bar{x}), \bar{x} - x \rangle + \langle \mu, g(x) \rangle \leqslant 0 \qquad \forall x \in X \tag{22}$$

where the inequality must be strictly verified if $\lambda = 0$.

What has been said for problem (11), and in particular the duality scheme, can now be extended, with obvious changes, to (19). When φ is differentiable it is well known that, by setting $F(x) = \nabla\varphi(x)$ and under suitable assumptions, the stationary points of problem (11) are the solutions of (19). Hence,

(21) or (22) are a further way to state (local) optimality by means of a theorem of alternative. For instance, when φ is a quadratic form (not necessarily convex), g is affine and $X = \mathbb{R}^n$, to state whether (22) is true, i.e. whether $\bar{x}$ is a stationary point of (11), amounts to solve a linear program.

4. APPLICATIONS TO VECTOR EXTREMUM PROBLEMS

Consider real-valued functions $\varphi : X \to \mathbb{R}^h$, $\theta : X \to \mathbb{R}^k$; and the following vector extremum problem[3]

$$\text{(23)} \qquad \nu\text{-min } \varphi(x), \qquad x \in \mathscr{R} \triangleq \{x \in X : \theta(x) \geqslant 0\}$$

Theorems 1 and 2 may be used to state a necessary and sufficient (vector) optimality condition for problem (23). To this end set

$$f_i(x) = \varphi_i(\hat{x}) - \varphi_i(x), \qquad g(x) = \begin{pmatrix} \varphi(\hat{x}) - \varphi(x) \\ \theta(x) \end{pmatrix}$$

and consider the system

$$\text{(24)} \qquad \begin{array}{ll} f_i(x) > 0 & \text{for at least one } i = 1, \ldots, h \\ g(x) \geqslant 0, & x \in X \end{array}$$

It is easy to show that $\hat{x} \in \mathscr{R}$ is (vector) optimal for (23) iff (24) is impossible. For every fixed i identify (24) with (1) at $\ell = 1$. Then, Theorem 1 (or 2) may be read this way: $\hat{x} \in \mathscr{R}$ is (vector) optimal for (23) iff (2) (or (3)) holds for every i=1,...,h. To show an application of this proposition, consider the particular case in which φ is linear, θ affine and $X = \mathbb{R}^n$; set

$$\varphi(x) = Dx, \qquad \theta(x) = Ax - b$$

where D, A, b are of orders $h \times n$, $k \times n$, $k \times 1$, respectively. Denote by D_i the ith row of D and set $y = D\hat{x}$. Now Theorem 3 may be applied. We have that $\hat{x} \in \mathscr{R}$ is an optimal solution of (23) iff there exist non-negative vector ω^i, μ^i, such that, for every $i = 1, \ldots, h$, either

(3) ν-min is the notation for finding all $\hat{x} \in \mathscr{R}$ for which there exists no $x \in \mathscr{R}$, such that $\varphi(x) \leqslant \varphi(\hat{x})$ and $\varphi(x) \neq \varphi(\hat{x})$; analogous definition for ν-max. Note that, at $h = 1$, (23) and (11) coincide.

(25*a*) $$y_i - \langle D_i, x\rangle + \langle \omega^i, y - Dx\rangle + \langle \mu^i, Ax - b\rangle \leqslant 0, \qquad \forall x \in \mathbb{R}^n$$

or

(25*b*) $$\langle \omega^i, y - Dx\rangle + \langle \mu^i, Ax - b\rangle < 0, \qquad \forall x \in \mathbb{R}^n$$

Without any loss of generality assume $\mathscr{R} \neq \emptyset$, so that $g(x) \geqslant 0$ is possible; this implies that (25*b*) is impossible. Hence, (25) is equivalent to (25*a*). Note that (25*a*) is possible iff system

(26) $$\mu A - \omega D = D_i, \quad y_i + \langle \omega, y\rangle \leqslant \langle \mu, b\rangle, \quad \omega \geqslant 0, \quad \mu \geqslant 0$$

is possible for every $i = 1,\ldots,h$. Denote by $\mathscr{Y}$ the set of y, such that system (26) is possible for every $i = 1,\ldots, h$. One is led to consider the following vector extremum problem

(27) $$v - \max(y), \qquad y \in \mathscr{Y}$$

which is called *dual* to (23), according to what has been recently stated [4] [17, 34]. The main duality property, namely

(28)
$$\{x \in \mathscr{R}\ ; y \in \mathscr{Y}\ \} \Rightarrow \{\ Dx \leqslant y; Dx \neq y \text{ is impossible}\ \}$$
$$\{x \in \mathscr{R}\ ; y \in \mathscr{Y}\ ; Dx = y\ \} \Rightarrow \{x, y \text{ are optimal for (23), (27), respectively}\}$$

is a straightforward consequence of Theorem 3.

Of course, when $h = 1$ (23) becomes an ordinary linear program, (27) its dual, and (28) the well known duality theorem.

Remark 14 A duality theory for (23), which has been developed in [16, 17] for the linear case, may be discussed under more general assumptions by means of the preceding approach.

Remark 15 (24) may be equivalently written as

(29) $$f(x) \triangleq \langle a, \varphi(\hat{x}) - \varphi(x)\rangle > 0, \qquad g(x) \geqslant 0, \qquad x \in X$$

where a is an h-vector whose elements equal 1, and g is defined as above. Identify (29) with (1); then from Theorem 1 (or 2) we obtain that $\hat{x} \in \mathscr{R}$ is (vector) optimal for (23), iff (2) (or (3)) holds. In the linear case, (4) becomes: there exist non-negative ω, μ such that either

(4) In [17] y is replaced by Ub, U being an $h \times n$ matrix of (dual) unkowns. Duality for (23) is contained also in [16, 33].

$$\langle a, y - Dx\rangle + \langle \omega, y - Dx\rangle + \langle \mu, Ax - b\rangle \leqslant 0, \quad \forall\, x \in \mathbb{R}^n$$

or

$$\langle \omega, y - Dx\rangle + \langle \mu, Ax - b\rangle < 0, \quad \forall x \in \mathbb{R}^n$$

As it is not restrictive to assume the latter inequality to be impossible, we have that $\hat{x}$ is (vector) optimal for (23) in the linear case iff the system

$$-\omega D + \mu A = aD, \quad -\langle \omega, y\rangle + \langle \mu, b\rangle \geqslant \langle a, y\rangle, \quad \omega \geqslant 0, \quad \mu \geqslant 0$$

is possible, or equivalently iff

$$0 \geqslant \min \langle A\hat{x} - b, \mu\rangle$$

subject to

$$-\omega D + \mu A = aD, \quad \omega \geqslant 0, \quad \mu \geqslant 0$$

Remark 16 The concept of variational inequalities can be introduced in the field of vector extremum problems too. Consider real-valued functions $F_i : X \to \mathbb{R}^n$, $i = 1,\ldots, h$, and the problem which consists of finding an $\hat{x} \in \mathscr{R}$, such that there exists no $x \in \mathscr{R}$ for which

(30) $\langle F_i(\hat{x}), x - \hat{x}\rangle \leqslant 0, i = 1,\ldots h$ and $\langle F_i(\hat{x}), x - \hat{x}\rangle \neq 0$ for at least one i

This problem, which may be called a *vector variational inequality*, becomes (19) at $h = 1$. Note that $\hat{x} \in \mathscr{R}$ is a solution of (30), iff system

$$(31)\quad \begin{aligned} &f_i(x) = \langle F_i(\hat{x}), \hat{x} - x\rangle > 0 \quad \text{for at least one} \quad i = 1, \ldots, h \\ &f(x) \geqslant 0, \quad \theta(x) \geqslant 0, \quad x \in X; \quad (f = (f_1, \ldots, f_h)) \end{aligned}$$

is impossible. Now (31) can be identified with (24) to obtain analogous results.

5. CONCAVE AUXILIARY FUNCTION FOR QUADRATIC PROGRAMS

Consider a quadratic strictly concave program, i.e. in (11) set

$$(32)\quad \varphi(x) = \langle c, x\rangle + \tfrac{1}{2} \langle x, Cx\rangle, \quad g(x) = Ax - b$$

where C is an $n \times n$ symmetric and negative-definite matrix, A an $m \times n$

matrix, b an $m \times 1$ vector and c an $1 \times n$ vector, all with real elements; X is assumed to be compact. Consider the auxiliary function

$$\Delta(\varphi, g; p, q) = \varphi - p \sum_{i=1}^{m} [1 - \exp(-qg_i)]$$

where p and q are scalars[5].

Theorem 4
Let (32) hold. Then $\forall\ q \in\ ]0, +\infty[$, there exists $\bar{p} > 0$, such that $\forall\ p \in\ ]0, \bar{p}[$ $\Delta(\varphi(x), g(x); p, q)$ is strictly concave.

Proof: There exist $n \times n$ matrices K and D, the former orthogonal and the latter diagonal[6] such that $DK^{-1} CKD = -I$. Set $D^{-1} K^{-1} x = z$; $cKD = d$. Hence, we have

$$\bar{\varphi}(z) \triangleq \varphi(KDz) = \langle d, z \rangle - \frac{1}{2} \langle z, z \rangle$$

$$\bar{g}(z) \triangleq g(KDz) = AKDz - b$$

To show the strict concavity of $\Delta(\varphi, g; p, q)$ is equivalent to showing the strict concavity of $\Delta(\bar{\varphi}, \bar{g}; p, q)$. Consider q as fixed and let $\hat{H}(z), H(z)$ be the Hessian matrices of $\Delta(\bar{\varphi}, \bar{g}; p, q)$ and $\Sigma_{i=1}^{m} \exp[-q\bar{g}_i(z)]$, respectively. We have $\hat{H}(z) = pH(z) - I$. $H(z)$ is continuous and because of a well known property, the same is also true for all of its eigenvalues, say $\rho_i(z)$. Set $Z = \{z \in \mathbb{R}^n : KDz \in X\}$. If X is compact, then Z is compact. Hence, as q is fixed, $\rho_i(z)$ is bounded, so that

$$0 < \tilde{\rho} \triangleq \max_i \sup_{z \in Z} |\rho_i(z)| < +\infty$$

Let $\bar{z} \in Z$; v is an eigenvalue of $\hat{H}(\bar{z})$ iff $\det[\hat{H}(\bar{z}) - vI] = 0$, i.e. $\det[pH(\bar{z}) - I - vI] = 0$. It follows that $\forall\ \bar{z} \in Z$, v is an eigenvalue of $\hat{H}(\bar{z})$ iff $\omega = (1+v)/p$ is an eigenvalue of $H(\bar{z})$.

Thus, if p is such that $p < 1/\tilde{\rho}$, we have either

$$v = p\omega - 1 < (\omega/\tilde{\rho}) - 1 \leqslant 0, \qquad \text{if } \omega > 0$$

or

$$v = p\omega - 1, \qquad \text{if } \omega \leqslant 0$$

(5) We have set $\mu_i = p$ and $\alpha_i = q$ in Γ_1.

(6) The non-zero entries of D are the reciprocal of the square roots of the absolute values of the eigenvalues of C. I denotes the identity matrix of order n.

In each case we have $v < 0$. This completes the proof.

Remark 17 In a certain sense, the 'simplest' kind of (weak) S-function necessary to get the convexity or, at least, the strict concavity of the auxiliary function may be regarded as an index of the computational complexity of the given problem in the sense of [18, 19]. For instance, as the penalty approach (14) is based on a polynomial weak S-function, which does not guarantee anything for the auxiliary function in the quadratic concave programs, it is natural to expect computational difficulties when these programs are handled by the classical penalty approach (see Example 4 for an instance of a concave auxiliary function).

6. DECOMPOSITION OF CONVEX QUADRATIC PROGRAMS

Now it will be shown how a theorem of alternative may be used to decompose an extremum problem. To this end, by means of Theorem 3, a procedure for solving convex quadratic programs will now be devised. For the sake of simplicity, the strictly convex case will be considered as the first one; then the variants necessary to handle the general convex case will be described.

Consider again problem (11) in case (32); assume now C to be positive definite, $m > n$, and that A has full rank. The dual of (11) is(7)

$$\text{(33)} \qquad \begin{aligned} &\max\ [\psi(x, \lambda) \triangleq -\tfrac{1}{2}\langle x, Cx\rangle + \langle b, \lambda\rangle] \\ &\text{s.t.}\quad Cx - A^T\lambda + c = 0, \qquad \lambda \geqslant 0 \end{aligned}$$

When x is regarded as a parameter, (33) becomes a linear program

$$L(x): \quad -\tfrac{1}{2}\langle x, Cx\rangle + \max_{\substack{A^T\lambda = Cx + c \\ \lambda \geqslant 0}} \langle b, \lambda\rangle$$

whose dual is

$$D(x): \quad -\tfrac{1}{2}\langle x, Cx\rangle + \min_{A\omega \geqslant b} \langle Cx + c, \omega\rangle$$

(7) T as superscript denotes transpose. Note that, by eliminating x, (33) becomes a sign-constrained quadratic convex problem in the unknown λ. This transformation may be useful when $m - n$ is small.

The following solving procedure can now be devised[8].

Step I: Linear subproblem Consider any $\bar{x} \in \mathbb{R}^n$, such that $L(\bar{x})$ has finite extremum; if no such $\bar{x}$ exists, then (11) has no feasible solutions. Let $\bar{\lambda}$ and $\bar{\omega}$ be optimal basic solutions of $L(\bar{x})$ and $D(\bar{x})$, respectively; and let $\bar{k} = \psi(\bar{x}, \bar{\lambda})$ be the (common) value of their extrema. Execute step II.

There exists a partition of A^T into a nonsingular $n \times n$ matrix B and a $n \times (m-n)$ matrix N, and corresponding partitions of λ into subvectors[9] λ_B, λ_N and of b into subvectors b_B, b_N, such that

$$A^T\lambda = B\lambda_B + N\lambda_N \; ; \; \bar{\lambda} = \begin{pmatrix} \lambda_B = \bar{\lambda}_B \triangleq B^{-1}(C\bar{x} + c) \\ \lambda_N = \bar{\lambda}_N \triangleq 0 \end{pmatrix}$$

Set

$$q(\lambda_B) = -\tfrac{1}{2} < \lambda_B, B^T C^{-1} B\lambda_B > + < (c^T C^{-1} + \bar{\omega}^T) B, \lambda_B >$$

By applying Theorem 3 to (11), we obtain that every x, which statisfies the inequality $-\frac{1}{2}\langle x, Cx\rangle + \langle Cx + c, \bar{\omega}\rangle \leq \bar{k}$, can be cut off, in searching for an x, such that the extremum of $L(x)$ be greater than $\bar{k}$. By using equalities $x = C^{-1}(B\lambda_B - c)$ and $\langle b, \bar{\lambda}\rangle = \langle \bar{\omega}, B\bar{\lambda}_B\rangle$, we obtain $-\frac{1}{2}\langle x, Cx\rangle + \langle Cx + c, \bar{\omega}\rangle - \bar{k} = q(\lambda_B) - q(\bar{\lambda}_B)$. It follows that, if the maximum of the strictly concave problem

$$\text{(34)} \qquad \max\,[q(\lambda_B) - q(\bar{\lambda}_B)], \text{ s.t. } \lambda_B \geq 0,$$

(which is obviously non-negative) is positive, then we can improve the current value $\bar{k}$ of ψ; while, if it is zero, no improvement is possible with the current basis B.

Step II: quadratic subproblem. Solve problem (34), and let $\hat{\lambda}_B$, $\hat{q}$ be an optimal solution and the corresponding maximum, respectively. Set $\hat{x} = C^{-1}(B\hat{\lambda}_B - c)$. Execute step III.

We want to state if $\hat{x}, \hat{\lambda}_B, \hat{\lambda}_N = 0$ form an optimal solution of (33). Denote by A_s the s-th column of A^T, by α_{is} the i-th element of column $B^{-1}A_s$; and by J_B, J_N the sets of indices of basic and nonbasic variables (namely, the elements of λ_B and λ_N), respectively. It is easy to show that no improvement is possible for ψ, iff $b_N - NC^{-1}(B\hat{\lambda}_B - c) \leq 0$. Then, we have the following step.

Step III: first optimality condition. If $b_j - \langle A_j, C^{-1}(B\hat{\lambda}_B - c)\rangle \leq 0, \forall\, j \in J_N$, then $\hat{x}$ is an optimal solution of (11). Otherwise, let $s \in J_N$ be such that

(8) The statements of the algorithm are contained in the preliminary paragraph of each step; comments and proof follow in subsequent paragraphs.

(9) Capital (small) letters denote subvectors (elements, respectively), when applied as indices.

$$s \in \arg\max \{b_j - \langle A_j, C^{-1}(B\hat{\lambda}_B - c)\rangle, j \in J_N\}$$

and let $r \in J_B$ be such that $\hat{\lambda}_r = 0$ or, if $\hat{\lambda}_i > 0 \quad \forall i \in J_B$, such that

$$r \in \arg\min \{\frac{\hat{\lambda}_i}{\alpha_{is}} : \alpha_{is} > 0, \ i \in J_B\}$$

(an ordering rule must be adopted when there is not unicity). Call $\widetilde{B}$ the basis obtained from B by replacing A_r with A_s. Execute step IV.

Kuhn-Tucker condition can be applied to justify the definition of r. Now denote by ω^i, $i \in I$, all the basic solutions (or rays) of $D(x)$; and denote by $\overline{I}$ any subset of I. Then, it is easy to show that, if ω is restricted to the points ω^i, $i \in I$, then the supremum of the extrema of problems $D(x)$ still equals the extremum of (33). It follows that the extremum of problem

$$(35) \qquad \max(\theta) \quad \text{s.t.} \quad \begin{array}{l} \theta \leqslant -\frac{1}{2}\langle x, Cx\rangle + \langle x, C\omega^i\rangle + \langle c, \omega^i\rangle, \qquad i \in I \\ (\theta, x) \in \mathbb{R}^{n+1}, \end{array}$$

equals the extremum of (33); while it is merely an upper bound of it if I is replaced by any subset $\overline{I}$.

Step IV: Second optimality condition Define $\overline{I}$ as the set of indices of optimal solutions of $D(x)$ met so far. Solve (35) with $\overline{I}$ in place of I. Let k^* be its maximum. If $\overline{k} = k^*$, then $\overline{x} = C^{-1}(B\overline{\lambda}_B - c)$ is an optimal solution of (11). Otherwise, execute Step II.

Note that $k^* - \overline{k}$ is an upper bound for the error we make, if we stop the algorithm at $x = \overline{x}$. Example 5 shows an application of this algorithm.

Remark 18. It is easy to prove that, by executing the preceding steps a finite number of times, an optimal solution of (11) is obtained. In fact, as the number of bases of $L(x)$ is finite, the increase of the objective function of $L(x)$ implies that no basis can be met more than once. On the other hand, we remain on the same value only a finite number of times, i.e. when the minimum of (34) is zero. Numerical experience has shown that the utility of the present method increases with respect to $m - n$.

Remark 19 Assume now C to be merely positive-semidefinite. A one-to-one correspondence between λ_B and x is not ensured. Hence, we cannot obtain the subproblem (34) with only the non-negativity constraints. It follows that there is no convenience in changing the unknown x into λ_B. Thus, if we set

$$r(x) = \frac{1}{2}\langle x, Cx\rangle - \langle C\overline{\omega}, x\rangle - \langle c, \overline{\omega}\rangle$$

it is easy to show that, instead of (34), we now meet the subproblem

$$(36) \qquad \min\,[r(x) - r(\overline{x})] \qquad \text{s.t.} \qquad B^{-1}Cx + B^{-1}c \geqslant 0$$

Hence, problem (11) is reduced to a finite sequence of quadratic convex subproblems having just n constraints (instead of strictly convex sign-constrained subproblems). However, by partitioning x into two subvectors (say x' and x'') in such a way to get a positive-definite principal submatrix in correspondence of one of them (say x'), one can decompose (11) into strictly convex and convex subproblems. In fact, one can fix x'', solve the resulting (strictly convex) subproblem, and hence look for a new x'' which improves the current value (this is achieved by solving a convex suproblem).

Remark 20 Let the rank of A be less than n. In the strictly convex case the only variant is that the dimension of the basis B is less than n. In the convex case we again have a reduction in the size of the basis, and moreover in (36) x must satisfy additional constraints to ensure the feasibility of $A^T\lambda = Cx + c$.

Remark 21 When $m = n$ and A has full rank, the preceding decomposition algorithm leads to solving (34) (where $A^T = B$) just once, so that it is neither useful nor useless. When $m = n$ and A has not full rank or when $m < n$, a transformation like the one described in Remark 19 enables one to apply the preceding algorithm. However, it should be clear that such an algorithm shows its utility when $m > n$, and that it grows with $m - n$.

Remark 22 Of course, special forms of A and C matrices may simplify the algorithm. For instance, when

$$A = \begin{pmatrix} \overline{A} \\ -\overline{A} \end{pmatrix}, \qquad b = \begin{pmatrix} b' \\ -b'' \end{pmatrix}$$

we have the so-called interval problem, in the sense that (if $b_i' < b_i''$) between every pair of constraints one at most is binding at an optimal solution. This means that at least $m/2$ of the multipliers are zero, so that in the preceding algorithm only $\overline{A}$ must be recorded. Another instance is $C = I$, $c = 0$; in this case the method finds the minimum distance between the origin of $\mathbb{R}^n$ and the polyhedron R. A further particular case is φ bilinear. Moreover, it can be extended to solve (11), where again $g(x) = Ax - b$, while $\varphi(x) = \langle c, x\rangle + \Phi(x)$ and Φ is convex. An instance may be given by a class of generalized quadratic forms [32]:

$$\Phi(x) = \psi^k(x) + \psi^{k-1}(x)\,\ell(x) + \sum_{s=2}^{2k-2} \tau_{2k-s}(x)$$

where ψ is a quadratic form, ℓ a linear form, τ_{2k-s} a form whose degree is $2k - s$, and where the third term must not be written if $k = 1$.

In these cases $Cx + c$ must be replaced by $\nabla\varphi(x)$. Of course, the possibility of solving the equation $\nabla \varphi(x)=\gamma$ for every given vector γ enables one to reduce (11) to a sequence of subproblems like (34). Otherwise, (11) can be reduced to a sequence of subproblems like (36).

Remark 23 As a convex quadratic program can be equivalently formulated as a linear complementarity system (or as a linear variational inequality), the preceding decomposition algorithm can be translated in terms of a linear complementarity system (variational inequality). This should lead to defining, for this system (inequality), an algorithm which, at every step, handles a system of smaller size (a simpler inequality, i.e. an inequality on $\mathbb{R}_+^n$, instead of any polyhedron). It should be useful to know a class of such systems (inequalities), not necessarily coming from a convex quadratic program, which can be solved by means of this approach.

Remark 24 The decomposition of an extremum problem may be obtained in several ways. For instance, with obvious notation, consider the problem

$$\text{(37)} \qquad \begin{aligned} &\min\,(\langle c, x\rangle + \langle d, y\rangle) \qquad \text{s.t.} \qquad Ax + \Delta y = b \\ &x \in \mathbb{R}_+^n, \qquad y \in \mathbb{R}_+^m \end{aligned}$$

where Δ is a diagonal matrix, whose principal entries are $\delta_i = 0, \pm 1$, $i = 1,\dots,m$. In some applications, where the vector b may assume more than one value, Δy has been interpreted as a change of b and $\langle d, y\rangle$ as a penalty implied by this change. In this order of ideas, an optimal solution (x, y) of (37) and the corresponding minimum $\bar{k}$ being known at $\Delta = \bar{\Delta}$ and $d = \bar{d}$, one may look for a different change Δ and a penalty d, which imply a minimum of (37) less than $\bar{k}$. If $\bar{\lambda}$ denotes an optimal solution of the dual of (37) at $\Delta = \bar{\Delta}$, $d = \bar{d}$, then a straightforward application of the Corollary 3 leads one to state that every y, Δ, d, which satisfy $\langle d, y\rangle \leqslant 0$ and $\bar{\lambda}\Delta \leqslant d$ or $\bar{\lambda}_i \delta_i \leqslant d_i$, $i = 1, \dots, m$, are such that the minimum of (37) is greater than or equal to $\bar{k}$, so that it may be cut off. Then, a necessary and sufficient condition to have a minimum less than $\bar{k}$ is that, for at least one $i = 1, \dots, m$, we have $d_i < \bar{\lambda}_i$ if $\delta_i = 1$, or $d_i < -\bar{\lambda}_i$ if $\delta_i = -1$, or merely $d_i < 0$ if $\delta_i = 0$.

Remark 25 Along the lines of Remark 22 another instance is the following (with obvious notation) maximum problem

$$\text{(38)} \qquad \max \varphi(x) \qquad \text{s.t.} \qquad Ax \leqslant b$$

where $\varphi(x) = \min\{\langle D_i, x\rangle, i = 1, \dots, h\}$. If D denotes the $h \times n$ matrix whose rows are the vectors D_i, and a is a column vector of unit entries, then (38) may be equivalently formulated as a linear program:

$$\max\,(\rho) \qquad \text{s.t.} \qquad \rho a \leqslant Dx\;,\; Ax \leqslant b, \qquad \rho \in \mathbb{R}$$

whose dual may be written as

(39) $$\min \langle b, \mu \rangle \quad \text{s.t.} \quad A^T \mu = D^T \lambda, \quad \mu \geqslant 0$$

where the vector λ is regarded as a parameter, satisfying $\lambda \geqslant 0$, $\langle a, \lambda \rangle = 1$. Instead of solving (38) it is equivalent to solve (39) for every λ. The preceding decomposition approach can now be applied to (39). If $\bar{\mu}$ is an optimal solution of (39) at a given $\lambda = \bar{\lambda}$ and $\bar{k}$ the corresponding minimum, and if $\bar{y}$ is an optimal solution of the dual of (39) (again at $\lambda = \bar{\lambda}$), it is easy to show that to every λ, such that

$$\langle D\bar{y}, \lambda \rangle \geqslant \bar{k}, \quad \langle a, \lambda \rangle = 1, \quad \lambda \geqslant 0$$

there corresponds a problem (39), which has a minimum greater than or equal to $\bar{k}$.

Remark 26 The decomposition approach of this section may be generalized to handle some classes of non-convex problems. To see this, consider the problem

$$\min \left(\langle c, x \rangle + \tfrac{1}{2} \langle x, Cx \rangle \right) \quad \text{s.t.} \quad x \in X = \bigcup_{t \in \mathscr{T}} X(t)$$

where t is a vector parameter of domain $\mathscr{T}$, $A(t)$ is an $m \times n$ matrix and $b(t)$ an m-vector, whose elements are functions of t, C is positive semidefinite, and where

$$X(t) = \{x \in \mathbb{R}^n : A(t)x \geqslant b(t)\}$$

Denote by Q the above problem and by $Q(t)$ the subproblem obtained by replacing X with $X(t)$ in Q. $Q(t)$ is a convex problem, while Q is not necessarily so. The dual of $Q(t)$, call it $D(t)$, is

$$\max \left(-\tfrac{1}{2} \langle x, Cx \rangle + \langle b(t), \lambda \rangle \right) \quad \text{s.t.} \quad c + Cx - A(t)^T \lambda = 0, \quad \lambda \geqslant 0$$

A decomposition procedure for solving Q is briefly the following. Fix any $\bar{t} \in \mathscr{T}$, and solve $Q(\bar{t})$. Let $\bar{x}$ be an optimal solution of $Q(\bar{t})$ and $\bar{k}$ the corresponding mimimum. Thus, an optimal solution $(\tilde{x}, \tilde{\lambda})$ of $D(\bar{t})$ is easily found. In searching for a new $\hat{t}$ such that the minimum of $Q(\hat{t})$ is less than $\bar{k}$, one can disregard every t which satisfies

$$-\tfrac{1}{2} \langle \tilde{x}, C\tilde{x} \rangle + \langle b(t), \tilde{\lambda} \rangle \geqslant \bar{k}, \quad c + C\tilde{x} - A(t)^T \tilde{\lambda} = 0$$

This cut may receive a simpler form, if a special Q is given. For instance, this happens when

$$C = 0, \quad c = \begin{pmatrix} c^1 \\ c^2 \end{pmatrix}, \quad x = \begin{pmatrix} x^1 \\ x^2 \end{pmatrix}$$

$$A(t) = \begin{pmatrix} A_1 & A_2 \\ I & 0 \\ 0 & I \\ 0 & -I \end{pmatrix}, \qquad b(t) = \begin{pmatrix} b \\ 0 \\ \ell(t) \\ -u(t) \end{pmatrix}, \qquad \lambda = \begin{pmatrix} \lambda^1 \\ \lambda^2 \\ \lambda^3 \\ \lambda^4 \end{pmatrix}$$

In this particular case, which has already been studied in the literature, and which embraces some known kinds of programs like linear integer and complementarity ones, the preceding cut becomes

$$\langle \tilde{\lambda}^3, \ell(t)\rangle - \langle \tilde{\lambda}^4, u(t)\rangle \geqslant \bar{k} - \langle b, \tilde{\lambda}^1\rangle$$

In such a case the preceding approach is essentially the one proposed in [1] as a dual fathoming procedure for mixed-integer programs.

7. DECOMPOSITION OF NON-CONVEX QUADRATIC PROGRAMS

A theorem of alternative or duality may be useful in decomposing non-convex quadratic programs too. To this end consider again problem (11) in case (32), and assume now C to be negative-definite(10). A first way of decomposing (11) consists in noting that (11) is equivalent to a linear complementarity problem:

(40*a*) $$\min \tfrac{1}{2} (\langle c, x\rangle + \langle b, \lambda\rangle)$$

subject to

(40*b*) $$c + Cx - A^T\lambda = 0, \qquad Ax \geqslant b, \qquad \lambda \geqslant 0, \qquad \langle Ax - b, \lambda\rangle = 0$$

In fact, when (40*b*) holds, we have

$$\varphi(x) = \langle c, x\rangle + \tfrac{1}{2}\langle x, Cx\rangle = \tfrac{1}{2}(\langle c, x\rangle + \langle Ax, \lambda\rangle) = \tfrac{1}{2}(\langle c, x\rangle + \langle b, \lambda\rangle)$$

Problem (40) can be decomposed by means of a theorem of alternative as shown in the next section.

A second way consists in noting that (11), when it has optimal solutions, is equivalent to

(10) A non-convex quadratic program can be decomposed into subproblems which are either convex or strictly concave. Hence, the non-convex case is easily handled when the strictly concave does. However, the former of the two following approaches does not require any assumption on C.

(41) $$\min_{\delta \in \Delta} \ [\max (\langle c, x\rangle + \tfrac{1}{2}\langle x, Cx\rangle)]$$

subject to

$$b - \rho(a - \delta) \leqslant Ax \leqslant b + \rho\delta$$

where $\delta = (\delta_1, \ldots, \delta_m)^T$ denotes a binary m-vector, $a = (1, \ldots, 1)^T$, ρ is large enough and real, and Δ is the set of all δ, such that $Ax \geqslant b$ remains feasible when we reverse the inequality sign in its ith inequality iff $\delta_i = 0$. x is an optimal solution of the maximization subproblem of (41) iff there exist m-vectors λ^- and λ^+, such that

(42) $$\begin{gathered} -Cx - c - A^T\lambda^- + A^T\lambda^+ = 0, \qquad b - \rho(a - \delta) \leqslant Ax \leqslant b + \rho\delta \\ \lambda^- \geqslant 0, \qquad \lambda^+ \geqslant 0 \\ \langle Ax - b + \rho(a - \delta), \lambda^-\rangle = \langle b + \rho\delta - Ax, \lambda^+\rangle = 0 \end{gathered}$$

the corresponding maximum equals

(43) $$\tfrac{1}{2}(\langle c, x\rangle + \langle -b + \rho(a - \delta), \lambda^-\rangle + \langle b + \rho\delta, \lambda^+\rangle)$$

As $x = C^{-1}(-c - A^T\lambda^- + A^T\lambda^+)$, by setting

$$\bar{A} = \begin{pmatrix} -AC^{-1}A^T & AC^{-1}A^T \\ AC^{-1}A^T & -AC^{-1}A^T \end{pmatrix}, \qquad c(\delta) = \begin{pmatrix} -AC^{-1}c - b + \rho(a - \delta) \\ AC^{-1}c + b + \rho\delta \end{pmatrix}, \qquad \lambda = \begin{pmatrix} \lambda^- \\ \lambda^+ \end{pmatrix}$$

system (42) and (43) become

(44) $$\bar{A}\lambda + c(\delta) \geqslant 0, \qquad \lambda \geqslant 0, \qquad \langle \bar{A}\lambda + c(\delta), \lambda\rangle = 0$$

(45) $$-\tfrac{1}{2}\langle c, C^{-1}c\rangle + \tfrac{1}{2}\langle c(\delta), \lambda\rangle$$

respectively.

Now, assume any $\bar{\delta} \in \Delta$ to be given. At $\delta = \bar{\delta}$ solve the maximization subproblem of (41) or equivalently system (44), which has a unique solution; let $\bar{x}$ be an optimal solution of the former, $\bar{\lambda}$ a solution of the latter, and $\bar{k}$ the corresponding value of (45). In order to find a $\hat{\delta}$, such that at $\delta = \hat{\delta}$ the maximum of (41) is less than $\bar{k}$, we have now to find a solution of (44) such that (45) is less than $\bar{k}$. This requires a parametric analysis of the unique solution, say $\lambda(\delta)$, of the linear complementarity system (44), in order to define a set of δ, such that

$$-\frac{1}{2}\left(\langle c, C^{-1}c\rangle + \langle c(\delta), \lambda(\delta)\rangle\right) \geq \bar{k}$$

This set can be cut off in searching for $\hat{\delta}$.

Remark 27 Recently a condition has been stated under which a feasible solution of a complementarity system can be obtained by solving a linear program [6, 23]. Unfortunately, such a condition is not fulfilled by system (44); in fact, it is easy to show that, in the present case, (3) of [23] becomes impossible. The above condition is easily extended to state equivalence between a complementarity problem (like (40)) and a linear program. Unfortunately, neither is this condition fulfilled by problem (40).

8. DECOMPOSITION OF A CLASS OF COMPLEMENTARITY PROBLEMS

A class of complementarity problems, which arise in the field of structural engineering [12, 13], may be decomposed by means of Theorem 3.

Consider the complementarity problem (with obvious notation):

$$\Phi_0 = \min\left(\sum_{i=1}^{3}\langle c^i, x^i\rangle + \frac{1}{2}\sum_{i,j=1}^{3}\langle x^i, C_{ij}x^j\rangle\right) \tag{46a}$$

subject to

$$\sum_{i=1}^{3} A_i x^i \geq b, \qquad x^i \geq 0, \qquad i = 1, 2, 3 \tag{46b}$$

and

$$\langle x^1, x^2\rangle = 0 \tag{46c}$$

where the matrix $C = (C_{ij}; i, j = 1, 2, 3)$ is assumed to be positive-semidefinite, where A_1, A_2, A_3 are of orders $m \times n, m \times n, m \times N$, respectively. Denote by $\delta = (\delta_1, \dots, \delta_n)$ a binary vector, let ρ be large enough and real, and set $c^1(\delta) = c^1 + \rho\delta$, $c^2(\delta) = c^2 + \rho(a - \delta)$, where $a = (1, \dots, 1)$. Consider the following convex quadratic program

$$\Phi(\delta) = \min\left(\sum_{i=1}^{2}\langle c^i(\delta), x^i\rangle + \langle c^3, x^3\rangle + \frac{1}{2}\sum_{i,j=1}^{3}\langle x^i, C_{ij}x^j\rangle\right) \tag{47a}$$

subject to

(47*b*) $$\sum_{i=1}^{3} A_i x^i \geqslant b, \qquad x^i \geqslant 0, \qquad i = 1, 2, 3$$

where δ is a parameter; (47) will be denoted by $Q(\delta)$, and its feasible region by R; assume $R \neq \emptyset$. Let Δ denote the set of all δ, such that (47*b*) is feasible if one sets $x_j^1 = 0$ when $\delta_j = 1$, and $x_j^2 = 0$ when $\delta_j = 0$. It is easy to show that an optimal solution of $Q(\delta)$ is a feasible solution of (46) iff $\delta \in \Delta$.

Set $x = (x^1, x^2, x^3)$, $A = (A_1, A_2, A_3)$ and denote by $F(x; \delta)$ the objective function in (47*a*); by means of the preceding statement it is easy to prove the following equality

(48) $$\Phi_0 = \min_{\delta \in \Delta} \quad \min_{x \in R} F(x; \delta) \quad = \min_{\delta \in \Delta} \Phi(\delta)$$

which enables one to reduce (46) to a finite sequence of convex quadratic programs. Of course, the number of such subproblems may be far too large. A theorem of alternative may be used to reduce the number of subproblems to be solved.

Consider any $\overline{\delta} \in \Delta$, and let $\overline{x} = (\overline{x}^i, i = 1, 2, 3)$ denote an optimal solution of $Q(\overline{\delta})$, so that $\Phi(\overline{\delta}) = F(\overline{x}; \overline{\delta})$. According to Theorem 3 (where we set $\ell = 1$, $f(x) = \Phi(\overline{\delta}) - F(x; \delta)$; $g(x) \geqslant 0$ denotes (47*b*)), $\overline{x}$ is an optimal solution of (46), iff there exists $\mu \in \mathbb{R}_+^m$ (depending on δ), such that [11]

(49) $$\Phi(\overline{\delta}) - F(x; \delta) + \langle \mu, Ax - \beta \rangle \leqslant 0, \qquad \forall x \in \mathbb{R}^{2n+N}, \qquad \forall \delta \in \Delta$$

Denote by Δ' the set of $\delta \in \Delta$ for which there exists $\mu(\delta)$ satisfying (49). The set Δ' may be cut off; in other words, it is not necessary to solve the subproblems $Q(\delta)$, $\delta \in \Delta'$. Hence, if $\Delta - \Delta' = \emptyset$, we have finished; otherwise, we choose $\delta \in \Delta - \Delta'$ and repeat the procedure.

As the determination of Δ' may be expensive, it is useful to remark that a subset is easily obtained. Let $\overline{\mu}$ be a vector of Kuhn-Tucker multipliers associated with $\overline{x}$. Because of condition (49), such a subset is

$$\Delta'' = \left\{ \delta \in \Delta : \Phi(\overline{\delta}) \leqslant \min_{x \in \mathbb{R}^{2n+N}} \left[F(x; \delta) - \langle \overline{\mu}, g(x) \rangle \right] \right\}$$

Of course, $\Delta'' \neq \emptyset$, as certainly $\overline{\delta} \in \Delta''$.

(11) As (46*ab*) is convex, a similar condition may be obtained by using the dual of (46*ab*); these two conditions enable one to cut off sets (of subproblems) which are not necessarily equal. We can assume $\lambda = 1$.

$\Phi(\overline{\delta})$ is of course an upper bound for Φ_0. If a lower bound is available, then their being equal is another useful optimality condition. To this end assume, without loss of generality, ρ to be an upper bound for every element of a feasible (or merely optimal) solution of (46). Then, it is easily noted that, instead of solving the sequence $\{Q(\delta), \delta \in \Delta\}$, it is equivalent to solving the sequence $\{Q'(\delta), \delta \in \Delta\}$, where $Q'(\delta)$ is the following problem

(50*a*) $$\Psi(\delta) = \min\left(\sum_{i=1}^{3} \langle c^i, x^i\rangle + \frac{1}{2} \sum_{i,j=1}^{3} \langle x^i, C_{ij}x^j\rangle\right)$$

subject to

(50*b*) $$\sum_{i=1}^{3} A_i x^i \geqslant b$$

and

(50*c*) $$0 \leqslant x^1 \leqslant \rho(a - \delta), \qquad 0 \leqslant x^2 \leqslant \rho\delta, \qquad x^3 \geqslant 0$$

Let a be a vector with unit entries; consider the dual of (50):

(51*a*) $$\max\left(-\frac{1}{2}\langle x, Cx\rangle + \langle b, \mu\rangle + \rho\langle \delta - a, \eta\rangle + \rho\langle -\delta, \xi\rangle\right)$$

subject to

(51*b*) $$Cx - A^T\mu + \begin{pmatrix} \eta \\ 0 \\ 0 \end{pmatrix} + \begin{pmatrix} 0 \\ \xi \\ 0 \end{pmatrix} + \begin{pmatrix} c^1 \\ c^2 \\ c^3 \end{pmatrix} \geqslant 0$$

and

(51*c*) $$\mu \geqslant 0, \qquad \eta \geqslant 0, \qquad \xi \geqslant 0$$

Denote by $S(\delta)$ and S the feasible regions of (50) and (51), respectively; by $t(x)$ and $T(x, \mu, \eta, \xi; \delta)$ the objective functions in (50*a*) and (51*a*), respectively, and by $\overline{S}$ any subset of S. We have

$$\Phi_0 = \min_{\delta \in \Delta} \min_{x \in S(\delta)} t(x) = \min_{\delta \in \Delta} \max_{(x, \mu, \eta, \xi) \in S} T(x, \mu, \eta, \xi; \delta)$$

$$\geq \min_{\delta \in \Delta} \max_{(x, \mu, \eta, \xi) \in \overline{S}} T(x, \mu, \eta, \xi; \delta)$$

It follows that

$$\text{(52)} \qquad \begin{aligned} &\min (\phi) \quad \text{s.t.} \quad \phi \geq T(x, \mu, \eta, \xi; \delta) \\ &\phi \in \mathbb{R}, \qquad \delta \text{ binary} \end{aligned}$$

equals the minimum of (50) if the inequalities are written for every $(x, \mu, \eta, \xi) \in S$, or is a lower bound for it if they are written merely for every $(x, \mu, \eta, \xi) \in \overline{S}$.

As at every step of the outlined procedure we get vectors $\overline{x}$, $\overline{\mu}$, and as it is easy to determine corresponding vectors $\overline{\eta}$, $\overline{\xi}$ (without solving (51)), we dispose of an element of S. Thus, at every step we can define a subset $\overline{S}$ and get a lower bound for Φ_0, by solving (52).

Remark 28 The condition stated in [22], and mentioned in Remark 27, may be extended to state equivalence between a convex quadratic complementarity problem (like (46)) and a convex quadratic program. Of course, when such a condition is fulfilled, the preceding decomposition procedure is useless. Unfortunately, in the above-mentioned engineering problem [12, 13] such a condition is not satisfied.

Remark 29 The above described solving procedure for (46) can be improved in the following way. Introduce the function

$$\pi(x_i^1, x_i^2) = x_i^1 - \max \{0, x_i^1 - x_i^2\}$$

and note that it is concave. If r denotes a large enough real, then it is easy to show that (46) is equivalent to minimize the function

$$G(x) \triangleq \sum_{i=1}^{3} \langle c^i, x^i \rangle + \sum_{i,j=1}^{3} \langle x^i, C_{ij} x^j \rangle + r \sum_{i=1}^{n} \pi(x_i^1, x_i^2)$$

under the constraints (46*b*). When G is concave (this happens if $C = (C_{ij})$ is negative semi-definite; in particular $C_{ij} = 0$), denote by $\hat{x} = (\hat{x}^i, i = 1, 2, 3)$ any local minimum (it is not restrictive to assume that be a vertex of (46*b*)) of the above problem, and by $x^s(t) = \hat{x} + \omega^s t$, $t \geq 0$, the parametric equation of the s-th extremal ray of polyhedron (46*b*) incident to $\hat{x}$; where ω^s is a suitable $(2n + N)$-vector. Denote by t_s the positive root of equation $G(x^s(t)) =$

$= G(\hat{x})$; set $t_s = +\infty$ if such root does not exist. Let $\langle c, x \rangle = \gamma$ represent the hyperplane passing through the (proper or improper) points $x^s(t_s)$; it is not restrictive to assume that $\langle c, \hat{x} \rangle < \gamma$. Hence, the halfspace $\langle c, x \rangle < \gamma$ does not contain any point x, such that $G(x) < G(\hat{x})$. It follows that $\langle c, x \rangle \geqslant \gamma$ can be used to reduce the polyhedron (46*b*), as well as to define a cutting plane solving algorithm for (46) through the above concave problem.

When the objective function in (46*a*) is convex a possible approach might consist in applying the above cutting plane approach to the (linear) complementarity problem (which contains the parameter δ both in the objective function and in the right-hand side terms) coming from the Kuhn-Tucker condition associated to the (convex) parametric program (47).

9. NUMERICAL EXAMPLES

Example 1

Consider the problem

$$\min(\text{sen } x) \qquad \text{s.t.} \qquad x \geqslant 0 \tag{53}$$

$$x \in X \triangleq [-\pi/3, \pi/3]$$

which is problem (11) at $m = n = 1$, $\varphi(x) = \text{sen } x$, $g(x) = x$. We want to prove that $\hat{x} = 0$ is an optimal solution of (53). As $f(x) = \varphi(\hat{x}) - \varphi(x) = -\text{ sen } x$ is not concave, Theorem 3 cannot be applied. However, as $1 - \exp(-\alpha \text{ sen } x)$ is concave on X, if $\alpha \geqslant 2\sqrt{3}$, we set $\alpha = 4$ and replace the optimality condition for $\hat{x}$, namely the impossibility of system

$$f(x) = -\text{sen } x > 0, \qquad g(x) = x \geqslant 0, \qquad x \in X \tag{54}$$

with the impossibility of system

$$1 - \exp(4 \text{ sen } x) > 0, \qquad x \geqslant 0, \qquad x \in X \tag{55}$$

Now Theorem 3 can be applied and the optimality of $\hat{x}$ shown by simply using a Γ_2 function. Note that the set E of Remark 1 is not convex in case (54), while it becomes convex in case (55).

Example 2
Consider the problem

$$\text{(56)}\qquad \begin{array}{ll} \min\ (\varphi(x) = -x_1^2 - x_2^2) \quad \text{s.t.} \quad 2x_1 + x_2 \leqslant 1 \\ x = (x_1, x_2) \in X \end{array}$$

where $X = \{x \in \mathbb{R}_+^2 : x_1 \leqslant 2; x_2 \leqslant 2\}$. If we set $G = \Gamma_2$ in (2), as we can set $\lambda = 1$, problem (16) becomes

$$\text{(57)}\qquad \max \psi(\mu), \qquad \mu \in \mathbb{R}_+$$

where

$$\psi(\mu) = \min_{x \in X} [-x_1^2 - x_2^2 - \mu(1 - 2x_1 - x_2)] = \begin{cases} 5\mu - 8 & \text{if } 0 \leqslant \mu \leqslant 1 \\ \mu - 4 & \text{if } 1 \leqslant \mu \leqslant 2 \\ -\mu & \text{if } 2 \leqslant \mu \end{cases}$$

It follows that (17*a*) is satisfied, while (17*b*) is not, as (57) equals - 2 and (56) equals - 1. In other words, weak alternative and hence weak duality hold. The difference between (56) and (57), namely 1, is the *duality gap.*

Example 3
Consider again problem (56) and replace G with Γ_1 in (2). Problem (16) now becomes

$$\text{(58)}\qquad \max \psi(\mu, \alpha) \qquad (\mu, \alpha) \in \mathbb{R}_+^2$$

where

$$\psi(\mu, \alpha) = \min_{x \in X} \{-x_1^2 - x_2^2 - \mu[1 - \exp(-\alpha(1 - 2x_1 - x_2))]\}$$

Let us simply verify that now (17) holds. To this end, set $t = 2x_1 + x_2$, so that

$$\psi(\mu, \alpha) = \min_t \Delta(t; \mu, \alpha) = \begin{cases} \alpha \exp[\alpha(t-1)] - \mu - t^2 & \text{if } 0 \leqslant t \leqslant 2 \\ \alpha \exp[\alpha(t-1)] - \mu - \frac{1}{4} t^2 + t - 5 & \text{if } 2 \leqslant t \leqslant 4 \end{cases}$$

As $\Delta(1; \mu, \alpha) = -1$, $\forall (\mu, \alpha) \in \mathbb{R}_+^2$, it follows that $\psi(\mu, \alpha) \leqslant -1$, $\forall (\mu, \alpha) \in \mathbb{R}_+^2$, so that (17*a*) holds. Some calculations allow us to state that

$$\min_t \Delta(t; 2e^{-2}, 2) = -1$$

and it follows that (17*b*) holds too; thus, ($\mu = 2e^{-2}$, $\alpha = 2$) is an optimal solution of (58).

The duality gap is now avoided, but we have to handle more complicated functions.

Example 4
Consider again problem (58). We have now

$$\Delta = -x_1^2 - x_2^2 - p\ \{1 - \exp[-q(1 - 2x_1 - x_2)]\}$$

Its Hessian matrix is

$$pq^2 \exp[q(2x_1 + x_2 - 1)] \begin{pmatrix} 4 & 2 \\ 2 & 1 \end{pmatrix} - \begin{pmatrix} 2 & 0 \\ 0 & 2 \end{pmatrix}$$

and is negative-definite iff $10pq^2 \exp[q(2x_1 + x_2 - 1)] < 4$. If we take account of the fact that $2x_1 + x_2 - 1 \leqslant 5$ on X, this happens if $p < [2/(5q^2)] \cdot \exp(-5q)$.

Example 5
Consider problem (11) in case (32), where $m = 3$, $n = 2$, and

$$c = \begin{pmatrix} -5 \\ 10 \end{pmatrix}, \qquad C = \begin{pmatrix} 2 & 0 \\ 0 & 10 \end{pmatrix}, \qquad A = \begin{pmatrix} 2 & 1 \\ 1 & 2 \\ 1 & 1 \\ 1 & 4 \end{pmatrix}, \qquad b = \begin{pmatrix} 1 \\ 1 \\ 1 \\ -2 \end{pmatrix}$$

Problem $L(x)$ of section 6 becomes

$$-x_1^2 - 5x_2^2 + \max(\lambda_1 + \lambda_2 + \lambda_3 - 2\lambda_4)$$

subject to

$$2\lambda_1 + \lambda_2 + \lambda_3 + \lambda_4 = 2x_1 - 5, \qquad \lambda_1 + 2\lambda_2 + \lambda_3 + 4\lambda_4 = 10x_2 + 10,$$

$$\lambda_j \geqslant 0, \qquad j = 1, 2, 3, 4$$

Set $x=\bar{x}=(5/2, -1)$ (i.e. consider the free minimum point of the objective function as starting position). Remark that $L(x)$ may be equivalently written in the following way:

$$-x_1^2 - 5x_2^2 + 10x_2 + 10 + \max(-\lambda_2 - 6\lambda_4)$$

$$\text{s.t.} \quad \begin{aligned} \lambda_1 - \lambda_2 \qquad - 3\lambda_4 &= 2x_1 - 10x_2 - 15 \\ 3\lambda_2 + \lambda_3 + 7\lambda_4 &= -2x_1 + 20x_2 + 25 \\ \lambda_j \geqslant 0, j &= 1, 2, 3, 4 \end{aligned}$$

which shows the optimal solution $\bar{\lambda}=(0, 0, 0, 0)$ at $x=\bar{x}$. An optimal solution of $D(\bar{x})$ is $\bar{\omega}=(0, 1)$, the (common) value of extrema of $L(\bar{x})$ and $D(\bar{x})$ is $\bar{k}=-45/4$. Problem (34) becomes max $[-(21/20)\lambda_1^2-(3/10)\lambda_3^2-(11/10)\lambda_1\lambda_3-3\lambda_1-(1/2)\lambda_3-45/4]$; s.t. $\lambda_1 \geqslant 0$; $\lambda_3 \geqslant 0$; and has $\hat{\lambda}_B=(\hat{\lambda}_1=\hat{\lambda}_3=0)$ as optimal solution and $\hat{q}=-45/4$ as maximum. The corresponding x-vector is $\hat{x}=C^{-1}(B\hat{\lambda}_B-c)=(\hat{x}_1=5/2, \hat{x}_2=-1)$. As $b_2-\langle A_2, C^{-1}(B\hat{\lambda}_B-c)\rangle=1/2$, at step III we cannot stop, and hence we find $s=2$ and $r=1$; $\widetilde{B}$ is now defined by columns 2 and 3 of A^T. At step IV problem (35) becomes

$$\max(\theta), \qquad \text{s.t.} \qquad \theta \leqslant -x_1^2 - 5x_2^2 + 10x_2 + 10$$

and its maximum is $k^*= 15$. As $\bar{k} \neq k^*$, we go to step II, where problem (34) becomes

$$\max(-\frac{9}{20}\lambda_2^2 - \frac{3}{10}\lambda_3^2 - \frac{7}{10}\lambda_2\lambda_3 + \frac{1}{2}\lambda_2 - \frac{1}{2}\lambda_3 - \frac{45}{4}); \text{ s.t. } \lambda_2 \geqslant 0; \ \lambda_3 \geqslant 0$$

and has $\widetilde{\lambda}_B=(\widetilde{\lambda}_2=5/9, \widetilde{\lambda}_3=0)$ as optimal solution and $\widetilde{q}=-100/9$ as maximum. The corresponding x-vector is $\widetilde{x}=C^{-1}(\widetilde{B}\widetilde{\lambda}_B-c)=(\widetilde{x}_1=25/9, \widetilde{x}_2=-8/9)$. At step III the optimality condition is fulfilled and hence $\widetilde{x}$ is the optimal solution.

REFERENCES

[1] J. F. Benders. 'Partitioning procedures for solving mixed-variables programming problems'. *Numer. Math.*, **4** (1962), 238-52.

[2] J. Borwein. 'Multivalued convexity and optimization: a unified approach to inequality and equality constraints'. *Math. Program.*, **13**, no. 2 (1977), 183-99.

[3] A Cambini, L. Martein, and L. Pellegrini. 'A decomposition algorithm for a particular class of nonlinear programs'. Presented at *AFCET-SMF Meeting*, Palaiseau, (Paris), September 1978.

[4] G. Castellani, and F. Giannessi. 'Decomposition of mathematical programs by means of theorems of alternative for linear and nonlinear systems'. *Proc. IX Int. Symp. on Mathematical Programming*, Hungarian Academy of Science, Budapest, 1979.

[5] R. W. Cottle. 'Complementarity and variational problems'. *Symposia Mathematica* (Istituto Nazionale Alta Matematica), vol. XIX, Academic Press, New York (1976), pp. 177-208.

[6] R. W. Cottle, and J. S. Pang. 'On solving linear complementarity problems as linear programs'. *Math. Program. Stud.*, 7 (1978), 88-107.

[7] J. Edmonds and R. Giles. 'A min-max relation for submodular functions on graphs'. In P. L. Hammer, E. L. Johnson, B. H. Korte and G. L. Nemhauser (eds): *Studies in integer programming*. Annals of Discrete Mathematics (1977), V. 1.

[8] K. Florek. 'Concerning dual system of linear relations (I)'. *Colloq. Math.*, vol. XX, fasc. 1 (1969). 143-51.

[9] A. M. Geoffrion. 'Generalized Benders decomposition'. *Proc. Symp. of Nonlinear Programming*, Mathematics Research Center, Univ. of Wisconsin, Madison, May 1970.

[10] F. Giannessi, and F. Niccolucci. 'Connections between nonlinear and integer programming problems'. *Symposia Mathematica* (Istituto Nazionale Alta Matematica), vol. XIX, Academic Press, New York (1976), pp. 161-76.

[11] F. Giannessi, and B. Nicoletti. 'The Crew scheduling problem: a travelling salesman approach'. In N. Christofides (ed.): *Combinatorial Optimization.* John Wiley, New York (1979).

[12] F. Giannessi, L. Jurina, and G. Maier. 'Optimal excavation profile for a pipeline freely resting on the sea floor'. Engineering Structures, I.P.C. Science and Technology Press, Guildford (U.K.), vol. 1, (1979), pp. 81-91.

[13] F. Giannessi, and G. Maier. 'A quadratic complementarity problem related to the optimal design of pipelines resting on a rough seabottom'. *Engineering Structures, I.P.C. Science and Technology Press*, Guildford (U.K.) To appear.

[14] P. E. Gill, and W. Murray. 'Numerically stable methods for quadratic programming'. *Math. Program.*, **14**, no. 3 (1978), 349-72.

[15] F. G. Gould. 'Nonlinear pricing: applications to concave programming'. *Oper. Res.*, **19**, no. 4 (1971), 1026-35.

[16] E. L. Hannan. 'Using duality theory for identification of primal efficient points and for sensitivity analysis in multiple objective linear programming'. *J. Oper. Res. Soc.*, **29**, no. 7 (1978), 643-9.

[17] H. Isermann. 'On some relations between a dual pair of multiple objective linear programs'. *Z. Oper. Res.*, **22** (1978), 33-41.

[18] R. M. Karp. 'Reducibility among combinatorial problems'. In *Complexity of Computer Computations*. R. E. Miller and J. W. Thatcher (eds), Plenum Press, New York (1972), pp. 85-104.

[19] R. M. Karp. 'On the computational complexity of combinatorial problems'. *Networks*, **5** (1975), 45-68.

[20] J. F. Kosma. 'Diophantische Approximationen'. Ergeb. der Math. u. ihrer Grenzgeb., Bd. IV, Heft 4, Springer-Verlag (1974).

[21] R. Lehmann, and W. Oettli. 'The theorem of the alternative, the key-theorem, and the vector-maximum problem'. *Math. Program.*, 8, no. 3 (1975), 332-44.

[22] O. L. Mangasarian. '*Nonlinear Programming*'. McGraw-Hill, New York (1969).

[23] O. L. Mangasarian. 'Linear complementarity problems solvable by a single linear program'. *Math. Program.*, **10**, no. 2 (1976), 263-70.

[24] L. McLinden. 'Duality theorems and theorems of the alternative'. *Proc. Am. Math. Soc.*, **53**, no. 1 (1975), 172-5.

[25] G. Maier. 'Mathematical programming methods in structural Analysis'. In *Variational methods in Engineering*, C. A. Bubbia and H. Tottenham (eds.), Proc. Inter. Conference, Southampton Univ. Press (U. K.), vol. II, 8/1 (1973).

[26] G. Maier, F. Andreuzzi, F. Giannessi, L. Jurina, and F. Taddei. 'Unilateral contact, elastoplasticity and complementarity with reference to offshore pipeline design'. Inter. Conference on Finite Elements in Nonlinear Mechanics, Stuttgart 1978. Published in Computer Methods in Applied Mech. and Engineering, vol 17/18, pp. 469-95 (1979).

[27] M. J. D. Powell. 'Algorithms for non linear constraints that use Lagrangian functions', *Math. Program.*, **14**, no. 2 (1978), 224-48.

[28] D. A. Pyne. 'On interior and convexity conditions, development of dual problems and saddlepoint optimality criteria in abstract mathematical programming'. *Math. Program.*, **8**, no. 2 (1975), 125-33.

[29] R. T. Rockafellar. 'Penalty methods and augmented Lagrangians in nonlinear programming'. *Proc. 5th IFIP Conf. on Optimization Techniques*, Rome, 1973, Springer-Verlag, Berlin (1973), pp. 418-25.

[30] R. T. Rockafellar. 'Solving a nonlinear programming problem by way of a dual problem'. *Symposia Mathematica* (Istituto Nazionale Alta Matematica), vol XIX, Academic Press, New York (1976), pp. 135-60.

[31] R. T. Rockafellar. 'Augmented Lagrangians and applications of the proximal point algorithm in convex programming'. *Math. Oper. Res.*, **1**, no. 2 (1976), 97-116.

[32] B. A. Rosina. 'Nuove considerazioni sulla famiglia delle quadriche generalizzate'. *Ann. Univ. Ferrara*, sez. VII (Matematica), vol. V, N. 1 (1956).

[33] E. E. Rosinger. 'Duality and alternative in multiobjective optimization'. *Proc. Am. Math. Soc.*, **64**, no. 2 (1977), 307-12.

[34] P. D. Scott. 'A quadratic programming dual algorithm for minimax control'. *IEEE Trans. Autom. Control*, **AC-20** (1975), 434-5.

[35] V. A. Sposito, and H. T. David. 'Saddlepoint optimality criteria of nonlinear programming problems over cones without differentiability'. *SIAM J. Appl. Math.*, **20** (1971), 698-702.

[36] R. M. Van Slyke, and R. J. B. Wets. 'A duality theory for abstract programs with applications to optimal control theory'. *J. Math. Anal. & Appl.* **22** (1968), 679-706.

Chapter 13

An Existence Result for the Global Newton Method

F. J. Gould and C. P. Schmidt

For a continuously differentiable function $f: \mathbb{R}^n \to \mathbb{R}^n$ the non-linear equations $f(x) = 0$ are considered. Homotopy methods are used to constructively prove a theorem on the existence of solutions.

In all of the following discussion f is a twice continuously differentiable function from $\mathbb{R}^n$ to $\mathbb{R}^n$. The following existence theorem is due to Smale [5].

Theorem 1

Suppose there is a bounded open connected set C, with ∂C connected and smooth, such that

(i) $\det f'(x) \neq 0 \quad \forall\, x \in \partial C$

(ii) at each $x \in \partial C$ the Newton direction $-(f'(x))^{-1} f(x)$ is transversal (not tangent) to ∂C

(iii) for some $x^0 \in \partial C$, 0 is a regular value of the homotopy $H: \mathbb{R}^{n+1} \to \mathbb{R}^n$ defined as $H(x, t) = f(x) - (1 - t) f(x^0)$, $x \in \mathbb{R}^n$, $t \in \mathbb{R}$ (i.e. rank $H'(x,t) = n$ on $H^{-1}(0)$).

Then f has a zero in C.

As Smale points out, the assumption that $\det f'(x) \neq 0$ on ∂C can be relaxed. The following approach gives a new extension of Theorem 1.

Definition Given any bounded open connected set C, with ∂C smooth, and any two points $Z_1 \in \partial C$, $Z_2 \in \partial C$, the Newton directions at Z_1, Z_2 are said *to agree* (or to be *in agreement*) with respect to C if the following conditions are satisfied

(i) f' is non-singular at Z_1 and Z_2;

(ii) the Newton directions at Z_1 and Z_2 are transversal to ∂C;

(iii) these directions both point into C or both point out of C.

The Newton directions at Z_1, Z_2 are said *to disagree* with respect to C if (i) and (ii) above hold and one of the directions points into C and the other out of C.

Theorem 2

Suppose there is a bounded open connected set C, with ∂C smooth, such that (i) 0 is a regular value of $H : \mathbb{R}^{n+1} \times \partial C \to \mathbb{R}^n$ defined as $H(x, t, w) = f(x) - (1 - t) f(w), x \in \mathbb{R}^n, t \in \mathbb{R}, w \in \partial C$;
(ii) $S = \{x \in \partial C\colon \det f'(x) = 0\}$ has measure zero in ∂C;
(iii) at all $x \in \partial C - S$ where $\det f'(x)$ has the same sign, the Newton directions at Z_1 and Z_2 disagree. If $Z_1, Z_2 \in \partial C - S$ and sgn det $f'(Z_1)$=-sgn det $f'(Z_2)$, then the Newton directions at Z_1 and Z_2 disagree.

Then f has a zero in $\overline{C}$. If it is assumed that 0 is a regular value of f, then f has a zero in C.

Proof: The plan of the proof is as follows.

(*a*) Let $H_{x^0}(x, t) = H(x, t, x^0)$. Choose an $x^0 \in \partial C$ such that $\det f'(x^0) \neq 0$ and let Γ_1 be the connected component of $H_{x^0}^{-1}(0)$ containing $(x^0, 0)$. Follow this component into $C \times \mathbb{R}^1$ and show that it intersects $\partial C \times \mathbb{R}^1$ at some (x^1, t_1). Let $\Gamma_{1,\,p}$ denote the projection of Γ_1 into $\mathbb{R}^n$.

(*b*) If $\det f'(x^1) \neq 0$, a result of Garcia and Gould [2] implies f has a zero in $\Gamma_{1,\,p} \cap C$.

(*c*) If $\det f'(x^1) = 0$, then either x^1 is a zero of f (in $\overline{C}$) or it is possible to pick x^2 near x^1 so that $f'(x^2)$ is non-singular. Let Γ_2 be the component of $H_{x_2}^{-1}(0)$ containing $(x^2, 0)$. Follow this component into $C \times \mathbb{R}^1$, and let (x^3, t_3) be the first point at which Γ_2 intersects $\partial C \times \mathbb{R}^1$. By choosing x^2 close enough to x^1, x^3 will be close enough to x^0 to ensure that $\det f'(x^3) \neq 0$. Using a previous argument, f has a zero in $\Gamma_{2,\,p} \cap C$. Since f is continuous, there is also a zero in $\Gamma_{1,\,p} \cap C$.

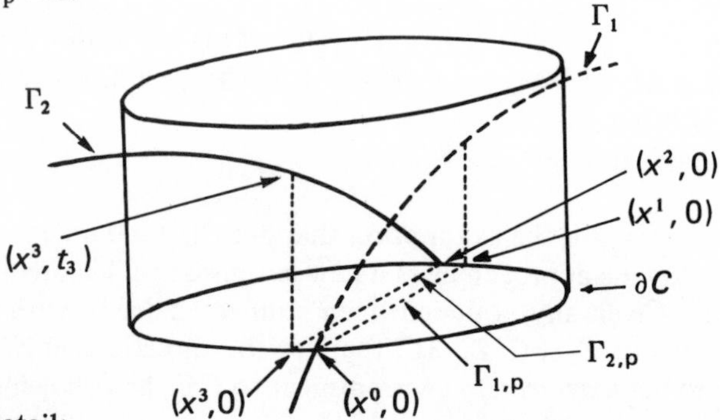

In more detail:

1. For almost every $x^0 \in \partial C$, $f'(x^0)$ is nonsingular and 0 is a regular value of $H_{x^0}(x, t)$. Choose one such x^0. Let Γ_1 be the component of $H_{x^0}^{-1}(0)$ containing $(x^0, 0)$. Let $(x(\theta), \tau(\theta))$ denote a parametrization of Γ_1. By differentiating $H(x(\theta), \tau(\theta)) = 0$ with respect to θ, it can be shown that

(1) $$x' = \frac{\tau'}{1-\tau}\,[-(f'(x))^{-1}f(x)]$$

This equation and the transversality of the Newton direction at x^0 imply that Γ_1 must cross $\partial C \times \mathbb{R}^1$ at $(x^0, 0)$. Noting that, for some i, $f_i(x^0) \neq 0$ (otherwise $f(x^0) = 0$ and the theorem is proved) the equation

$$\tau(\theta) = 1 - f_i(x(\theta))/f_i(x^0)$$

implies that $t = \tau(\theta)$ is bounded on $\Gamma_1 \cap \bar{C} \times \mathbb{R}^1$. Thus we may assume that $\Gamma_1 \cap \bar{C} \times \mathbb{R}^1 \subset \bar{C} \times (-\bar{t}, \bar{t})$. Since Γ_1 is not a loop in $\bar{C} \times \mathbb{R}^1$ (again by transversality) and $\bar{C} \times [-t, t]$ is compact, Γ_1 must intersect $\partial(\bar{C} \times [-\bar{t}, \bar{t}])$ after entering $C \times \mathbb{R}^1$ at $(x^0, 0)$. Let (x^1, t_1) be the first such point. It must be true that $x^1 \in \partial C$ (since $-\bar{t} < t_1 < \bar{t}$).

2.If $\det f'(x^1) \neq 0$, there are two cases to consider.

(a) Suppose sgn det $f'(x^0) =$ sgn det $f'(x^1)$. Suppose $(x^0, 0) = (x(\theta_0), \tau(\theta_0))$ and $(x^1, t_1) = (x(0), \tau(0))$. Then since the Newton directions at x^0 and x^1 must agree with respect to C (by assumption (iii) of Theorem 2), it must be true that $\tau'(0)/(1 - t_1)$ has sign apposite to $\tau'(\theta_0)/(1 - 0)$. However, a result of Garcia and Gould [2] says that since sgn det $f'(x^0) =$ sgn det $f'(x^1)$, it must be true that $\tau'(0)$ and $\tau'(\theta_0)$ have the same sign. Therefore, $1 - t_1 < 0$. By the intermediate value theorem there exists $(x^*, 1) \in \Gamma_1 \cap C \times \mathbb{R}^1$ such that $f(x^*) = 0$.

(b) Suppose sgn det $f'(x^0) = -$ sgn det $f'(x^1)$. Then again from the result of [2] it must be that $\tau'(0)$ and $\tau'(\theta_0)$ have opposite signs, but $\tau'(0)/(1 - t_1)$ and $\tau'(\theta_0)/(1 - 0)$ have the same sign because the Newton directions disagree (by assumption (iii). Therefore $1 - t_1 < 0$ and again f has a zero in C.

3. To complete the proof, it only remains to handle the case $\det f'(x^1) = 0$. If $f(x^1) = 0$, then f has a zero in $\bar{C}$ and the theorem is proved. Suppose that $f(x^1) \neq 0$, as must be if 0 is a regular value of f. Let $H^i_{x^0}$ be the Jacobian matrix of H_{x^0} with the ith column deleted. Garcia and Zangwill [3] have shown that Γ_1 is characterized by

(2a)
$$x_i' = (-1)^{i+1} \det H^i_{x^0}, \qquad i = 1, \ldots, n+1$$
$$(x(0), x_{n+1}(0)) = (x^1, t_1)$$

or

(2b)
$$x_i' = (-1)^{i} \det H^i_{x^0}, \qquad i = 1, \ldots, n+1$$
$$(x(0), x_{n+1}(0)) = (x^1, t_1)$$

The system chosen is the one which implies that $(x'(0), x_{n+1}(0))$ points into $C \times \mathbb{R}^1$ at (x^1, t_1). Assume that (2a) characterizes Γ_1.

4. Since none of the expressions det $H^i_{x^0}$ depends on x_{n+1}, because of the form of the Newton homotopy,

$$(3) \qquad \begin{aligned} x_i' &= (-1)^{i+1} \det H^i_{x^0}, \qquad i = 1, \ldots, n \\ x(0) &= x^1 \end{aligned}$$

describes the projection of Γ_1 into $\mathbb{R}^n$, denoted $\Gamma_{1,p}$.

5. Since

$$f(x^1) = (1 - t_1) f(x^0) = (1 - x^1_{n+1}) f(x^0)$$

then

$$\det H^i_{x^1} = (1 - x^1_{n+1}) \det H^i_{x^0}, \qquad \text{for} \quad i = 1, \ldots, n$$

Note that $x^1_{n+1} \neq 1$ because $f(x^1) \neq 0$ and hence (3) is equivalent to

$$(4) \qquad \begin{aligned} x_i' &= \frac{(-1)^{i+1}}{(1 - x^1_{n+1})} \det H^i_{x^1}, \qquad i = 1, \ldots, n \\ x(0) &= x^1 \end{aligned}$$

6. Consider the system

$$(5) \qquad \begin{aligned} x_i' &= (-1)^{i+1} \det H^i_{x^1}, \qquad i = 1, \ldots, n \\ x(0) &= x^1 \end{aligned}$$

This system also describes $\Gamma_{1,p}$. Let $x^0 = x(\theta_0)$ and define $\hat{x}$ as $x(\hat{\theta})$, $\hat{\theta} > \theta_0$ and such that $x(\theta) \notin \overline{C}$, $\theta \in (\theta_0, \hat{\theta}]$.

7. Pick an x^2 near x^1 so that $f'(x^2)$ is non-singular and 0 is a regular value of $H_{x^2}(x,t)$. Let Γ_2 be the component of $H^{-1}_{x^2}(0)$ containing $(x^2, 0)$. Consider the solution to

$$(6) \qquad \begin{aligned} y_i' &= (-1)^{i+1} \det H^i_{x^2}, \qquad i = 1, \ldots, n \\ y(0) &= x^2 \end{aligned}$$

The solution of (6) describes the projection of Γ_2 into $\mathbb{R}^n$. Since $x'(0)$ points into C, $y'(0)$ will also point into C if x^2 is close enough to x^1. By a theorem on differential equations (see p. 173 of [4]) $y(\theta)$ can be made ar-

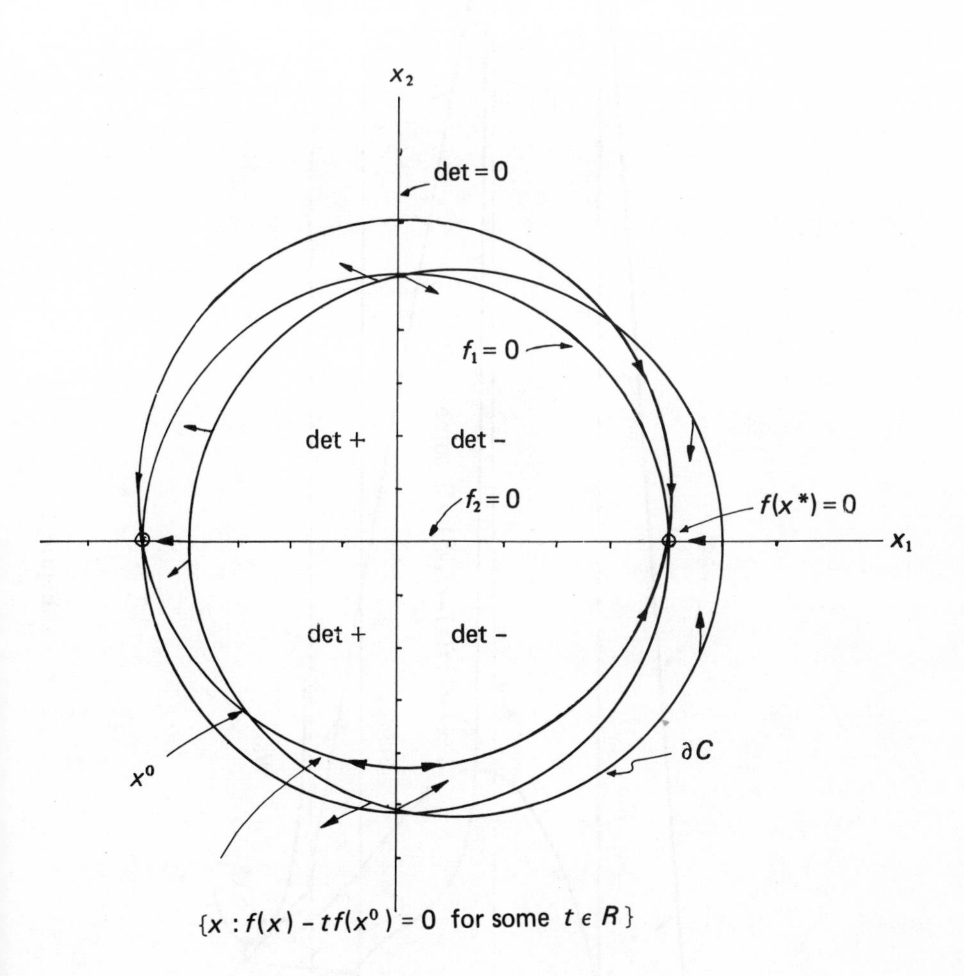
x_2
det = 0
$f_1 = 0$
det +
det -
$f_2 = 0$
$f(x^*) = 0$
x_1
det +
det -
∂C
x^0
$\{x : f(x) - tf(x^0) = 0 \text{ for some } t \in R\}$

Figure 1

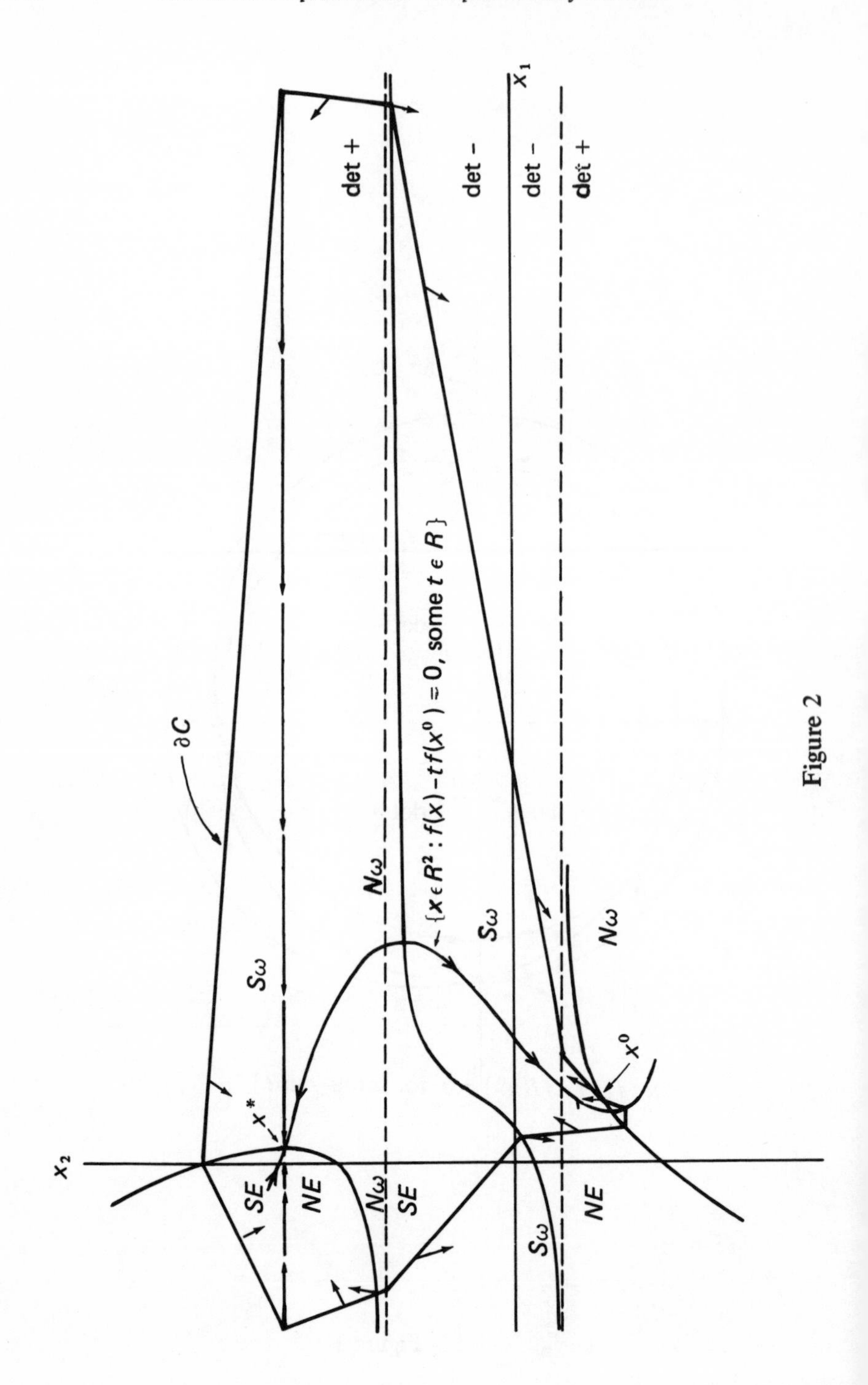
det +
det -
det -
det +
x_1
x_2
∂C
$N\omega$
$S\omega$
$S\omega$
$N\omega$
$S\omega$
$N\omega$
$S\omega$
SE
NE
SE
NE
x^*
x^0
$\{x \in R^2 : f(x) - tf(x^0) = 0, \text{ some } t \in R\}$

Figure 2

bitrarily close to $x(\theta)$, for $\theta \in [0, \hat{\theta}]$, where $x(\theta)$ is the solution of (5), by making x^2 arbitrarily close to x^1. Thus we can make $y(\theta)$ sufficiently close to $x(\theta)$, for $\theta \in [0, \hat{\theta}]$, that $y(\theta)$ intersects ∂C for the first time (after leaving x^2) at some point, say $x^3 = y(\theta_1)$, which is close enough to x^0 that det $f'(x^3) \neq 0$.

8. The theorem now follows by applying the argument of part 2 above to the path $y(\theta), \theta \in [0, \theta_1]$.

A simple illustration of the theorem is provided by the following problem

$$
\begin{aligned}
f_1(x_1, x_2) &= -(x_1^2 + x_2^2) + 25 \\
f_2(x_1, x_2) &= x_2 \\
x^0 &= (-2.5, -3.5) \\
f(5, 0) &= 0 \\
\det f'(x_1, x_2) &= 0 \Leftrightarrow x_1 = 0
\end{aligned}
$$

Figure 1 shows a set C postulated in Theorem 2. This figure also shows that det $f'(x)$ cannot be everywhere non-zero on the boundary of any open bounded connected set C containing the zero (5.0, 0) and with $x^0 \in \partial C$. Consequently, the assumptions of Theorem 1 cannot be satisfied.

Another illustration is provided by a problem due to Freudenstein and Roth [1]:

$$
\begin{aligned}
f_1(x_1, x_2) &= -13 + x_1 - 2x_2 + 5x_2^2 - x_2^3 = 0 \\
f_2(x_1, x_2) &= -29 + x_1 - 14x_2 + x_2^2 + x_2^3 = 0 \\
x^0 &= (15, -2) \\
f(5.0, 4.0) &= 0 \\
\det f'(x_1, x_2) &= 0 \Leftrightarrow x_2 = 2.23 \text{ or } x_2 = -0.897
\end{aligned}
$$

Figure 2 shows a set C postulated in Theorem 2. This figure also shows that det $f'(x)$ cannot be everywhere non-zero on the boundary of any open bounded connected set C containing the zero (5.0, 4.0) and with $x^0 \in \partial C$.

REFERENCES

[1] F. Freudenstein, and B. Roth, 'Numerical solutions of systems of nonlinear equations'. *J. Assoc. Comput. Mach.*, **10** (1963), 550-6.

[2] C. B. Garcia, and F. J. Gould. 'A theorem on homotopy paths'. *Math. Oper. Res.* **3** (1978), 282-9.

[3] C.B. Garcia, and W.I. Zangwill. 'On a new approach to homotopy and degree theory'. *Report No. 7813,* Center for Mathematical Studies in Business and Economics, University of Chicago, February, 1978.

[4] W. Hirsch, and S. Smale. *'Differential Equations, Dynamical Systems, and Linear Algebra'.* Academic Press, New York (1974).

[5] S. Smale. 'Convergent process of Price Adjustment and global Newton methods'. *J. Math. Econ.* **3** (1976), 107-20.

Chapter 14

The Smoothness of Free Boundaries Governed by Elliptic Equations and Systems

D. Kinderlehrer

1. INTRODUCTION

The object of this note is to illustrate the use of hodograph transforms to determine the smoothness of free boundaries. The problems considered involve functions defined on one or both sides of a free boundary satisfying elliptic equations and overdetermined boundary conditions. All considerations are local and the free boundary is assumed C^1 initially. This report is a summary of the papers [8, 9] written with Louis Nirenberg and Joel Spruck (cf. also [4, 7]).

Free-boundary problems arise naturally in the study of variational problems where constraints are present, for example, in the theories of variational inequalities or non-linear eigenvalue problems, in addition to many other situations. Among the question we consider are:

(i) a plasma surrounded by vacuum; the boundary of the plasma region being free;

(ii) two membranes, one above the other, subjected to forces; the boundary of the region of coincidence being free;

(iii) three soap films which meet along a curve at constant angles of $2\pi/3$; the curve being free; and

(iv) a thin plate constrained by an obstacle; the boundary of the region of concidence being free.

Our approach to these problems has been unified by the introduction of partial hodograph mappings and their associated Legendre transforms. Certain reflection mappings are employed when the question at hand requires study of functions defined on both sides of the free boundary. The properties of these transformations are discussed in section 2.

To understand the origin of a free-boundary problem in this context, let us recall the most prominent case, the obstacle problem of variational inequalities.

Example 1
Let $G \subset \mathbb{R}^n$ be open, connected, with smooth boundary ∂G and let $\psi \in C^2(\overline{G})$ satisfy

$$\max_G \psi > 0, \quad \text{and} \quad \psi < 0, \quad \text{on } \partial G$$

Set

$$\mathbb{K} = \{v \in H_0^1(G) : v \geqslant \psi \ \text{ in } \ G\}$$

and let $f \in C^1(\overline{G})$ be given. Suppose that

$$(1) \qquad u \in \mathbb{K}: \quad \int_G u_{x_j}(v-u)_{x_j} dx \geqslant \int_G f(v-u)dx, \quad \text{for } v \in \mathbb{K}$$

We know by the theorem of Lewy and Stampacchia [13] that $u \in C^1(\overline{G})$. Defining $I = \{x \in G : u(x) = \psi(x)\}$ we easily check that

$$(2) \qquad -\Delta u = f, \quad \text{in } G - I$$

$$(3) \qquad u = \psi, \quad u_{x_i} = \psi_{x_i} \qquad \text{on } \partial I$$

The conditions (3) constitute Cauchy data for (2) on ∂I. Since the Cauchy problem for Δ is not well posed even for a smooth initial surface, the mere existence of a solution suggests restrictive conditions on the set ∂I. It is a free boundary.

With Example 1, it is natural to consider the problem of two membranes. This was formulated by Stampacchia and studied in a series of papers by Vergara-Caffarelli [18, 19, 20]. It remains among our most enticing questions, for despite the ease with which it may be formulated, the analysis of the free boundary requires the full strength of our theory. Indeed, the reflection mappings and unusual choices of weights we use in the sequel were devised in response to the difficulties we encountered here. Even so, only the case related to linear equations is within our grasp.

We describe a simple example which displays the significant features. With $G \subset \mathbb{R}^n$ as before, consider the following:

Example 2
Let $G \subset \mathbb{R}^n$ be a bounded domain with smooth boundary ∂G. Suppose we are given $\lambda_1, \lambda_2 \in \mathbb{R}$, $f_1, f_2 \in C^1(\overline{G})$, and $\varphi^1, \varphi^2 \in C^1(\overline{G})$ satisfying $\varphi^1 < \varphi^2$ on ∂G. Set

$$\mathbb{K} = \{(v^1,v^2) : v^1 \leq v^2 \text{ in } G, v^j = \varphi^j \text{ on } \partial G, v^j \in H^1(G), j = 1,2\}$$

and consider

$$(u^1, u^2) \in \mathbb{K} :$$

(4) $$\int_G \{u^j_{x_i}(v^j - u^j)_{x_i} + \lambda_j(v^j - u^j)\}dx \geq \int_G f_j(v^j - u^j)dx, \quad \text{for } (v^1, v^2) \in \mathbb{K}$$

Here the summation convention is in force on i, $1 \leqq i \leq n$, and $j = 1,2$. Under appropriate hypotheses about λ_1, λ_2, the form

$$a(u,v) = \int_G \{u^j_{x_i} v^j_{x_i} + \lambda_j u^j v^j\}\, dx \quad u = (u^1,u^2), \quad v = (v^1,v^2)$$

is coercive on $H^1_0(G) \times H^1_0(G)$, in which case (4) has a unique solution. Assuming this, $u^j \in H^{2,s}(G)$, $j = 1,2$, according to [18] so we may write the complementarity relations

(5)
$$\begin{aligned}
&(-\Delta u^1 + \lambda_1 u^1 - f_1)(u^1 - u^2) = 0, \quad \text{in } G\\
&(-\Delta u^2 + \lambda_2 u^2 - f_2)(u^1 - u^2) = 0, \quad \text{in } G\\
&-\Delta u^1 + \lambda_1 u^1 - f_1 \geq 0, \quad \text{in } G\\
&-\Delta u^2 + \lambda_2 u^2 - f_2 \geq 0, \quad \text{in } G\\
&u^2 - u^1 \geq 0, \quad \text{in } G
\end{aligned}$$

Furthermore, for any $\zeta \in H^1_0(G)$, $(v^1, v^2) = (u^1 + \zeta, u^2 + \zeta) \in \mathbb{K}$ from which we derive the relation

(6) $$-\Delta(u^1 + u^2) + \lambda_1 u^1 + \lambda_2 u^2 - (f_1 + f_2) = 0, \quad \text{in } G$$

With $I = \{x \in G : u^1(x) = u^2(x)\}$, we write analogously to (2), (3), that

$$u^j \in C^{1,\lambda}(\overline{G}) \cap H^{2,s}(G)$$

and

(7) $$-\Delta u^1 + \lambda_1 u^1 = f_1, \quad -\Delta u^2 + \lambda_2 u^2 = f_2, \quad \text{in } G - I$$

(8) $$u^1 = u^2, \quad -\Delta u^1 + \frac{1}{2}(\lambda_1 + \lambda_2)u^1 = \frac{1}{2}(f_1 + f_2) \quad \text{in } I$$

Although the system for three functions u^1, u^2 in $G - I$ and u^1 in I is overdetermined, this was not obvious to us. Note in particular that the smoothness of the u^j means that

$$u^1_{x_i} = u^2_{x_i}, \quad \text{in } I, \quad 1 \leqslant i \leqslant n \tag{9}$$

2. HODOGRAPH MAPPINGS

The purpose of a hodograph mapping is to transform the given free boundary into a portion of hyperplane, albeit at the expense of changing the equation to a highly non-linear one. When the order of the equation is more than two, a single new equation may not suffice and a system, usually overdetermined, must be analyzed.

Let Γ be a C^1 manifold in $\mathbb{R}^n$ containing the origin 0 which separates a neighbourhood of the origin into two neighbourhoods Ω^+ and Ω^-. Say that the normal to Γ at the origin is in the positive x_n direction and points into Ω^+. Our hodograph mappings are defined in terms of a function $u(x)$, $x \in \Omega^+ \cup \Gamma$, satisfying

$$\partial_n^s u = 0, \quad \text{for} \quad s \leqslant p \quad \text{and} \quad \partial_n^{p+1} u \neq 0, \quad \text{on } \Gamma \tag{10}$$

Define the mapping

$$\begin{aligned} x \to y &= (x_1, \ldots, x_{n-1}, u) = (x', u), & p = 0 \\ x \to y &= (x_1, \ldots, x_{n-1}, -\partial_n^p u) = (x', -\partial_n^p u), & p \geqslant 1 \end{aligned} \tag{11}$$

which is locally 1:1, say 1:1, in $\Omega^+ \cup \Gamma$ and maps Ω^+ onto a domain $U \subset \{y_n > 0\}$ and Γ onto $S \subset \{y_n = 0\}$, say.

For a new dependent variable we choose a Legendre transform

$$\begin{aligned} \psi(y) &= x_n, & \text{if } p = 0 \\ v(y) &= x_n y_n + \partial_n^{p-1} u(x), & \text{if } p \geqslant 1 \end{aligned} \tag{12}$$

When $p > 1$ this is usually insufficient to describe the original equation which will be defined in Ω^+ and we shall have recourse to additional unknowns. These will be discussed later.

Suppose that $p \geqslant 1$. Then, with $1 \leqslant \sigma \leqslant n-1$,

$$\begin{aligned} dv &= x_n \, dy_n + y_n \, dx_n + \sum_\sigma \partial_\sigma \partial_n^{p-1} u \, dx_\sigma + \partial_n^p u \, dx_n \\ &= x_n \, dy_n + \sum_\sigma \partial_\sigma \partial_n^{p-1} u \, dy_\sigma \end{aligned}$$

or

$$(13)\qquad \begin{aligned} \nu_\sigma(y) &\equiv \frac{\partial}{\partial y_\sigma}\nu = \partial_\sigma \partial_n^{p-1} u, \qquad 1 \leqslant \sigma \leqslant n-1 \\ \nu_n(y) &= x_n \end{aligned}$$

In particular

$$\Gamma : x_n = \nu_n(x', 0), \qquad (x', 0) \in S$$

gives a parametrization of Γ. In this way, the smoothness of Γ may be interpreted as a question about the boundary behaviour of ν. In particular, the inverse of (11) is given by

$$(14)\qquad g^+ : U \cup S \to \Omega^+ \cup \Gamma \qquad g^+(y) = (y', \nu_n(y))$$

Passing to the case $p = 0$, we obtain

$$\begin{aligned} \mathrm{d}y_n = \mathrm{d}u &= \Sigma\, u_\sigma\, \mathrm{d}x_\sigma + u_n\, \mathrm{d}x_n \\ &= \Sigma\, u_\sigma\, \mathrm{d}y_\sigma + u_n\, \mathrm{d}\psi \end{aligned}$$

so

$$(15)\qquad \psi_\sigma = -\, u_\sigma / u_n, \qquad \psi_n = 1/u_n$$

In this case, the inverse of (11) is simply

$$(16)\qquad g^+(y) = (y', \psi(y))$$

Equations (14), (16) suggest simple reflection mappings defined in $U \cup S$ to $U^- \cup \Gamma$ on the opposite side of Γ defined by

$$(17)\qquad g^-(y) = \begin{cases} (y', \psi(y) - Cy_n), & \text{if } p = 0 \\ (y', \nu_n(y) - Cy_n), & \text{if } p > 1 \end{cases}$$

where C is any constant larger than $\psi_n(y)$ or $\nu_{nn}(y)$ respectively.
Hence (17) agrees with (14) when $y \in S$ and $p > 1$ and similarly agrees with (16) when $y \in S$, $p = 0$.

Restricting ourselves to $p = 0$, 1, it is easily seen that the x derivatives of u are expressed in terms of the y derivatives of ψ or ν by the formulae

$$\frac{\partial}{\partial x_\sigma} = \frac{\partial}{\partial y_\sigma} - \frac{\psi_\sigma}{\psi_n}\frac{\partial}{\partial y_n}, \qquad \frac{\partial}{\partial x_n} = \frac{1}{\psi_n}\frac{\partial}{\partial y_n}, \qquad p = 0, \sigma < n$$

$$\frac{\partial}{\partial x_\sigma} = \frac{\partial}{\partial y_\sigma} - \frac{v_{n\sigma}}{v_{nn}}\frac{\partial}{\partial y_n}, \qquad \frac{\partial}{\partial x_n} = \frac{1}{v_{nn}}\frac{\partial}{\partial y_n}, \qquad p = 1, \sigma < n$$

Observe that if u satisfies a second-order equation in Ω^+ then ψ or v will satisfy a second-order equation in U and, in fact, ellipticity is preserved ([5], Lemma 2.1). For example, $+\Delta u = -f$ in Ω^+ is transformed to

$$(18)\quad \Delta u \equiv -\frac{\psi_{nn}}{\psi_{n^3}} + \sum_{\sigma<n}\left[-\left(\frac{\psi_\sigma}{\psi_n}\right)_\sigma + \frac{\psi_\sigma}{\psi_n}\left(\frac{\psi_\sigma}{\psi_n}\right)_n\right] = -f(y', \psi), \quad \text{in } U, \text{ for } p = 0$$

$$(19)\qquad \Delta u \equiv -\frac{1}{v_{nn}} + \sum_{\sigma<n}\left(v_{\sigma\sigma} - \frac{v_{n\sigma}^2}{v_{nn}}\right) = -f(y', v_n), \quad \text{in } U, \text{ for } p = 1$$

Again, an elementary calculation shows that the transformation laws associated with the reflections (17) are given by

$$\frac{\partial}{\partial x_\sigma} = \frac{\partial}{\partial y_\sigma} - \frac{\psi_\sigma}{\psi_n - C}\frac{\partial}{\partial y_n}, \qquad \frac{\partial}{\partial x_n} = \frac{1}{\psi_n - C}\frac{\partial}{\partial y_n}, \quad \sigma<n,\ p=0$$

$$\frac{\partial}{\partial x_\sigma} = \frac{\partial}{\partial y_\sigma} - \frac{v_{n\sigma}}{v_{nn} - C}\frac{\partial}{\partial y_n}, \qquad \frac{\partial}{\partial x_n} = \frac{1}{v_{nn} - C}\frac{\partial}{\partial y_n}, \quad \sigma<n,\ p=1$$

Given a function $w(x)$, $x \in \Omega^-$, we can pull it back to a function $\varphi(y)$ by $\varphi(y) = w(g^-(y))$ and then calculate its derivatives. For example, if $\Delta w = -f$ in Ω^-, then

$$(20)\quad \begin{aligned} &\frac{1}{\psi_n - C}\left(\frac{1}{\psi_n - C}\varphi_n\right)_n + \\ &\sum_{\sigma<n}\left[\left(\varphi_\sigma - \frac{\psi_\sigma}{\psi_n - C}\varphi_n\right)_\sigma - \frac{\psi_\sigma}{\psi_n - C}\left(\varphi_\sigma - \frac{\psi_\sigma}{\psi_n - C}\varphi_n\right)_n\right] = \\ &= -f(g^-(y)), \qquad \text{in } U, \text{ for } p = 0 \end{aligned}$$

$$(21)\quad \begin{aligned} &\frac{1}{v_{nn} - C}\left(\frac{\varphi_n}{v_{nn} - C}\right)_n + \sum_{\sigma<n}\left[\left(\varphi_\sigma - \frac{v_{n\sigma}}{v_{nn} - C}\varphi_n\right) - \right. \\ &\left. - \frac{v_{n\sigma}}{v_{nn} - C}\left(\varphi_\sigma - \frac{v_{n\sigma}}{v_{nn} - C}\right)_n\right] = -f(y', v_n - Cy_n), \qquad \text{in } U, \text{ for } p = 1 \end{aligned}$$

We now have a coupled system for ψ, φ or v, φ. Note especially that (21) is third order in v although only second order in φ.

3. THE OBSTACLE PROBLEM: A SINGLE SECOND-ORDER EQUATION

This is the simplest problem. Suppose that $u \in C^2(\Omega^+ \cup \Gamma)$ and satisfies

$$\Delta u = a, \qquad \text{in } \Omega^+$$
$$u = 0, \quad u_{x_i} = 0, \qquad \text{on } \Gamma \tag{22}$$

where $a(x)$ is smooth and $a(x) > 0$ in $\overline{\Omega}^+$. Introduce the change of variables (11) with $p = 1$ and the Legendre transform

$$v(y) = x_n y_n + u(x), \qquad x \in \Omega^+, y \in U$$

To verify condition (10), notice that $u_{x_j} = 0$ on Γ and x_σ tangential to Γ at $x = 0$ imply that

$$u_{x_i x_\sigma}(0) = 0, \qquad 1 \leqslant i \leqslant n, \; 1 \leqslant \sigma \leqslant n\text{-}1$$

so

$$u_{x_n x_n}(0) = a(0) > 0$$

We may assume that $u_{x_n x_n} > 0$ in $\Omega^+ \cup \Gamma$. In view of (19), v is a solution of

$$\frac{1}{v_{nn}} + \sum_{\sigma < n} \left(v_{\sigma\sigma} - \frac{v_{n\sigma}^2}{v_{nn}} \right) = a(y', v_n), \qquad \text{in } U$$
$$v = 0, \qquad \text{on } S \tag{23}$$

The boundary condition holds because $u = 0$ on Γ and $y_n = 0$ on S. According to well known theorems about elliptic equations [1, 14, 5], we have

$$a \in C^{m,\lambda}(\Omega^+ \cup \Gamma) \Rightarrow v \in C^{m+2,\lambda}(U \cup S), \qquad m = 0, 1, 2, \ldots$$
$$a \text{ analytic} \Rightarrow v \text{ analytic} \tag{24}$$

In particular

$$\Gamma : x_n = \frac{\partial v}{\partial y_n}(y', 0), \qquad (y', 0) \in S$$

is a smooth parametrization of Γ.

4. THE CONFINED PLASMA

For the formulation of this problem, the reader is referred to [16, 17, 10]. In the last reference there is a discussion of the topological nature of the plasma set and the separatrix, which in our case is Γ.

Let $u \in C^2(\Omega^+ \cup \Gamma \cup \Omega^-)$ and

$$\begin{aligned} \Delta u &= 0, && \text{in } \Omega^+ \\ \Delta u + \lambda u &= 0, && \text{in } \Omega^- \\ u = 0 \text{ and } \frac{\partial u}{\partial \nu} &\neq 0, && \text{in } \Gamma \end{aligned} \tag{25}$$

We may suppose that $u > 0$ in Ω^+. Introduce the transformation (11) for $p = 0$ in all of $\Omega^+ \cup \Gamma \cup \Omega^-$, and set

$$\psi(y) = x_n$$

Then with $\widetilde{U}$ the image of Ω^- and U the image of Ω^+, $\psi \in C^2(U \cup S \cup \widetilde{U})$ and satisfies

$$\begin{aligned} -\frac{\psi_{nn}}{\psi_n^3} + \sum_{\sigma<n}\left[-\left(\frac{\psi_\sigma}{\psi_n}\right)_\sigma + \frac{\psi_\sigma}{\psi_n}\left(\frac{\psi_\sigma}{\psi_n}\right)_n\right] &= 0, && \text{in } U \\ -\frac{\psi_{nn}}{\psi_n^3} + \sum_{\sigma<n}\left[-\left(\frac{\psi_\sigma}{\psi_n}\right)_\sigma + \frac{\psi_\sigma}{\psi_n}\left(\frac{\psi_\sigma}{\psi_n}\right)_n\right] + \lambda y_n &= 0, && \text{in } \widetilde{U} \end{aligned} \tag{26}$$

in view of (18). Now define

$$\varphi(y) = \psi(y', -y_n), \qquad \text{for } y \in U$$

The differentiability of ψ then provides us with boundary conditions

$$\psi - \varphi = 0, \quad \psi_n + \varphi_n = 0, \qquad \text{on } S \tag{27}$$

whereas from (26) we obtain the system

$$\left.\begin{aligned} -\frac{\psi_{nn}}{\psi_n^3} + \sum_{\sigma<n}\left[-\left(\frac{\psi_\sigma}{\psi_n}\right)_\sigma + \frac{\psi_\sigma}{\psi_n}\left(\frac{\psi_\sigma}{\psi_n}\right)_n\right] &= 0 \\ -\frac{\varphi_{nn}}{\varphi_n^3} + \sum_{\sigma<n}\left[-\left(\frac{\varphi_\sigma}{\varphi_n}\right)_\sigma + \frac{\varphi_\sigma}{\varphi_n}\left(\frac{\varphi_\sigma}{\varphi_n}\right)_n\right] + \lambda y_n &= 0 \end{aligned}\right\} \text{ in } U \tag{28}$$

The system (27), (28) is a (non-linear) elliptic system with coercive boundary data. Hence from [1], the solution ψ, φ is analytic in $U \cup S$. Since

$$\Gamma : x_n = \psi(y', 0), \qquad (y', 0) \in S$$

the separatrix Γ is also analytic.

To resolve this question it was not necessary to use the reflection (17); however, for more complicated situations it is. An an example, the reader may apply the method (see [8], Theorem 3.2) to study.

$$(29) \qquad \begin{aligned} \Delta u^+ &= 0, && \text{in } \Omega^+ \\ \Delta u^- &= 0, && \text{in } \Omega^- \\ u^- = u^+ &= 0, && \\ & && \text{on } \Gamma \\ f(u_\upsilon^+, u_\upsilon^-) &= 0, && \end{aligned}$$

where $(u_\upsilon^+)^2 + (u_\upsilon^-)^2 \neq 0$ and $f_{u_\upsilon^+} u_\upsilon^+ - f_{u_\upsilon^-} u_\upsilon^- \neq 0$ on Γ.

5. THE DOUBLE MEMBRANE

As in the previous example, we shall try to write (7), (8) as an elliptic system in U with coercive boundary data on S. There is no loss in generality in setting $\lambda_1 = 0$, $\lambda_2 = \lambda$. Near a suitably smooth and small portion Γ of ∂I we may rewrite (7), (8) as

$$(30) \qquad \Delta u^1 = -f_1, \quad \Delta u^2 + \lambda u^2 = -f_2, \qquad \text{in } \Omega^+$$

$$(31) \qquad \Delta u^3 + \frac{1}{2}\lambda u^3 = -\frac{1}{2}(f_1 + f_2), \qquad \text{in } \Omega^-$$

$$(32) \qquad u^1 = u^2 = u^3, \quad \frac{\partial u^1}{\partial \upsilon} = \frac{\partial u^2}{\partial \upsilon} = \frac{\partial u^3}{\partial \upsilon} \qquad \text{on } \Gamma$$

Here we have set $u^3 = u^1 = u^2$ in Ω^-. It is not immediately evident that (30), (31), (32) is overdetermined. In fact, neglecting the equation in Ω^- temporarily gives the problem

$$\begin{aligned} \Delta u^1 &= -f_1, \\ \Delta u^2 + \lambda u^2 &= -f_2, \end{aligned} \qquad \text{in } \Omega^+$$

$$\begin{aligned} u^1 &= u^2 \\ \frac{\partial u^1}{\partial \upsilon} &= \frac{\partial u^2}{\partial \upsilon}, \end{aligned} \qquad \text{on } \Gamma$$

which generally is coercive and hence solvable when Γ has only a limited degree of smoothness. We must utilize the information given in Ω^-.

To assist us, introduce $w = u^1 - u^2$ and $w' = u^1 + u^2$ in Ω^+ and $w' = 2u^1 = 2u^2 = 2u^3$ in Ω^-. Then

$$(33) \qquad \begin{aligned} &(\Delta + \tfrac{1}{2}\lambda)w + \tfrac{1}{2}\lambda w' = -(f_1 - f_2), && \text{in } \Omega^+ \\ &\tfrac{1}{2}\lambda w + (\Delta + \tfrac{1}{2}\lambda)w' = -(f_1 + f_2) && \text{in } \Omega^+ \cup \Gamma \cup \Omega^- \\ &w = w_{x_i} = 0, && \text{in } \Gamma \cup \Omega^-, \ 1 \leq i \leq n \end{aligned}$$

From the second equation

$$\Delta w' = -(f_1 + f_2) - \tfrac{1}{2}\lambda(w' + w) \in H^{2,s}(\Omega^+ \cup \Gamma \cup \Omega^-), \qquad s < \infty$$

The elliptic regularity theory applied here informs us that

$$w' \in H^{4,s}(\Omega^+ \cup \Gamma \cup \Omega^-) \subset C^{3,\alpha}(\Omega^+ \cup \Gamma \cup \Omega^-), \quad 1 \leq s < \infty, 0 \leq \alpha < 1$$

Imposing the non-degeneracy criterion

$$(34) \qquad f_1(0) - f_2(0) - \lambda u^1(0) = f_1(0) - f_2(0) - \tfrac{1}{2}\lambda(w(0) + w'(0)) \neq 0$$

we see that w may be interpreted as the solution of

$$(35) \qquad \begin{aligned} &\Delta w = a(x,w) = f_1 - f_2 - \tfrac{1}{2}\lambda(w' + w), \quad \text{in } \Omega^+ \\ &w = w_i = 0, \quad \text{on } \Gamma, \ 1 \leq i \leq n \end{aligned}$$

with $a(x,w)$ analytic in w and $C^{3,\alpha}$ in x assuming that $f_1(x)$, $f_2(x)$ are at least $C^{3,\alpha}$. Moreover, $a(0,0) \neq 0$, It now follows from (24) for $m = 0$ that

$$w \in C^{5,\alpha}(\Omega^+ \cup \Gamma), \quad \text{and} \quad \Gamma \in C^{4,\alpha}$$

Unfortunately, this technique cannot be iterated to obtain improved regularity. Instead we turn to our reflection technique described in section 2. Assuming the non-degeneracy condition (34), introduce the new coordinates

$$y = (x', -w_n), \qquad w_j = w_{x_j}$$

which we assume map $\Omega^+ \cup \Gamma$ onto $U \cup S$, as usual. Set

$$v(y) = x_n y_n + w(x), \quad x \in \Omega^+ \cup \Gamma, \ y \in U \cup S \tag{36}$$

and define two additional dependent variables

$$\begin{aligned} \varphi^+(y) &= w'(g^+(y)), \quad y \in U \cup S \\ \varphi^-(y) &= w'(g^-(y)) \end{aligned} \tag{37}$$

where g^+ and g^- are given by (14) and (17) for $p = 1$.

Reserving, for the moment, our calculation of the various equations, we determine the boundary conditions obeyed by v, φ^+, φ^- on $y_n = 0$. As usual for a Legendre transform, $v = 0$ on $y_N = 0$. Since w' is continuous across Γ, $\varphi^+ = \varphi^-$ on $y_N = 0$. Finally, since $\partial w'/\partial x_n$ is continuous across Γ, $\varphi_n^+/v_{nn} = w'_n = \varphi_n^-/v_{nn} - C$ on $y_n = 0$. Summarizing,

$$\left.\begin{aligned} v &= 0 \\ \varphi^+ - \varphi^- &= 0, \quad \text{on } S \\ \frac{1}{v_{nn}}\varphi_n^+ - \frac{1}{v_{nn} - C}\varphi_n^- &= 0 \end{aligned}\right. \tag{38}$$

For ease of presentation, we now assume that $f_1(x)$, $f_2(x)$ are analytic. The first equation of (33) transforms in a familiar manner. From (19) we have that

$$-\frac{1}{v_{nn}} + \sum_{\sigma<n}\left(v_{\sigma\sigma} - \frac{v_{n\sigma}^2}{v_{nn}}\right) = F_1(v_n, v, \varphi^+, y), \quad \text{in } U \tag{39}$$

where

$$F_1(v_n, v, \varphi^+, y) = f_2(y', v_n) - f_1(y', v_n) - \tfrac{1}{2}\lambda(-y_n v_n + v + \varphi^+)$$

is an analytic function of its arguments. For φ^+ we obtain equation (21) with $C = 0$, namely

$$
\begin{aligned}
&\varphi_n^+ \left\{ \sum_{\sigma<n} \left[\frac{1}{v_{nn}} \left(\frac{1}{v_{nn}}\right)_\sigma - \left(\frac{v_{\sigma n}}{v_{nn}}\right)_\sigma \right] + \frac{1}{v_{nn}} \left(\frac{1}{v_{nn}}\right)_n \right\} + \\
(40) \qquad &+ \sum_{\sigma<n} \left[\varphi_{\sigma\sigma}^+ - 2 \frac{v_{n\sigma}}{v_{nn}} \varphi_{n\sigma} + \left(\frac{v_{n\sigma}}{v_{nn}}\right)^2 \varphi_{nn}^+ \right] + \left(\frac{1}{v_{nn}}\right)^2 \varphi_{nn}^+ = \\
&= F_2(v_n, v, \varphi^+, y), \qquad \text{in } U
\end{aligned}
$$

where

$$F_2(v_n, v, \varphi^+, y) = -[f_1(y', v_n) + f_2(y', v_n)] - \frac{1}{2}\lambda(-y_n v_n + v + \varphi^+)$$

For $x \in \Omega^-$, note that $w(x) = 0$. Consequently

$$
\begin{aligned}
&\varphi_n^- \left\{ \sum_{\sigma<n} \left[\frac{1}{v_{nn}-C} \left(\frac{1}{v_{nn}-C}\right) - \left(\frac{v_{\sigma n}}{v_{nn}-C}\right)_\sigma \right] + \frac{1}{v_{nn}-C} \left(\frac{1}{v_{nn}-C}\right)_n \right\} + \\
(41) \quad &\sum_{\sigma<n} \left[\varphi_{\sigma\sigma}^- - 2 \frac{v_{n\sigma}}{v_{nn}-C} \varphi_{n\sigma}^- + \left(\frac{v_{n\sigma}}{v_{nn}-C}\right)^2 \varphi_{nn}^- \right] + \left(\frac{1}{v_{nn}-C}\right)^2 \varphi_{nn}^- = \\
&= F_3(v_n, \varphi^-, y)
\end{aligned}
$$

where

$$F_3(v_n, \varphi^-, y) = -f_1(y', v_n) + f_2(y', v_n)) - \frac{1}{2}\lambda\varphi^-$$

We assign the weights $s_1 = -2$ to (39) and $s_2 = s_3 = 0$ to (40) and (41), $t_v = 4$, $t_+ = 2$, and $t_- = 2$ to v, φ^+, and φ^- and $r_1 = -4$, $r_2 = -2$, $r_3 = -1$, respectively to the boundary conditions of (28); see [1, 14, 11] for an explanation of weights.

To linearize these equations at $y = 0$, we assume that $v_{nn}(0) = a > 0$. Note that $v_{n\sigma}(0) = 0$, $1 \leq \sigma \leq n-1$, The terms of order 3 in v do not occur in the linerarized equations, nor do the terms of order 2 in v in the last boundary condition occur. For variations $\bar{v}$, $\bar{\varphi}^+$, $\bar{\varphi}^-$, we find the system

$$
(42) \qquad
\begin{aligned}
&\Sigma\, \bar{v}_{\sigma\sigma} + \frac{1}{a^2} \bar{v}_{nn} + \frac{1}{2} \lambda \varphi^+ = 0 \\
&\Sigma\, \bar{\varphi}_{\sigma\sigma}^+ + \frac{1}{a^2} \bar{\varphi}_{nn}^+ = 0 \qquad\qquad \text{in } y_n > 0 \\
&\Sigma\, \bar{\varphi}_{\sigma\sigma}^+ + \frac{1}{(C-a)^2} \bar{\varphi}_{nn}^+ = 0
\end{aligned}
$$

$$
(43) \qquad
\begin{aligned}
&\bar{v} = 0, \\
&\bar{\varphi}^+ - \bar{\varphi}^- = 0 \qquad\qquad \text{on } y_n = 0 \\
&\frac{1}{a} \bar{\varphi}^+ + \frac{1}{C-a} \bar{\varphi}_n^- = 0
\end{aligned}
$$

It is easy to see that this system is elliptic and has no bounded exponential solutions, keeping in mind that $C - a > 0$. Hence (39), (40), (41) is an elliptic system with the coercive boundary conditions (38). It follows that $\nu, \varphi^+, \varphi^-$ are analytic in $U \cup S$, and so is Γ.

6. THE SMOOTHNESS OF THE LIQUID EDGE

Consider a configuration of three minimal surfaces in $\mathbb{R}^3$ meeting along a $C^{1,\alpha}$ curve γ, the liquid edge, at equal angles of $2\pi/3$. Such a configuration represents one of the stable singularities that soap films can form [15]. With the origin of a coordinate system on this curve, let us represent the three surfaces as graphs over the tangent plane to one of them at the origin and denote by Γ the projection of γ in this plane. Two of the height functions u^1, u^2 will be defined on one side Ω^+ of Γ, while the third u^3 will be defined on the opposite side Ω^- of Γ. We obtain the following system for the u^j, $j = 1,2,3$.

$$
(44)\qquad
\begin{aligned}
&Au^1 = Au^2 = 0 && \text{in } \Omega^+ \\
&Au^3 = 0 && \text{in } \Omega^- \\
&u^1 = u^2 = u^3 && \\
&\frac{\nabla u^j \nabla u^3 + 1}{(1 + |\nabla u^j|^2)^{1/2}(1 + |\nabla u^3|^2)^{1/2}} = \frac{1}{2} && j = 1,2 \text{ on } \Gamma \\
&\nabla u^1 \neq \nabla u^2 &&
\end{aligned}
$$

where

$$Au = \frac{\partial}{\partial x_i}[u_{x_i} / (1 + |\nabla u|^2)^{1/2}] \text{ is the minimal surface operator.}$$

By combining a zeroth-order hodograph transform and a reflection, Γ is shown to be analytic.

We are able to prove more general results, for example:

If three distinct minimal hypersurfaces in $\mathbb{R}^{n+1}$ meet at constant angles on the $n - 1$ surface γ, then γ is analytic.

7. HIGHER ORDER EQUATIONS: THE DEFLECTION OF A THIN PLATE

Again we let $G \subset \mathbb{R}^n$ denote a bounded domain with smooth boundary.

Set

(45) $$\mathbb{K} = \{v \in H^2(G) \cap H_0^1(G) : v \geqslant \psi \text{ in } G\}$$

where

$$\max_G \psi \geqslant 0, \quad \text{and} \quad \psi < 0, \quad \text{on } \partial G$$

and ψ is a smooth function. Let $u \in \mathbb{K}$ satisfy

(46) $$\int_G \Delta u \Delta(v - u) dx \geqslant \int_G f(v - u) dx, \qquad \text{for } v \in \mathbb{K}$$

where f is a given smooth function. It is known that this problem has a unique solution. Moreover [3], $u \in H^3(G)$, and if $n = 2$ [2], $u \in C^2(\overline{G})$. Assuming that $u \in C^2(G)$ we derive G into the sets

$$I = \{x \in G : u(x) = \psi(x)\}, \qquad \text{and} \qquad G - I$$

From (46) we derive the conditions

(47) $$\begin{aligned} &\Delta^2 u = f && \text{in } G - I \\ &D^\alpha u = D^\alpha \psi && \text{in } I \text{ for } |\alpha| \leqslant 1 \\ &D^\alpha u \in C(G) && \text{for } |\alpha| \leqslant 2 \end{aligned}$$

Two situations of interest may 'typically' occur. We describe these in terms of Ω^+, Γ, and Ω^-. The first, more obvious, case is that

(48) $$\begin{aligned} &\Delta^2 u = f && \text{in } \Omega^+ \\ &D^\alpha u = D^\alpha \psi && \text{in } \Omega^- \cup \Gamma \text{ for } |\alpha| \leqslant 2 \end{aligned}$$

However, I may also contain a hypersurface Γ of dimension $(n - 1)$, namely

(49) $$\begin{aligned} &\Delta^2 u = f && \text{in } \Omega^+ \cup \Omega^- \\ &D^\alpha u = D^\alpha \psi && \text{on } \Gamma \text{ for } |\alpha| \leqslant 1 \\ &D^\alpha u \in C(\Omega^+ \cup \Gamma \cup \Omega^-) && \text{for } |\alpha| \leqslant 2 \end{aligned}$$

Since $u \notin H^4(G)$, one cannot conclude that $\Delta^2 u = f$ in the sense of distributions in $\Omega^+ \cup \Gamma \cup \Omega^-$. So (48) and (49) are higher-order versions of the obstacle problem (section 3) and the plasma problems (section 4) respectively.

First we describe the result we have in mind which is pertinent to (48). A more general theorem is given in [9] Theorem 4.3.

Theorem

Let $u \in C^{2m}(\Omega^+ \cup \Gamma) \cap C^{p+1}(\Omega^+ \cup \Gamma)$ satisfy the elliptic equation

$$F(x, u, \ldots, \nabla^{2m} u) = 0 \qquad \text{in } \Omega^+ \tag{50}$$

and the boundary conditions

$$D^\alpha u = 0, \qquad \text{for } |\alpha| \leqslant p$$

(51) and

$$\left(\frac{\partial}{\partial x_n}\right)^{p+1} u \neq 0, \qquad \text{on } \Gamma$$

where $p \geqslant m$. If F is analytic in all its arguments, then Γ is analytic.

The proof of this theorem requires the study of an overdetermined system in the hodograph space. We give a brief description of this system referring to [9] for its analysis. The case of interest is $m \leqslant p \leqslant 2m - 1$. When $p \geqslant 2m$, the result follows easily from the implicit function theorem applied to

$$\left(\frac{\partial}{\partial x_n}\right)^{p-2m} F(x, 0, \ldots, 0) = 0 \qquad \text{on } \Gamma$$

Introduce the hodograph mapping of order p (11) and the Legendre transform (12). In addition, we select as new unknown functions almost all the derivatives of u of order less than or equal to p. For $\alpha = (\alpha_1, \ldots, \alpha_r)$, $\alpha_i < n$, set

$$\begin{aligned} w^{\alpha,k}(y) &= D_{\alpha_1} \ldots D_{\alpha_r} D_n^k u(x), \qquad k \leqslant p-1, r+k < p \\ w^{\alpha,k}(y) &= D_{\alpha_1} \ldots D_{\alpha_r} D_n^{p-r} u(x), \qquad r \geqslant 2 \end{aligned} \tag{52}$$

where $D_\ell = \partial / \partial x_\ell$, $\ell = 1, \ldots, n$. Define the length of α, $|\alpha| = r$, for the multi-index $\alpha = (\alpha_1, \ldots, \alpha_r)$. We must point out that this use of the multi-index α differs from the usual one.

The functions $w^{\alpha,k}$ are symmetric in the indices $(\alpha_1, \ldots, \alpha_r)$. Note that we have not included derivatives of order p of the form $D_n^p u$ or $D_\sigma D_n^{p-1} u$, $\sigma = 1, \ldots, (n-1)$. This first is $-y_n$ and the others are v_σ.

In F we replace x by (x', v_n) and derivative of order inferior to p by the corresponding $w^{\alpha,k}$. Consider the derivative $D^\alpha D_n^k u$, $|\alpha| + k > p$. If $k \geqslant p-1$, then, according to (13) we may represent this derivative in terms of v and we replace it by this expression in F. If $k < p - 1$, we write $D^\alpha D_n^k u = D^\beta D^\gamma D_n^k u$ where $|\gamma|$ is as large as possible consistent with $|\gamma| + k \leqslant p$. We may then express this in terms of $w^{\alpha,k}$ and v. The choice is not unique, but this will not matter. So (50) assumes the form

$$(53) \quad \widetilde{F}(y', v_n, \ldots, \nabla^{2m-p+1} v, \ldots, \nabla^{2m-p} w^{\alpha,k}, \ldots, w^{\beta,k}) = 0 \quad \text{in } U$$

where derivatives of v occur to order $2m - p+1$; derivatives of $w^{\alpha,k}$, $|\alpha|+k=p$, to order $2m - p$; and $w^{\beta,k}$ appear to order zero if $|\beta|+k<p$.

In addition, x being derivatives of u, the $w^{\alpha,k}$ satisfy a number of compatibility relations. Since there are many of these relations the new system is overdetermined [1, 14, 9]. We then appeal to 'known' regularity theorems.

The new system is elliptic and its boundary conditions are covering. From this, analyticity of the solution may be deduced, whence it follows that Γ is analytic.

The case of (49) may be resolved by using our reflection mapping in conjunction with the construction given above.

8. AN EXTENSION OF A THEOREM OF HANS LEWY

Preliminary to our study of higher-order reflection problems like (49) we returned to a question of Hans Lewy regarding elliptic equations which share Cauchy data on a hyperplane. Since the methods we employed to generalize Lewy's result [12] were useful in resolving (49), we give a brief description of the theorem we proved.

Let L be an elliptic operator of order $2m$ with analytic coefficients in a neighbourhood of the origin in $\mathbb{R}^n$. Let $u, v \in C^{2m}$ in $x_n \geqslant 0$ near $x=0$ satisfy

$$(54) \qquad Lu = f_1, \qquad Lv + \lambda v = f_2$$

where λ, f_1, f_2 are analytic in a full neighbourhood of the origin and $\lambda(0) \neq 0$. Suppose that for some ℓ, $0 \leqslant \ell < m$, we have on $x_n = 0$

$$(55) \qquad \left(\frac{\partial}{\partial x_n}\right)^j u = \left(\frac{\partial}{\partial x_n}\right)^j v = 0 \qquad j \leqslant \ell - 1$$

$$(56) \qquad \left(\frac{\partial}{\partial x_n}\right)^j (u - v) = 0 \qquad j = \ell, \ldots, 2m - 1$$

Then u, v are analytic in $x_n > 0$ near $x=0$.

Observe that (55), (56) constitute $2m$ conditions. For $\ell=0$, (31) is vacuous. We treated this case in [3]. When $\ell > 0$, the idea of the proof remains the same, but verifying the coerciveness of the boundary conditions involves considerable work. The considerations are sufficiently delicate that they do not apply when L has complex coefficients. A proof of the theorem in this case would be of interest.

REFERENCES

[1] S. Agmon, A. Douglis, and L. Nirenberg. 'Estimates near the boundary for solutions of elliptic partial differential equations satisfying general boundary conditions, I, II.' *Commun. Pure & Appl. Math.*, **12** (1959), 623-727; **17** (1964), 35-92.

[2] L. Caffarelli, and A. Friedman. 'The obstacle problem for the biharmonic operator'. To appear.

[3] J. Frehse. 'On the regularity of the solution of the biharmonic varational inequality', *Manuscr. Math.*, **9** (1973), 91-103.

[4] D. Kinderlehrer. 'Variational inequalities and free boundary problems'. *Bull. Am. Math. Soc.*, **84** (1978), 7-26.

[5] D. Kinderlehrer, and L. Nirenberg. 'Regularity in free boundary value problems'. *Ann. Sc. Norm. Sup., Pisa,* ser. IV, vol. IV (1977), 373-91.

[6] D. Kinderlehrer, and L. Nirenberg. 'Analyticity at the boundary of solutions of nonlinear second-order parabolic equations'. *Commun. Pure & Appl. Math.*, **31** (1978), 283-338.

[7] D. Kinderlehrer, L. Nirenberg, and J. Spruck. 'Regularité dans les problèmes elliptiques à frontière libre'. *C.R. Acad. Sci, Paris,* **286** (1978), 1187-90.

[8] D. Kinderlehrer, L. Nirenberg, and J. Spruck. 'Regularity in elliptic free boundary problems' I.. *J. Analyse.* To appear.

[9] D. Kinderlehrer, L. Nirenberg, and J. Spruck. 'Regularity in elliptic free boundary problems, II'. *Ann. Scu. Norm. Sup., Pisa.* To appear.

[10] D. Kinderlehrer, and J. Spruck. ' The shape and smoothness of stable plasma configurations'. *Ann. Scu. Norm. Sup., Pisa.* **5** (1978), 131-48.

[11] D. Kinderlehrer, and G. Stampacchia. *'An Introduction to Variational Inequalities and its Applications'.* Academic Press, New York. To appear.

[12] H. Lewy. 'On the reflection laws of second order differential equations in two independent variables'. *Bull. Am. Math. Soc.*, **65** (1959), 37-58.

[13] H. Lewy, and G. Stampacchia. 'On the regularity of the solution of a variational inequalty.', *Commun. Pure & Appl. Math.*, **22** (1969), 153-88.

[14] C.B. Morrey, Jr. *'Multiple Integrals in the Calculus of Variations'.* Springer-Verlag, New York/Berlin (1966).

[15] J. Taylor. 'The structure of singularities in soap-bubble-like and soap-film-like minimal superfaces', *Ann. Math.*, **103** (1976), 486-539.

[16] R. Temam. 'A non-linear eigenvalue problem: the shape at equilibrium of a confined plasma'. *Arch. Ration. Mech. & Anal.*, **60** (1925), 51-73.

[17] R. Temam. 'Remarks on a free boundary problem arising in plasma physics'. *Commun. Partial Differ. Equations,* **2** (1977), 563-85.

[18] G. Vergara-Caffarelli. 'Regolaritá di un problema di disequazioni varazionali relativo a due membrane'. *Rend. Acad. Lincei,* **50** (1971), 659-62.

[19] G. Vergara-Caffarelli. 'Variational inequalities for two surfaces of constant mean curvature'. *Arch. Ration. Mech. & Anal.,* **56** (1974), 334-47.

[20] G. Vergara-Caffarelli. 'Superficie con curvatura media assegnata in L^P; Appiicazioni ad un problema di disequazioni variazionali'. *Boll. Unione Mat. Ital.,* **8** (1973), 261-77.

Chapter 15

A Survey of Complementarity Theory

C.E. Lemke

1. INTRODUCTION

This contribution is a descriptive summary of some of the developments over the past fifteen years in an area of mathematical programming which has often been described by the term 'complementary pivoting'. However, other descriptive terms have also been used such as 'simplicial approximation methods', 'fixed-point methods', and more recently 'continuation' or 'path-following methods'. Primarily, however, the focus has been on *constructive* approaches to some mathematical problems. The word 'constructive' has been used to draw attention to the 'constructive proof' nature of the methods developed. A name is needed that includes some quite recent additions of 'path-following' methods, which I would like to include in the discussion. I shall therefore describe the area under discussion by the term 'constructive approximation methods' (CAM), meant only for this summary.

After some preliminary material, I shall consider the 'linear complementarity problem' (LCP) and then, more generally, CAM.

Roughly, the general models which have received attention during the CAM development are *fixed-point problems, complementarity problems,* finding solutions to *systems of equations,* and some types of *variational inequality problems.* By now, each of these has its own areas of theory and application. In this summary, however, to provide a focus, we shall stress *systems of equations* as a general model. Not too much will be lost by this emphasis, since there are certain relevant equivalences between the four models mentioned, and also, in consideration of developments up to the present, this seems to be the trend.

Further, two other terms tend to describe the CAM activity, namely *piecewise linear systems,* and the *homotopy approach.* To a large degree, we will be concerned with finding zeros to a system of equations using the homotopy approach, often when the functions defining the system are piecewise linear (PL).

The scope of the CAM activity over the past fifteen years is evidenced primarily by more than 200 publications, representing the work of perhaps 100 persons concerned either with theory, models, applications, associated numerical analyses, generalizations or consolidations, or with computational aspects, inclusive of computer programs.

Only a few of these publications will be cited here, and then generally only when needed. However, I shall draw heavily on the bibliography contained in Eaves [9]; when only names are noted here, work referred to will most likely be cited there. I also call attention to three recent volumes: references [2, 4, 13], and, for history and a view of the future, to the excellent 'Introduction' [18].

By now, we more or less agree on what constitutes non-linear programming (NLP). I will say that it also includes linear programming (LP). In retrospect, over the past twenty-five years, there seems to be a parallel emerging that LCP is to CAM as LP is to NLP. Much of the activity we are addressing concerns LCP, namely the problem:

given the matrix $M(n \times n)$, and the column vector $q(n \times 1)$:

$$\text{(1)} \qquad \begin{array}{l} \text{writing } y := q + Mx, \text{ find } x \geqslant 0 \\ \text{such that } y \geqslant 0 \text{ and } y^T x = 0 \end{array}$$

We shall try to note some of the parallelism as we proceed.

2. SOME EQUIVALENCES

As the general model of focus:

given the column vector $F := [F_1, F_2, ..., F_n]$, where each F_i is a real-valued function of the column variable $x = [x_1, x_2, \ldots, x_n]$, of real variables x_i, and a region $D \subset \mathbb{R}^n$:

$$\text{(2)} \qquad \text{find } x \in D \text{ such that } F(x) = 0$$

(Note that we use square brackets to denote a column vector).

Generally, CAM has been concerned with finite-dimensional cases, but there have been exceptions in the following problems, and one can expect some focus on infinite-dimensional problems in the next few years.

In (2) F is assumed to be continuous, at least. The simple fact that $x \in D$ is a zero of F iff it is a fixed point of the map G defined by $G(x) := F(x) + x$ gives an equivalence of (2) and

(3) FP $\qquad$ find $x \in D$ such that $G(x) = x$

(Usually, $G:D \to D$.) This equivalence is sufficient for our purposes since, in a given case, stated assumptions on D and F make the differences essentially notational.

Two others which have accounted for much of the CAM activity are as follows:

given $f:\mathbb{R}^n \to \mathbb{R}^n$ (the complementarity problem):

(4) CP $\qquad$ writing $y := f(x)$, find $x \geqslant 0$ such that $y \geqslant 0$ and $x^T y = 0$ (equivalently, $x_i y_i = 0, i = 1, 2, \ldots, n$);

The variational inequality problem:

(5) VIP $\qquad$ find a 'stationary point' of the pair (g, D), that is find $x \in D$ such that for all $y \in D$, $(y - x)^T g(x) \geqslant 0$ (or $x \in D$ satisfies $x^T g(x) = \inf_D y^T g(x)$)

Of course, CP is LCP when f is affine i.e. $f(x) := q + Mx$.

In (5), in the case $D = \mathbb{R}^n_+ := \{x \in \mathbb{R}^n : x \geqslant 0\}$, it is readily seen [13] that (4) and (5) are equivalent. We shortly note a reverse result.

From the standpoint of analytic generalization, (5) seems to be the most generalizable (see, for example, [5]) when V_1 and V_2 are vector spaces over $\mathbb{R} := \mathbb{R}^1$, $g : V_1 \to V_2$, and a bilinear form $B(V_1, V_2)$ replaces the scalar product $x^T y$.

The CP (4) has many equivalents of the form (2). Given f in (4), the simplest of these, often used to apply theoretical results, is obtained upon defining F in (2) by:

$$F_i(x) := \min(x_i, f_i(x)) \tag{6}$$

Then, for $D = \mathbb{R}^n$, $F(x) = 0$ iff x is a solution to CP.

A more general transformation, displayed and used by Mangasarian (in [4]) is the following:

let $\theta : \mathbb{R} \to \mathbb{R}$ be strictly increasing with $\theta(0) = 0$; define $h(a, b) := \theta(a) + \theta(b) - \theta(|a - b|)$. Then

$$h(a, b) = 0 \text{ iff } a, b \geqslant 0, \text{ and } ab = 0 \tag{7}$$

Hence, defining F (for (2)) by $F_i(x) := h(x_i, f_i))$, again $F(x) = 0$ iff x is a solution to CP. Also, since

(8) $$(a + b) - |a - b| = 2 \min (a, b)$$

the first example corresponds to the case $\theta(t) := t$, and F is then continuous when f is continuous. Mangasarian notes that even for simple θ (e.g. $\theta(t) = t|t|$), F is differentiable when f is differentiable.

The Karush-Kuhn-Tucker (KKT) conditions, well known in mathematical programming, for differentiable case of NLP:

(9) $$\min_D \; h_0(x) \text{ where } D := \{x : h(x) \leqslant 0; x \geqslant 0\}$$

furnish the most well known (and, to date, probably the most applicable) example of the CP (4). Generalizing the classical method of Lagrange, they furnish necessary conditions for an x to yield a solution to (9). Briefly, writing $z := [x, u]$(a column), where the u_i are the 'Lagrange multipliers', if one forms the 'Lagrangian'

(10) $$L(z) := h_0(x) + u^T h(x)$$

a necessary condition that x yield a solution to (9) is that (with a sufficient 'constraint qualification' on D) for some $u \geqslant 0$:

(11) $$\begin{array}{ll} D_x L(z) \geqslant 0 & x^T D_x L(z) = 0 \\ \text{and} & \\ -D_u L(z) \geqslant 0 & -u^T D_u L(z) = 0 \end{array}$$

which has the form (4), with $f(z) = [D_x L(z), -D_u L(z)]$.

In the *convex* case, namely when h_0 and the h_i are convex (so that D is then a convex region), a necessary and sufficient condition that an $\bar{x}$ solve (9) is that $z = [\bar{x}, u] \geqslant 0$ satisfies (11) for some u.

The KKT conditions are 'geometrical', and also show that a solution $\bar{x}$ to (9) also solves

$$\min_D \; x^T (Dh_0(\bar{x}))$$

in which the functional to be minimized is linear, which further shows that $\bar{x}$ solves the VIP (5), where $g(x) := Dh_0(x)$. In this sense, VIP (5), for the region D, is more general since a 'general' g is not the derivative (gradient) of a scalar function.

But, on the other hand, as observed by Rockafellar (see Chapter XX of this book), consider (5) for the case of D which has the representation

$$D := \{x : x \geqslant 0,\ h(x) \leqslant 0\}$$

where the h_i are differentiable over D. If $\bar{x}$ is a stationary point, writing $c := g(\bar{x})$, $\bar{x}$ also solves the NLP

$$\min_D \ c^T x$$

whose KKT conditions are (11) upon replacing $h_0(x)$ by $c^T x = g(\bar{x})^T x$, so that, for some $u \geqslant 0$, $[\bar{x}, u]$ is a solution to the CP:

$$\text{(12)} \quad \begin{array}{ll} g(x) + \nabla h(x) u \geqslant 0 & -h(x) \geqslant 0 \\ \text{and} & \\ x^T(g(x) + \nabla h(x) u) = 0 & u^T h(x) = 0 \end{array}$$

And further, therefore, for such VIP (5) where the h_i are convex, $\bar{x}$ solves VIP iff for some $u \geqslant 0$, $[\bar{x}, u]$ solves (12).

3. THE LINEAR COMPLEMENTARITY PROBLEM

LCP, building on the works of Cottle, Dantzig, Kuhn, Tucker, and so many others, is more closely observable as a generalization of 'linear and quadratic programming' than CAM is of, say, NLP, because of the new thrusts of 'methods of simplicial approximation'.

The KKT conditions for the quadratic programming (QP) problem:

$$\text{(13)} \qquad \min c^T x + \tfrac{1}{2} x^T Q x \text{ where } y := b + Ax \geqslant 0, x \geqslant 0$$

(where $Q = Q^T$) may be expressed (as LCP) in the form

$$\text{(14)} \quad \begin{array}{l} \begin{pmatrix} y \\ v \end{pmatrix} = \begin{pmatrix} b \\ c \end{pmatrix} + \begin{pmatrix} 0 & A \\ -A^T & Q \end{pmatrix} \begin{pmatrix} u \\ x \end{pmatrix} \\ \begin{pmatrix} y \\ v \end{pmatrix}, \begin{pmatrix} u \\ x \end{pmatrix} \geqslant 0 \quad \begin{pmatrix} y \\ v \end{pmatrix}^T \begin{pmatrix} u \\ x \end{pmatrix} = 0 \end{array}$$

which of course gives the LCP for LP when $Q = 0$. In particular, for the convex QP (including LP), namely when Q is positive semi-definite (psd), any solution to (14) yields a solution to QP (and also to its dual problem).

In 1965 the author proposed a pivot scheme, which we shall call the 'complementarity pivot scheme' (CPS), essentially as a method for solving QP via the LCP (14), CPS is similar to the 'principal pivot' methods (PPM) of Cottle and Dantzig, which had been proposed prior to CPS for the matrix

M psd (such as the coefficient matrix in (14) when Q is psd), or when M is a P-matrix (that is, M has positive principal minors).

In turn, the CPS was modelled on the Lemke-Howson pivot scheme published in 1964 which furnished the first 'constructive' proof (by pivoting) of the existence of a Nash equilibrium point for a bi-matrix game. Later, Eaves pointed out that CPS itself could process the game case also.

As we shall note, there are applications of LCP other than to QP and to bimatrix games. Now we shall spend some time fitting LCP (and the CPS for it) into the CAM setting. In so doing, we assume some familiarity with the 'simplex method' (SM) of Dantzig for LP.

It is well known that the LP model (QP (13) with $Q=0$) has many applications. Historically, the growth of the LP area, and the NLP area as well, was in a large part due to the advent of SM as an efficient way of solving (or 'processing') LP problems. To some extent, there is a parallel with regard to the CPS for LCP. We develop this thought as a portion of this summary. Both SM and CPS generate a *sequence of pivots* on a linear system. In turn, the sequence of pivots, viewed geometrically, generates a PL path, comprised of a finite number of line segments. This generation of a PL path is also characteristic of the methods of simplicial approximation we are discussing and, further, one may consider each of the segments of the path as generated by a *pivot*. We briefly consider this basic computation of pivoting.

4. PIVOTING

Consider a linear system in the form

(15) $$y = b + Ax \qquad A \text{ rectangular (write } z := [y, x])$$

On the left side are the *dependent* or *basic* variables associated with the form, namely the y_i, and on the right are the *independent* or *non-basic* variables, namely the x_j. Associated with the form is the *basic solution* (bs), namely $z = [y, x] = [b, 0]$, obtained by setting all independent variables equal to 0.

A *pivot* (according to Gauss) entails an exchange of the roles of an independent and a dependent variable yielding an equivalent (i.e. having the same set of solutions) system of equations.

After a sequence of some k pivots from (15), one has the 'updated' system, which we write in the form

(16) $$y^k = b^k + A_k x^k \qquad (\text{write } z^k := [y^k, x^k])$$

where therefore z^k is some permutation (after the k exchanges) of z. The (r,s) pivot from (16) involves an exchange of the pair (y_r^k, x_s^k) followed by the 'standard' updating to an equivalent form. In z-space the (r,s) pivot is viewed as generating the line segment from the bs of the first form to that of the second. In this way, a finite number of different forms can be obtained. Looking at $y = b + Ax$ (or indeed, any form), there are precisely as many other forms as there are non-singular submatrices of A. In particular, therefore, if a sequence of pivots from (15) is generated so that no form *repeats*, that sequence must terminate. In the case of CPS, the linear system is $w = q + Mz$, and M is square. In both SM and CPS, to guarantee that no form repeats, the linear system is assumed to be *non-degenerate* (ND) (that is, for all forms, such as (16), for each j, we have $b_j^k \neq 0$). Then the *algorithm* in each case is such that no form repeats. Thus, the PL path generated does not intersect itself. Likewise, for the methods of simplicial approximation, essentially the same ND is assumed.

In LP one seeks to minimize a functional ($c^T x$ in (13)), and SM (like most methods of NLP) is a method of 'successive improvement' in $c^T x$, whereas the novelty of CPS is that it does not. LCP is not *per se* an 'optimization' problem. Rather (but roughly) one deals with 'equilibrium'-type settings (such as the KKT conditions, or the game setting generally).

To describe CPS briefly (while alluding to the more general 'homotopy approach'), prior to pivoting one has selected an 'artificial' column $d \geqslant 0$ for which for some $z_0 > 0$ one has $\bar{q} := q + z_0 d > 0$. In the generation of the PL path, by pivoting, using the 'augmented' linear system

$$w = (q + z_0 d) + Mz \tag{17}$$

the 'parameter' z_0 is systematically varied. Along the path the conditions (a) $\bar{q} \geqslant 0$, and (b) $z^T w = 0$ are maintained. A solution to LCP is therefore obtained for a point with $z_0 = 0$.

We note here, to expand upon later, that also in the methods of simplicial approximation used in generating solutions to 'homotopy systems' $H(x_0, x) = 0$ in the cases when H is PL, one generates PL paths of zeros of H (on which the parameter x_0 varies) quite like that which we have just discussed. (Then the linear system is 'huge', and a column is generated only when used.) In general, the PL mapping H is a (linear) approximation to the underlying H whose zeros are sought. This brief allusion to these methods accounts also for the use of the term 'complementary pivoting'. Shortly, we shall exhibit CPS itself as generating a zero of a system by this homotopy approach.

In LP, essentially all the methods in use utilize pivoting (essentially SM or a variant), as distinct from methods which are *iterative* (i.e. basically non-terminating) methods. To date this holds also for LCP.

Finally it might be recalled that SM 'processes' any LP problem – that is, for any data A, b, c, SM applied to LP terminates after a finite number of pivots and specifies which of the three *states* of LP applies: whether LP is infeasible (i.e. $y = b + Ax$, $[y,x] \geq 0$ has no solution); is feasible with a finite minimum; or is feasible with an infinite minimum.

CPS, on the other hand, is more 'matrix-dependent'. For given M and q it also terminates in a finite number of pivots, with three possibilities of termination: in a solution to LCP (with $z_0 = 0$); not in a solution with z_0 tending to infinity; or not in a solution with z_0 at a fixed value. However, for 'general' M, if termination is not in a solution, then nothing further regarding existence of a solution can be inferred.

This fact has given rise to much effort directed toward identifying *classes* of matrices M about which, on the one hand, one can make definite statements regarding existence (and/or uniqueness) of solutions, and on the other hand, classes which can be 'processed' by CPS: meaning an M such that one can say that CPS either, for a given q, terminates in a solution to LCP, or else determines that LCP is infeasible. Two associated classes have been defined:

(18)
$$Q := \{M : \text{for all } q, \text{ LCP has a solution}\},$$
and
$$Q_0 := \{M : \text{for each } q \text{ such that LCP is feasible there is a solution}\}$$

Over the years, quite substantial subclasses of Q and Q_0 have been identified. Worthwhile characterizations of Q and Q_0 themselves have yet to be found. Such characterizations would definitely be of great interest.

We note next some of the classes of matrices which have been identified and studied, and also give some of the facts relating to LCP and CPS. The developments to date in this area will be reported in a forthcoming paper [6].

5. SOME CLASSES OF MATRICES. EXISTENCE AND UNIQUENESS IN LCP

First there is the class (D_0), consisting of psd matrices M of all orders (the M are not restricted to be *symmetric*). Thus, $M \in (D_0)$ iff, for all x, $x^T M x \geq 0$. This class is the most prevalent in applications. Essentially all of the classes studied have defining properties which generalize those of (D_0). In particular, (D_0) includes $M = 0$; more generally M *anti-symmetric* (i.e. $M = -M^T$, which is true iff, for all x, $x^T M x = 0$, or iff, for all $x \geq 0$, $x^T M x = 0$). In particular, QP in the KKT form (14) yields an LCP where M is psd (and for LP – namely $Q = 0$ – M is anti-symmetric).

$(D_0) \subset Q_0$, and in fact, given $M \in (D_0)$, for all q such that LCP is ND and feasible, LCP has a *unique* solution.

Associated with (D_0) is its 'strong' subclass (D) (without subscript 0) of *positive definite* (pd) matrices M: $M \in (D_0)$, and $x^T Mx = 0$ iff $x = 0$. If $M \in (D)$ then $M \in Q$. (Recall the important use in NLP, for example: the quadratic function $x^T Mx$ is *convex* iff M is psd; is *strictly* convex iff M is pd.)

Generalizing a property of (D_0), (P_0) is the class of all M (called 'P_0-matrices') having non-negative *principal minors* (pm); $(P) \subset (P_0)$ is the (strong) class of matrices M ('P-matrices') with *positive* pm. In particular, $(D_0) \subset (P_0)$; $(D) \subset (P)$. The class (P) has the outstanding property (is characterized by): $M \in (P)$ iff, for each q, LCP has a *unique* solution. (Although $(P) \subset Q$, $(P_0) \not\subset Q_0$.)

If, for given M and q, LCP has a solution, then, of course, LCP is *feasible* (for some $z \geqslant 0; q + Mz \geqslant 0$); that is, feasibility is necessary for solvability of LCP. Note that the important classes Q and Q_0 are related by feasibility. In fact, clearly, if $M \in Q_0$, and for all q LCP is *feasible*, then already $M \in Q$. This relates to the following broad classes of matrices which, to great extent, form the basis for the study of classes of matrices we are discussing:

$$
(19) \qquad \begin{array}{l} S_0 : M \in (S_0) \text{ iff, for some } 0 \neq x \geqslant 0,\ Mx \geqslant 0 \\ S\ : M \in (S) \text{ iff, for some } \qquad x \geqslant 0,\ Mx > 0 \end{array}
$$

Clearly, $(S) \subset (S_0)$. (The notation is that used by Fiedler and Ptak.) For these classes M may be rectangular. To indicate the general nature of these classes, there is a classical theorem (Ville):

$$
(20) \qquad \text{for any } A; A \in (S_0) \text{ or } -A^T \in (S) \text{ (not both)}
$$

One sees easily that (S) has the characterization:

$$
(21) \qquad \begin{array}{l} A \in (S) \text{ iff, for all } b, y = b + Ax \text{ is feasible} \\ \text{(i.e. for some } x \geqslant 0, b + Ax \geqslant 0) \end{array}
$$

Hence, in particular, if $M \in (D_0)$ then $M \in Q$ iff $M \in (S)$. One might have $A \in (S_0)$, but $A^T \notin (S_0)$, for example. One can see that the 'complementary' condition in LCP is intimately related to the *principal sub-matrices* (psm) of M (obtained from M by deleting some rows and the same columns); formally there are $2^n - 1$ psm.

The following related basic classes have been introduced by Cottle and Dantzig and by Eaves:

$$
(22) \qquad \begin{array}{l} E_0 :\ M \in (E_0) \text{ iff, for all psm } \overline{M} \text{ of } M,\ \overline{M} \in (S_0) \\ E\ :\ M \in (E) \text{ iff, for all psm } \overline{M} \text{ Of } M,\ \overline{M} \in (S) \end{array}
$$

First, relative to LCP, these have the easily seen characterizations (Eaves):

(23)
$$M \in (E_0) \text{ iff, for all } q > 0, \text{ LCP has the unique solution } z = 0$$
$$M \in (E) \text{ iff, for all } q \geq 0, \text{ LCP has the unique solution } z = 0$$

One has $(E) \subset Q$. Most of the subclasses of Q studied are in (E); in particular (P_0), $(D_0) \subset (E_0)$, and (P), $(D) \subset (E)$. But (E) is not all of Q.

Generalizing the defining properties of (D_0) and (D):

(24)
$$C_0: \; M \in (C_0) \text{ iff, for all } x \geq 0, x^T M x \geq 0$$
$$C: \; M \in (C) \text{ iff, } M \in (C_0) \text{ and, for all } x \geq 0$$
$$x^T M x = 0 \text{ iff } x = 0$$

In particular, (C_0), the class of 'co-positive' matrices, includes non-negative matrices (i.e. $M \geq 0$), anti-symmetric matrices, and, like (D_0), is closed under non-negative linear combinations (i.e. M_1, $M_2 \in (C_0)$ and $a, b \geq 0$ implies $aM_1 + bM_2 \in (C_0)$). (C) is closed under non-negative, non-zero linear combinations. Moreover

(25)
$$(C_0) \subset (E_0) \quad \text{and} \quad (C) \subset (E)$$

Other less general, but nonetheless important, classes have been studied. We mention one, namely the class (Z) of 'Z-matrices': $M \in (Z)$ iff, for $i \neq j$, $M_{ij} \leq 0$. These, like the previous classes (excepting (E_0) and (E)) have more classical importance in other fields.

Relative to LCP, (Z) has the following distinction: generally if M is $n \times n$, the number of pivots taken in a pivot scheme, such as CPS, for solving (or processing) LCP is not obviously related to n. However, if $M \in (Z)$, there is a pivot scheme (due to Chandrasekaran) which, for any q, takes no more than n pivots to process (i.e. find a solution or decide that LCP for the given q is infeasible). Because of this, $(Z) \subset Q_0$.

Again, therefore, one can say that if $M \in (Z)$, then $M \in Q$ iff $M \in (S)$. However, in the case of this class (Z) more can be said, namely, that if $M \in (Z) \cap (S)$, then $M \in (P)$.

Thus, we define the subclasses of (Z):

(26)
$$K_0: \; M \in (K_0) \text{ iff } M \in (Z) \text{ and } M \in (S_0)$$
$$K: \; M \in (K) \text{ iff } M \in (Z) \text{ and } M \in (S)$$

(Again, the notation S_0, S, K_0, K is that used by Fiedler and Ptak.) (K) is the class of *Minkowski* or *Leontief* matrices. In particular, $(K_0) \subset (P_0)$, $(K) \subset (P)$, and if $M \in (K)$, one has $M^{-1} \geq 0$.

6. CLASSES OF MATRICES AND CPS

We have noted the three terminal states for LCP: (I) a final form giving a solution (with $z_0 = 0$); (II) a final form with z_0 going to infinity (a z_0-*ray*); and (III) a final form with z_0 fixed but the PL path going to infinity (a *secondary* ray). In particular, observe therefore that if, for a given pair M and q, it is known that (II) and (III) cannot occur, CPS must (with ND) yield a solution.

To begin, using the characterization (23), it is easy to see that if $M \in (E_0)$, then case (II) *cannot occur* (Eaves). Secondly, for the 'strong' class (E), which is in Q, in addition case (III) cannot occur. Hence, CPS yields a solution for an M in (E) (and *any* q). Thirdly, a large class, L_0 has been defined by Eaves (and extended by Garcia) with the property that $(L_0) \subset (E_0)$ and for $M \in (L_0)$ *if*, for a given q, case (III) does occur, then LCP is already *infeasible.* Hence, one may say that CPS processes any $M \in (L_0) \subset (Q_0)$. Eaves defined the following class:

(27) $\qquad E_1 : M \in (E_1)$ iff, when $z \geqslant 0$, $w := Mz \geqslant 0$ and $z^T w = 0$

(i.e. z is a solution to LCP for $q = 0$), then for some x, writing $y := -M^T x$ one has $0 \leqslant [y, x] \leqslant [w, z]$, and if $z \neq 0$, then $x \neq 0$.

The class (L_0) is $(L_0) := (E_0) \cap (E_1)$. In particular, $(E) \subset (L_0)$ and $(D_0) \subset (L_0)$. Finally, classes not in (L_0) can be processed by CPS. We note shortly, for example, that (Z) can be processed by CPS (Saigal).

7. SOME RELATIONS BETWEEN LCP AND LP

We note some results which tend to clarify distinctions between LP and the more general LCP.

First, since the KKT conditions for the LP problem ((13), with $Q = 0$) form an LCP problem with matrix $M \in (D_0)$, CPS can process the LP problem. When it does (and $d := e$, a column of ones) it is precisely the 'self-primal-dual method' discussed by Dantzig in his text 'Linear Programming and Extensions' which uses either simplex-method or dual-simplex-method iterations, depending upon the state. Some empirical results (due to Ravindran and others) seem to indicate that this technique applied to LP might be as efficient for general problems as the SM.

Secondly, there are large classes of matrices M (as yet incompletely identified), first and extensively studied by Mangasarian, then studied and augmented by Cottle and Pang, with the property that given M there is a readily available column p such that the solution to the LP:

$$\min p^T x \quad \text{where} \quad q + Mx \geqslant 0, \; x \geqslant 0$$

is also a solution to LCP with the same M and q.

Such classes are rather intimately related to the class (Z) which (thirdly) serves to yield the following relationship (Mohan) between CPS and SM: for CPS, again take $d := e$. Consider the SM applied to the LP:

$$\min z_0 \quad \text{where} \quad z_0 e + q + Mz \geqslant 0, \; z \geqslant 0 \tag{28}$$

Then SM and CPS go through exactly the same sequence of pivots in the processing!

Finally, of a different nature is the following recent result (an example of which is contained in the model of Dantzig and Manne [8], who were the first to apply it): if M is *co-positive*, and the LP:

$$\min q^T z \quad \text{where} \quad q + Mz \geqslant 0, \; z \geqslant 0 \tag{29}$$

has a *finite minimum*, then LCP defined by q and M has a solution which can be found by CPS.

Computer codes and computational experience, in particular with regard to CPS, are reported by Cottle [7].

8. MODELS AND APPLICATIONS OF LCP

As of the present writing, LCP has found application in the following areas among others:

(1) - LP and QP;
(2) - bimatrix and polymatrix games;
(3) - mechanics, plasticity, etc;
(4) - economic equilibrium;
(5) - approximate solutions to differential equations;
(6) - classification of square matrices; and
(7) - solutions to systems $F(x) = 0$ (n equations in n unknowns).

We mention the names of some of the more prominent workers in these areas, roughly in chronological order.

(1) - In a real sense, all of the proposed methods of QP, inclusive of LP, may be visualized in the context of the KKT. Emphasizing the roots of LCP, many of the following are concerned exclusively with QP: Dantzig, Wolfe, Beale, Zoutendijk, van de Panne, Whinston, Graves, Keller, Cottle, Cottle

and Dantzig, Sacher.
(2) Howson-Lemke, Howson, Eaves, Shapley.
(3) In this area, see the papers by Cottle [5], and Kaneko [12].
(4) Dantzig and Manne, Hansen, T. Evers, J.J.M. Engles, B.C. Eaves. It may be expected that additional such models of economic equilibrium will be developed.
(5) In particular, again see [5]. In many instances of this area of application, the LCP is based upon QP.
(6) A substantial proportion of the CAM effort has consisted of contributions in this area. The development to date is reported in the forthcoming paper [6], which unifies the contributions. Some of the main contributors are: Gale, Samelson-Thrall-Wesler, Cottle, Dantzig, Murty, Lemke, Eaves, Pang, Mangasarian, Chandrasekaran, Karamardian, Garcia, Saigal, Mohan, Periera, Evers. Of special note are the works on principal pivoting of A.W. Tucker, and the comprehensive papers of Fiedler and Ptak.
(7) One of the most recommended papers dealing with developments over the past fifteen years is a paper of Eaves [4] which develops a theory for solving general PL systems $F(x)=0$; and includes the methods of simplicial approximation. The paper builds upon a previous paper by Eaves and Scarf and is a forerunner of a book by Eaves and Saigal. In particular, a simple but potentially useful formulation of LCP as a PL system is given by Eaves, which we illustrate with an example:

Write $w = q + Mz$ as

$$(M, I)\begin{pmatrix} z \\ -w \end{pmatrix} + q = 0$$

Let O_r, for $r = 1, 2, \ldots, 2^n$ denote the orthants of $\mathbb{R}^n$ in some order. For each r let B_r denote the $n \times n$ matrix such that for $x \in \operatorname{int} O_r$:

$$(B_r)_i = \begin{cases} M_i & \text{iff} \quad x_i > 0 \\ I_i & \text{iff} \quad x_i < 0 \end{cases} \tag{30}$$

Define $F: \mathbb{R}^n \to \mathbb{R}^n$ by:

$$F(x) := B_r x + q \qquad \text{iff} \quad x \in O_r \tag{31}$$

Then x satisfies $F(x) = 0$ iff x solves LCP.

As a simple example of the homotopy approach, Eaves gives a picture of LCP as follows: if $d \geqslant 0$ and, for some $x_0 > 0$, $d + x_0 q > 0$, consider the simple homotopy $H: [0, \infty]\ \mathbb{R}^n \to \mathbb{R}^n$ given by:

$$H(x_0, x) := F(x) + x_0 d \tag{32}$$

If the linear system $w = (q + x_0 d) + Mz$ is *ND*, the PL path of zeros of H which contains a ray in the non-positive orthant of $\mathbb{R}^n$ corresponds exactly to the path followed by CPS (and this path may be visualized as a non-self-intersecting path in $\mathbb{R}^n$, parametrized by x_0). This is illustrated in Figure 1 for

$$q = \begin{pmatrix} -6 \\ -4 \end{pmatrix} \quad d = \begin{pmatrix} 1 \\ 1 \end{pmatrix}$$

and the *P*-matrix

$$M = \begin{pmatrix} 1 & 2 \\ 0 & 1 \end{pmatrix}$$

This example is actually the case $n = 2$ of the example used by Murty to illustrate the 'computational complexity' of CPS, which is not polynomial – the CPS takes $2^n - 1$ pivots; that is, the PL path pierces the interiors of all orthants in $\mathbb{R}^n$.

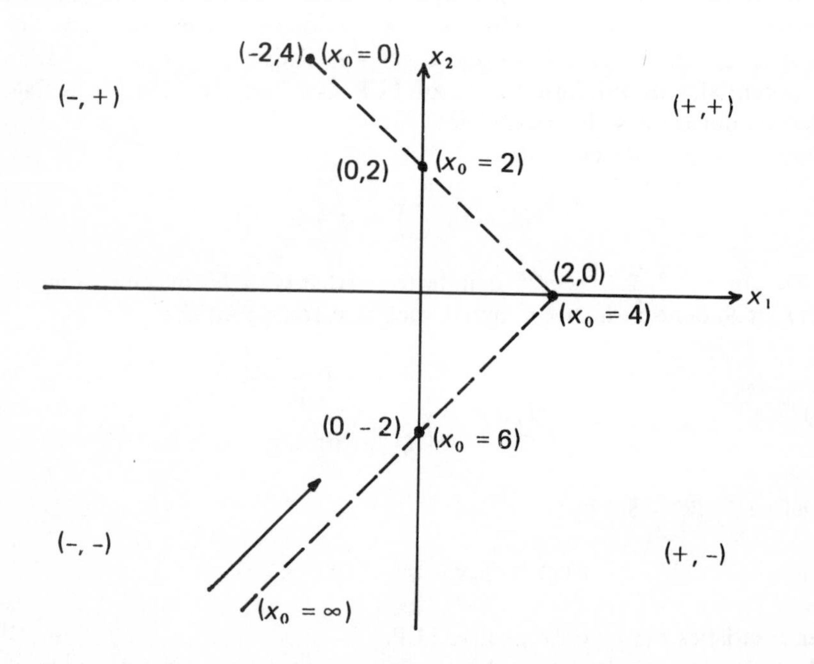

Figure 1

9. CAM DEVELOPMENTS

Even from the vantage point of the present, there is no doubt that the timely publication by H.E. Scarf in 1967 of his constructive proof of the Brouwer fixed-point theorem provided the greatest impetus for the overall developments in CAM. It is also quite possible that Scarf is one of the few who could have anticipated the subsequent effects of his work. I again refer to his version of developments since 1967 in the 'Introduction' of [13].

In his proof of the Sperner Lemma for the simplex, Scarf developed the notion of 'primitive set', together with a pivotal algorithm for proceeeding to a next primitive set. A sequence of primitive sets was thereby generated terminating in a 'completely-labelled' primitive set, which gave the desired approximation to a fixed point of the given function in the simplex. This initial use of the 'primitive-set' notion utilized the bimatrix-game setting, and the Howson-Lemke algorithm.

Following this, the process was 'refined', by Scarf and Hansen and by Kuhn, towards the process of generating a sequence of 'almost-completely labelled' members (simplices) of a standard simplicial subdivision of the simplex. This, in turn, has given rise to the many applications of the constructive simplicial approximation (or 'fixed-point') methods. Many of the early developments are described in Scarf's monograph *On The Computation of Economic Equilibria* (1973) written in collaboration with T. Hansen. Although almost all of these subsequent developments do not use Scarf's initial 'primitive set' concept, *per se* that concept is in some respects more general and will quite probably form the setting for some future developments.

10. SIMPLICIAL APPROXIMATION METHODS (SAM)

Uses of simplicial approximation subsequent to Scarf's initial contribution, peripheral developments (such as studies of appropriate simplicial subdivisions SS), and developments in related areas (such as the developments of CP), have been many and varied. We shall shortly identify these in some detail.

Basic to all of these uses is the following notion.

One is given an SS, call it K^n, in $\mathbb{R}^n$. Without loss of generality K^n is a subdivision of $\mathbb{R}^n$ itself, whose elements are simplices, each being the convex hull of some $(n+1)$ *vertices* of K^n. Let K^0 denote the set of all vertices V of K^n. Each member of K^n has $(n+1)$ *facets,* each facet being the convex hull of n of the $(n+1)$ vertices defining the simplex. K^{n-1} is the collection of all facets of K^n. Each member of K^{n-1} is a facet of exactly two members of K^n. Two members of K^n are *adjacent* iff they share (intersect in) a common facet. Thus, each member of K^n has $(n+1)$ *neighbours* (adjacent simplexes).

In each application a *sequence* of members of K^n is identified, such that adjacent members of the sequence are adjacent simplices, with no simplex appearing twice. Thus, adjacent members of the sequence differ in but one vertex. If S^{k-1}, S^k, S^{k+1} denote three consecutive members of the sequence, Figure 2 illustrates the implicit 'pivot' (or drop-add) process. In this sense, each pivot involves one 'new' vertex, so that the sequence may be considered as a sequence of vertices.

Given an SS K^n, let $L : K^0 \to \mathbb{R}^n$ be a function

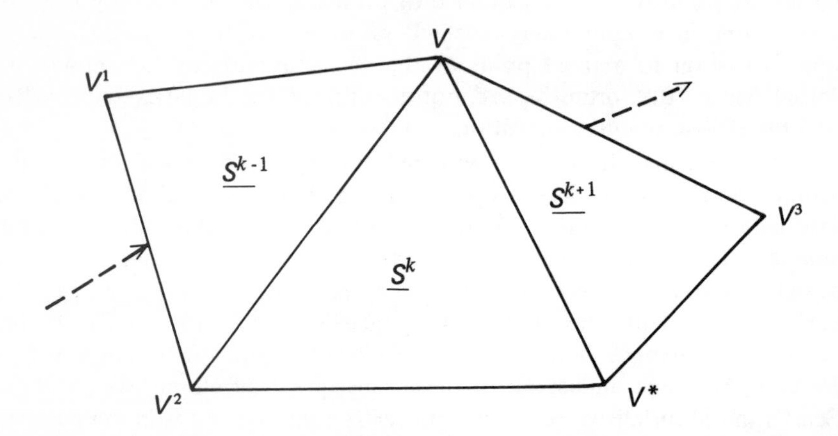

(Drop V^{-1} - Add V^* ; Drop V^2 - Add V^3)

Figure 2

defined on K^0, with values in $\mathbb{R}^n$. There is then a uniquely defined PL (continuous) function $L : \mathbb{R}^n \to \mathbb{R}^n$.

In each of the applications of SAM, one has such a function to generate the sequence of simplices. (L is the 'labelling' function.) Such an L generally has the property that it identifies a subset of K^n of 'permissible' simplices; each permissible simplex has at most two neighbours which are permissible. (In Figure 2 S^k has two permissible neighbours; namely S^{k-1} and S^{k+1}. S^{k+1} might have only S^k as a permissible neighbour.) Given K^n and an L, a sequence can be *started* at some permissible S^0 which has *one* permissible simplex. It is then characteristic in all of the SAM applications that a *unique* sequence is generated (which may be infinite) terminating if and only if a permissible simplex with one permissible neighbour is encountered. This is the feature inherent in the Howson-Lemke paper, subsequently utilized by Scarf.

From the computation point of view, one must be able to identify, and conveniently store, a current member of the sequence, such as S^k. Then, for the kth iteration, with V^* the vertex added on the previous iteration, one computes $L(V^*)$, subsequently determines the 'pivot pair' (V^2, V^3), and then 'updates' the stored S^k to S^{k+1}.

We discuss briefly each of the following main topics of development in the SAM area:
(1) refined approximations and the homotopy approach;
(2) algorithms;
(3) simplicial subdivisons (triangulations); and
(4) theoretical results and uses.

At the present time of writing, there is much indication that the general approach of SAM will find other, perhaps novel, applications, building upon the accumulated efforts to date. As we have noted, one may view most of the uses in terms of finding zeros of a system of equations but only for focus. To do proper justice, in a survey, to the scope of the CAM development *in toto* would require a much more comprehensive effort.

(1) Scarf's initial computation resulted in an approximation to a fixed point which, in particular, depended upon the fineness (mesh size) of the SS. In order to obtain an improved approximation, one would have to refine the SS (smaller simplices), and *restart* the process. For some time, people were concerned with this 'restart problem'. It is safe to say that this problem has been resolved, and not only with respect to approximating fixed-points over the simplex. There were two notable results in this direction, each of which in turn have had value well beyond just resolving the 'restart problem'. The two techniques involved have been called the 'sandwich method', and the 'homotopy approach'. Merrill is credited with initiating the sandwich method, and Eaves with initiating the homotopy approach. We shall develop the latter in some detail later. Roughly, in the sandwich method, using an SS one obtains an approximate solution to the given problem. Using a finer SS (generally a refinement of the previous), a next 'iteration', starting from the previous approximation, leads to an improved approximation. Eaves' initial paper utilizing the homotopy approach and concerned with fixed points over the simplex was, on the other hand, a method which 'continually' improved the approximation in the sense of 'moving through' a pre-assigned sequence of SS, with 'adjacent' SS's bounding an SS of one higher dimension. It was the first example, in SAM, of the (deliberate) generation of an infinite sequence of simplices which, in the limit, gave a fixed point. A sense in which each of these approaches is a special case of the other is discussed by Eaves [9]. Subsequently, others have utilized these ideas or provided variations. In this regard see the 'Introduction' in Scarf [18]. In particular, Eaves and Saigal developed a first extension of the Eaves result from the simplex to all of $\mathbb{R}^n$. Other contributors include Gould, Fischer and Tolle, Kuhn, Lüthi, Wilmuth, MacKinnon.

(2) Besides the more general algorithms included in (1) above, there have been algorithms (or methods) developed for specific problems, notably NLP, via the KKT conditions (several of quite recent vintage), or more generally the CP. But of course, in a broader sense, and in view of the fact that we are concerned with 'constructive' methods, the designation 'algorithm' may extend from a constructive proof for class of theorems, to a detailed, computer-based method, incorporating a specific type of SS, designed for a specific problem (such as finding zeros of a polynomial), and possibly combining, in hybrid fashion, SAM computations with Newton-like steps. Such variations and more have been described. We note some of the contributors.

In NLP and CP, of particular note are Gould, Fischer and Tolle, Kojima, Karamardian, Scarf and Hansen, Mangasarian, Merrill, Friedenfelds, Lüthi, P. Reiser, Habetler and Price, Garcia, Saigal.

For finding fixed points, and solving systems of equations, of particular note are Scarf and Hansen, Garcia, Garcia and Zangwill (general polynomial systems), Kuhn, Kojima (roots of a polynomial).

(3) There are reasons for studying different classes of SS. Whereas, in principle, for most 'theoretical' applications almost any SS may be used, in an actual application of, say, a SAM method (for example, for a computer program), there are practical considerations which require appropriate SS (for example, relating to 'storing' a current simplex, or to rates of convergence questions). Actually, it appears that there are very few kinds of SS that have practical value. In any case, it also appears that more work needs to be done in this area.

Some have, in particular, been contributed by Scarf and Hansen, Kuhn, Eaves, Saigal, Merrill, and others.

The survey by Saigal [17] is recommended here.

(4) Perhaps one of the most impressive contributions of the SAM approach, again starting with Scarf's initial results, concerns the 'constructive proof' aspect. Many of the classical results, e.g. dealing with fixed points or, more generally, with systems of equations, have been more simply proved, often generalized. Contributors in this regard include Scarf, Eaves, Garcia Karamardian, Wilson, Moré, Shapley, Garcia-Gould, Kojima, Kuhn, Kojima--Meggido, Garcia-Lemke-Lüthi, Charnes-Garcia-Lemke, Garcia-Zangwill.

11. SOLUTIONS TO SYSTEMS OF EQUATIONS

We have noted that most of the CAM development may be viewed in terms of finding solutions to a system $F(x)=0$ of n equations in n unknowns in a given region D in $\mathbb{R}^n$. We have also noted that most of the recent SAM effort in this regard stresses the 'homotopy approach'. It appears, moreover, that in this setting the overlap between the CAM accomplishments and the more classical areas of algebraic or differential topology will be more readily identified. In particular, there is much indication of this in the very recent 'path-following' approaches we have mentioned and which we will make more note of shortly. In any case, we now identify by an example the homotopy approach which is contained in a very recent paper [10].

If F is affine, $F(x) = b + Ax$, there is a unique solution iff the (constant) derivative $F'(x) = A$ is non-singular. In particular, then, the zero is *isolated.* More generally if, in a region D, at every zero of F, F' exists, is continuous, and is non-singular, then by the implicit-function theorem the zeros are isolated. Then, for example, if D is compact, it contains a finite number of zeros.

To locate zeros one sets up a homotopy $H : \mathbb{R}^{n+1} \to \mathbb{R}^n$.

$$(33) \qquad H(x_0, x) \quad \text{where} \quad H(0, x) =: F_0(x) \quad \text{and} \quad H(1, x) = F(x)$$

F_0 is some appropriately selected *start* function, with known zeros in D. x_0 is the associated 'parameter'. Two simple forms, which to date are the two most widely used, are (32) and

$$(34) \qquad H(x_0, x) := (1 - x_0) F_0(x) + x_0 F(x)$$

both linear in x_0. In almost all of the actual uses, F_0 is affine with a unique zero.

Let us write $D^* := [0,1] \times D$. Briefly, in D^*, with various choices or assumption, one visualizes the set of zeros of H as consisting of a finite number of non-intersecting curves, all of whose end-points are in bdry D^*. Thus, for $x_0 = 0$ they are the zeros of F_0, and for $x_0 = 1$ they are the (desired) zeros of F in D.

Generally, the SAM method is to approximate (constructively) one or more of these curves. For this one uses an SS of D^* and the PL approximation H^* which agrees with H on the vertex set of the SS, and is affine on each simplex of the SS. In particular, $F^*(x) := H^*(1, x)$ is the PL approximation to F in the SS of $\mathbb{R}^n$ induced by the SS of D^*. With the proper care, the zeros of H^* form PL paths – line segments in each simplex of the SS – whose end-points with $x_0 = 1$ approximate zeros of F in D.

In a more recent series of developments, one more directly 'follows a path' of zeros of H by some more classical (in any case different) approximating scheme. Such a scheme embodies the 'constructive' element of such an approach.

Returning to D^* and the zeros of H, as in (34), we consider a development which is essentially in [10].

Suppose that D is compact, so that D^* is also compact and that the derivative H' of H has rank n and is continuous at each point $x^* := (x_0, x)$ in D^*.

Visualize x^* on a path of zeros of H as a continuously differentiable function of a real parameter θ, so that $H(x^*(\theta)) = 0$ identically in θ. By the chain rule:

(35) $$H'(v^*)x^{*\prime}=0 \qquad (\text{where } x^{*\prime} := x^{*\prime}(\theta) := \mathrm{d}x^*(\theta)/\mathrm{d}\theta)$$

In the $n \times (n+1)$ matrix H' (of rank n) let H'_{-i} denote H' with its ith column deleted. The assumptions permit the assertion that the following system of $(n+1)$ ordinary differential equations:

(36) $$x_i^{*\prime} = (-1)^{i+1} \det H'_{-1} \qquad i = 0, 1, \ldots, n$$

satisfies (35), and hence its solution through some given zero of H is a 'solution path' of the set where $H = 0$. It is then seen rather simply that each zero of H in D^* lies on a path of solutions to (36), and that therefore the set of zeros of H consists of a finite number of non-intersecting curves with continuous tangents $x^{*\prime}$, all of whose end-points are in bdry D. This is a theorem in [10], and is illustrated in Figure 3.

To indicate how (36) comes about, for a fixed x^* write $A := H'(x^*)$. Since rank $A = n$, one has $Ac = 0$ for a column c such that $c^T c = 1$, and c is unique except for sign. Hence, $y = c$ is the unique solution to:

(37) $$\begin{pmatrix} c^T \\ A \end{pmatrix} y = \begin{pmatrix} 1 \\ 0 \end{pmatrix}$$

Hence (from the adjoint, or using Cramer's Rule):

(38) $$c_i = k(-1)^{i+1} \det H'_{-i} \qquad \text{where } \frac{1}{k} = \det \begin{pmatrix} c^T \\ A \end{pmatrix} = \pm [\det AA^T]^{1/2}$$

In particular, any continuous multiple of $c = c(x^*)$, such as defines the differential system in (36), gives a continuous measure of the tangent to a path through a given zero of H on which $H = 0$.

One objective of Garcia and Zangwill was to present a simplified development of many of the results of CAM *and* of more classical developments related to solving systems. In particular, they adopt the point of view that the situation illustrated in Figure 3 is a 'typical' one. That is, in particular, even though the above assumption of continuous differentiability is rather strong, they point out, using Sard's theorem and the Weierstrass approximation theorem, that their assumption holds 'with probability one. Again, in particular, an important element in the SAM development has been that in

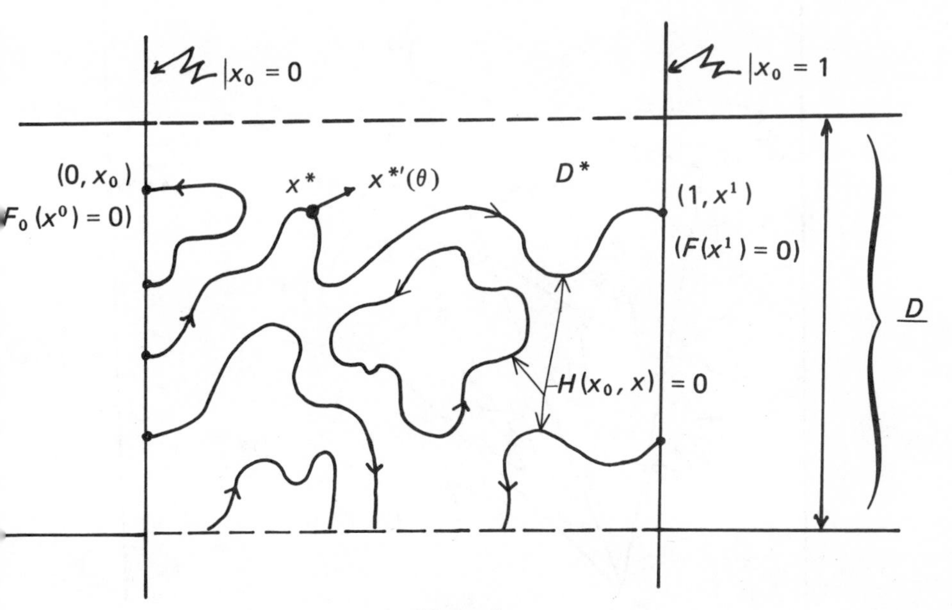

Figure 3

many instances assuming merely that F is continuous suffices to obtain results (for example, the Brouwer fixed-point theorem), and without reference to differentiability. It is nonetheless interesting and of value to retain the picture in Figure 3.

We next note the 'pivotal' aspect of a typical SAM application. Visualize the cylindrical region D^* as above, and an SS of D^* (more generally, and without loss of generality, of $[0,1] \times \mathbb{R}^n$). The homotopy H is now PL with respect to H (hence, determined by the vertices of the SS), and one is concerned with the zeros of H in D^*. With some mild assumptions of ND and 'regularity', the picture is quite like that in Figure 3. Again, with the zeros of $F_0(x) := H(0,x)$ known, it is first assumed that all zeros in D of $F_0(x)$ and of $F(x) := H(1,x)$ occur only in relative interiors of simplices in the $x_0 = 0, 1$ portions of bdry D^*. (Then it follows that H, in each such simplex containing a zero, has a non-singular derivative there, and hence that the zeros of F_0 and F in D are finite in number.) Then, from each such zero, a PL path of zeros of H may be initiated. We illustrate a local piece of such a PL path in Figure 4.

Corresponding to the simplex S^k is the abbreviated linear system

$$(39) \qquad \begin{pmatrix} 0 \\ 1 \end{pmatrix} = y_0 \begin{pmatrix} H(V^0) \\ 1 \end{pmatrix} + y_1 \begin{pmatrix} H(V^1) \\ 1 \end{pmatrix} + y_2 \begin{pmatrix} H(V^2) \\ 1 \end{pmatrix}, \quad \begin{pmatrix} y_0 \\ y_1 \\ y_2 \end{pmatrix} \geqslant 0$$

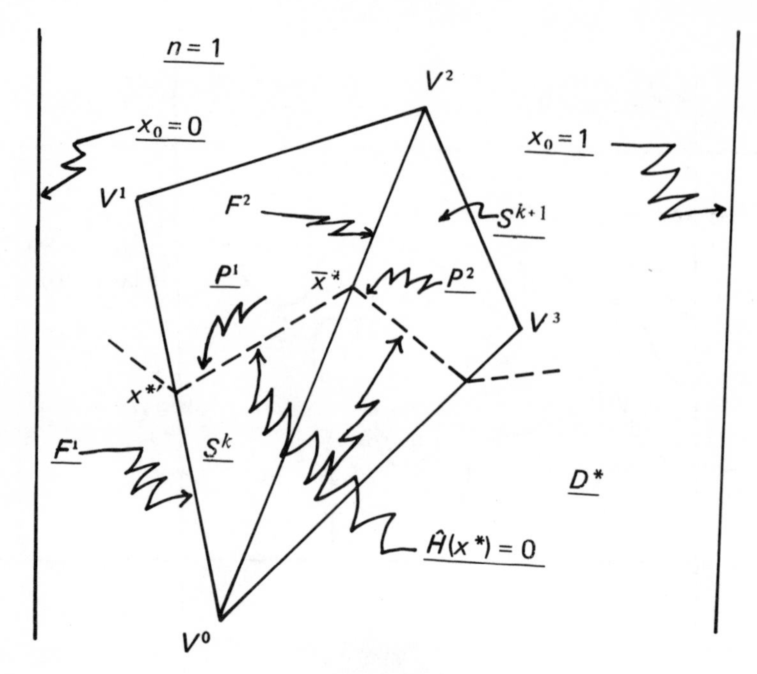

Figure 4

(Since $n = 1$, the values $H(V)$ are scalars.) The segment P^1 is the set of all (feasible) solutions to (39). We are supposing that, in the facet F^1 of S^k, $x^{*\prime} = \bar{y}_0 V^0 + \bar{y}_1 V^1$ is the only zero of H, and that $x^{*\prime}$ is in the relative interior of F^1 (so that $\bar{y}_0, \bar{y}_1 > 0$). Thus, $[y_0, y_1, y_2] = [\bar{y}_0, \bar{y}_1, 0]$ is the solution to (39) with $y_2 = 0$. Introducing y_2 as basic (increasing y_2 from 0), we find that y_1 becomes 0 at $\bar{x}^*$. The matrix

$$\begin{pmatrix} H(V^0) & H(V^1) \\ 1 & 1 \end{pmatrix}$$

is non-singular, and *by the pivot* process, the matrix

$$\begin{pmatrix} H(V^1) & H(V^2) \\ 1 & 1 \end{pmatrix}$$

is likewise non-singular. We make the (non-degeneracy) assumption that the linear system (39) is ND. Thus, at $y_1 = 0$, $[y_0, y_1, y_2] = [y_0', 0, y_2']$, where $y_0', y_2' > 0$, so that $\bar{x}^* := y_0' V^0 + y_2' V^2$ is in the relative interior of F^2, the facet of S^k opposite V^1.

Now the local knowledge of the SS yields (uniquely) the vertex V^3 (assumed to be in D^*), and geometrically we have the 'drop-add' operation: drop V^1 - add V^3 leading from S^k to S^{k+1}. Then the computed column

$[H(V^3), 1]$ is added to (39) (as 'next pivot column'), and the situation is exactly as with S^k. Note that, therefore, with the assumption that, as one proceeds, each segment of zeros of H generated has its end points in relative interiors of facets of the SS, so that a unique path of zeros is generated, the $n \times (n+1)$ matrix H' has rank n in each simplex used, so that all of the zeros in the sequence of simplices used comprise the path generated. In particular, therefore, since with the initial assumption, such a path may be initiated at any zero of F_0 or of F in D, there is a finite number of non-intersecting PL paths in D^* which contain the zeros of F_0 and F.

12. CONTINUATION METHODS

Of course, the problem of finding zeros of systems of equations has concerned us for many many years, as has the existence problem. It is safe to say that the SAM approach is yet another tool, and a valuable one, which adds much to the area. In particular, the SAM approach, being *constructive*, has attracted much attention. We again note that these methods derived from Scarf's constructive proof of the Brouwer theorem.

By 'continuation' methods, one means methods which trace a path of zeros of a homotopy $H(x_0, x)$, based on a function $F(x)$, whose zeros are sought. These have also been called 'Davidenko' methods (see, for example [3]) since they trace back at least to a paper by Davidenko entitled *On the Approximate Solution of a System of Nonlinear Equations* published in 1953.

In the past three years, a series of papers [1, 3, 14] indicate a development of 'constructive' methods which appears, on the one hand, to have a computational potential paralleling that of the SAM approach and, on the other hand, to indicate a larger area of fruitful (future) study which embraces both approaches. However, the extent of such a unification and the form it will take is a matter for the next few years to decide. It will no doubt have its bases in combinatorial and differential topology.

The series of papers in question was initiated by Kellogg, Li, and Yorke [14], which also concerned the Brouwer theorem with impressive computational results. We shall terminate this summary with a brief discussion of the approach, as it relates to our previous development.

Let $D \subset \mathbb{R}^n$ be a bounded, open, convex set (such as the interior of a simplex), so that $\overline{D}$, the closure of D, is compact. Let $G: \overline{D} \to D$ be continuous. Brouwer's theorem states the existence of a fixed point : for some x^* in D, $G(x^*) = x^*$. Let C denote the set of fixed point of G in D.

There is the 'classical' function, $H: (\overline{D} - C) \to (\text{bdry } D)$ defined as follows: for $x \in \overline{D} - C$, $H(x)$ is the point where the ray from $G(x)$ through x meets bdry D. This is illustrated in Figure 5 (D is the simplex). Note that, for $y \in$ bdry $D, H(y) = y$.

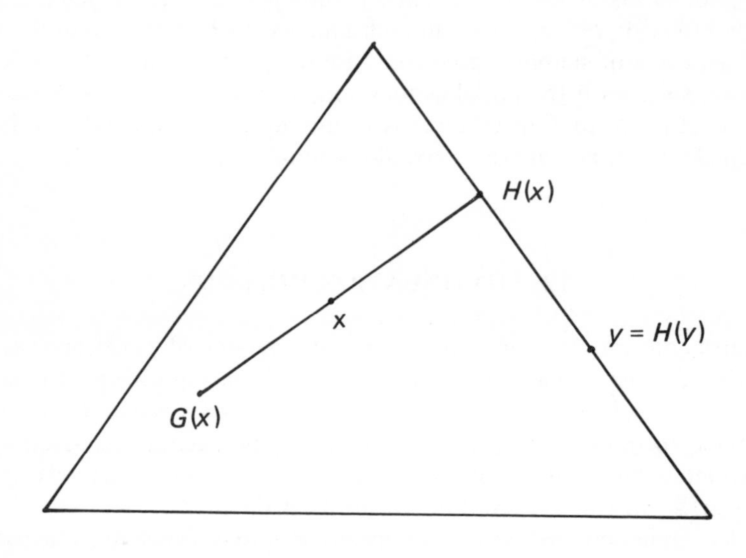

Figure 5

Thus, for a function $\theta : (\bar{D} - C) \to \mathbb{R}$,

$$H(x) = x + \theta(x)[G(x) - x] = [1 - \theta(x)]x + \theta(x)G(x) \qquad \text{where } \theta(x) < 0 \tag{40}$$

Then θ, hence H, is continuous. A classical *existence* proof is based on the fact that there is no continuous function $H : \bar{D} \to$ bdry D such that $H(y) = y$ when $y \in$ bdry D – which would contradict the assumption that $C = \emptyset$. In the following way, this proof was rendered 'constructive' [14].

In 1959, Pontryagin [16] studied the homotopy properties of maps $\phi : \bar{B}^n \to$ bdry B^n from a ball into its boundary. In particular, he used Sard's theorem (see below) and (quoting from [14]) it's corollary fact that for almost all $y \in$ bdry B^n the inverse image $\phi^{-1}(y)$ is a finite union of non-intersecting closed curves (such as we see in Figure 3).

Sard's theorem appears to be the source of the 'a.e.' (almost everywhere), or 'probability one' statements currently being used. We give a (quite simplified) version which is the one people have been using (e.g. Garcia-Zangwill [10]):

Theorem (Sard)
Let $D \subset \mathbb{R}^{n+1}$ be open and bounded. Let $f : \bar{D} \to \mathbb{R}^n$ have a continuous deriva-

tive f'. Then, if f has all second-order partial derivatives the set of all $y \in \mathbb{R}^n$ such that, for some point $x \in f^{-1}(y)$, rank $f'(x) < n$, has measure 0.

If rank $f'(x) = n$ for all $x \in f^{-1}(y)$, y is a *regular* value of f (which, in particular, is true when $f^{-1}(y) = \phi$). Thus, regularity holds almost everywhere.

In 1963, Hirsch published a paper [11] wherein, he very elegantly proved the fact noted after (40), and thereby the Brouwer theorem. With the assumption that no fixed points exist, he used the 'standard retraction' (40) and, employing SS and Sard's theorem, he investigated $H^{-1}(y)$, for a regular value y and, exhausting the possibilities, obtained a contradiction.

The above ideas were used in [14]. From a regular point $y^0 \in$ bdry D, that path of $H^{-1}(y^0)$ which contains y^0 is considered. Since H maps from $\mathbb{R}^n$ to an $(n-1)$-dimensional space, with the assumption that G has a continuous second derivative, by Sard's theorem the regularity of y^0 ensures that on that path rank $H'(x) = n-1$, so that the path is a non-self-intersecting curve with a continuous tangent. If z is on the path and $z \in$ bdry D, then $z = H(z) = H(y^0) = y^0$, but , with the rank condition, a piece of the path in some neighbourhood of y^0 is an arc. Hence, the path cannot lead back to bdry D, and it is argued that the path must contain a point of C, as desired.

From the computational point of view, the authors, set up a system of differential equations – *exactly* (36), but with $i = 0$ deleted, and, again, the 'constructive' aspect is embodied in the numerical 'path-following' to a point of C, starting at y^0.

It is of much interest (see Scarf's relevant remarkes [18]) to note how the authors relate their approach to the 'continuous' Newton method (compare the derivation of (36)).

From (40), writing $F(x) := G(x) - x$ we have

$$H'(x) = H' = (I + \theta F') + F\theta'^T = \theta(F' + \theta^{-1}I) + F\theta'^T \tag{41}$$

Since rank $H' = n-1$, a $v \neq 0$, unique with a multiple, satisfies $H'v = 0$. Now, away from y^0, $H(x) - x$ is bounded away from 0 and equals θF. Hence, as F approaches 0 along the path, θ approaches $-\infty$. Assuming $(F' + \theta^{-1}I)$ is non-singular (which is true if F' has no eigenvalues in $(1, \infty)$), with proper choice of the multiple, we may take the tangent $x' = v$ as

$$x' = -(F' + \theta^{-1}I)^{-1} F \tag{42}$$

This we visualize as the basis for an approximation scheme. Near a fixed point (a zero of F) θ^{-1} is small, so that the calculations are as those applied to

$$x' = -F'^{-1} F \tag{43}$$

(the Newton equations). For example, with 'step size 1', the kth Newton iteration would be

(44) $$x^{k+1} = x^k - [F'(x^k)]^{-1} F(x^k)$$

As with the SAM development, this 'path-following' technique has been extended in the applications to other, more general, problems, and apparently with some comparable numerical success. We have tried to indicate not only the differences, but the similarities of these two general 'constructive' approaches, and to suggest that there will be new developments in the near future embracing these two. Again, in this regard, the remarks of Scarf [18], especially as regards the conjunction with the Newton-like approaches, are to be noted.

REFERENCES

[1] J.C. Alexander, and J.A. Yorke. 'The Homotopy Continuation Method: Numerically Implementable Topological Procedures'. *Trans Am. Math. Soc.* (To appear).

[2] M.L. Balinski, and R.W. Cottle (Eds.) . 'Complementarity and fixed point problems'. *Math. Program. Study.*, 7 (1978).

[3] S.N. Chow, J. Mallet-Paret, and J.A. Yorke. 'Finding Zeroes of Maps: Homotopy Methods that are Constructive with Probability One'. (To appear).

[4] R.W. Cottle, and C.E. Lemke (Eds). *'Nonlinear Programming'. (Proc. Symp. Appl. Math., Am. Math. Soc. & SIAM).* Am. Math. Soc. (1976).

[5] R.W. Cottle. 'Complementarity and variational problems'. *Symposia Mathematica.* (Istituto Nazionale Alta Matematica), vol. XIX (1976).

[6] R.W. Cottle, and C.E. Lemke. 'Classes of Matrices and the Linear Complementarity Problem'. Tech. Rep., Dept of O.R., Stanford Univ. To appear.

[7] R.W. Cottle. 'Fundamentals of Quadratic Programming and Linear Complementarity'. *Proc. NATO Advanced Study Institute in Engineering Plasticity by Mathematical Programming.* To appear.

[8] G.B. Dantzig, and A.S. Manne. 'A Complementarity Algorithm for an Optimal Capital Path with Invariant Proportions'. *J. Econ. Theor.*, **9**, n. 3 (1974).

[9] B.C. Eaves. 'A Short Course in Solving Equations with Piecewise-linear Homotopies'. In *Nonlinear Programming. (SIAM-AMS Proc.*, vol. 9), Am. Math. Soc. (1976).

[10] C.B. Garcia, and W.I. Zangwill. 'On a New Approach to Homotopy and Degree Theory'. *Report,* Univ. of Chicago, 1978 To be published.

[11] M. Hirsch. 'A Proof of the Nonretractability of a Cell onto its Boundary'. *Proc. Am. Math. Soc.*, **14** (1963) 364-5.

[12] L. Kaneko. 'LCP with, an n x 2n P-matrix'. In [2].

[13] S. Karamardian (Ed. in collaboration with C.B. Garcia. 'Fixed-Points: Algorithms and Applications'. Academic Press, New York (1977).

[14] R.B. Kellogg, T.Y. Li and J. Yorke . 'A Constructive Proof of the Brouwer Fixed-point Theorem and Computational Results'. *SIAM J. Numer. Anal.*, **13**, n. 4 (1976).

[15] J.S. Pang, I. Kaneko, and W.P. Hallman. 'On the Solution of Some (Parametric) LCPs with Applications to Portfolio Analysis, Structural Engineering, and Graduation'. *MRC Tech. Summary Rep.*, Univ. of Wisconsin, WP 77-27, 1977.

[16] L.S. Pontryagin. 'Smooth Manifolds and Their Applications in Homotopy Theory'. *Am. Math. Soc.* **11** (1959). 1-114.

[17] R. Saigal. '*Fixed Point Computing Methods*'. Encyclopedia of Computer Science and Technology, vol. 8, Marcel Deker, Inc., New York (1977).

[18] H. E. Scarf. '*Fixed Points Algorithms and Applications*'. Academic Press, New York (1977).

Chapter 16

An Estimate of the Lipschitz Norm of Solutions of Variational Inequalities and Applications

P.L. Lions

1. INTRODUCTION

In this paper we study the Lipschitz character of solutions of variational inequalities of obstacle type, introduced by Stampacchia [11], and we give an application to a problem of equivalence of two variational inequalities, generalizing a result of Brezis and Sibony [2]. Let us give an example of our results.

Let $\mathcal{O}$ be a bounded domain in $\mathbb{R}^N$ with a smooth boundary Γ, and let A be the following operator:

$$A = -\sum_{i,j} a_{ij} \frac{\partial^2}{\partial x_i \partial x_j} + \sum_i b_i \frac{\partial}{\partial x_i} + c$$

where a_{ij}, b_i, c satisfy some smoothness conditions given in section 2. Then, if $f \in L^p(\mathcal{O})$ with $p > N$ and if $\psi \in W^{1,\infty}(\mathcal{O})$ such that $\psi|_\Gamma \geq 0$, the solution u of

$$(1) \quad \begin{cases} a(u, v-u) \geq (f, v-u) \quad \forall\, v \in H_0^1(\mathcal{O}) \quad v \leq \psi \text{ a.e. (almost everywhere)} \\ u \in H_0^1(\mathcal{O}) \quad u \leq \psi \text{ a.e.} \end{cases}$$

(where $a(u, v) = \langle Au, v\rangle_{H^{-1} \times H_0^1}$) satisfies $u \in W_0^{1,\infty}(\mathcal{O})$ and

$$(2) \qquad \|\nabla u\|_{L^\infty(\mathcal{O})} \leq \|\nabla \psi\|_{L^\infty(\mathcal{O})} + C\|f\|_{L^p(\mathcal{O})} + C\|\psi\|_{L^\infty(\mathcal{O})}$$

where C does not depend on ψ and f.

This kind of result was first obtained by H. Lewy and G. Stampacchia [6].

Remark 1 This result generalizes results given in [4, 5].

2. THE MAIN RESULT

We assume that the coefficients of the operator A satisfy

(3) $$a_{ij} = a_{ji},\ b_i,\ c \in W^{1,\infty}(\mathcal{O}) \qquad (1 \leqslant i, j \leqslant N)$$

(4) $$\exists \upsilon > 0 \text{ a.e.}, x \in \mathcal{O}, \forall \xi \in \mathbb{R}^N \qquad \sum_{i,j} a_{ij}(x)\, \xi_i \xi_j \geqslant \upsilon\, |\xi|^2$$

(5) $$c \geqslant 0 \text{ a.e.}$$

Then we have the following theorem

Theorem 1
Under assumptions (3)-(5) and if $f \in L^p(\mathcal{O})$ with $p > N$, if $\psi \in W^{1,\infty}(\mathcal{O})$ with $\psi|_\Gamma \geqslant 0$, the solution u of (1) satisfies $u \in W_0^{1,\infty}(\mathcal{O})$ and we have

(2) $$\|\nabla u\|_{L^\infty(\mathcal{O})} \leqslant \|\nabla \psi\|_{L^\infty(\mathcal{O})} + C\|f\|_{L^p(\mathcal{O})} + C\|\psi\|_{L^\infty(\mathcal{O})}$$

where C does not depend on ψ and f.

Remark 2 The existence (and uniqueness) of solution u of (1) comes from [11] in the case where $a(\cdot,\cdot)$ is coercive, i.e. $c(x) \geqslant \lambda > 0$ and λ is large enough. In the general case, this result can be deduced from [7] or [8].

Proof of Theorem 1: The proof will be divided into several steps:
(1) reduction to the case where $f = 0$;
(2) reduction to the case where ψ is smooth;
(3) estimate of u on the boundary;
(4) final estimate;
and (5) proof of a lemma.

Step 1 Let us introduce $\bar{u}$ as the solution of the following Dirichlet problem

(6) $$A\bar{u} = f \text{ a.e. in } \mathcal{O}$$
$$\bar{u} \in H^2(\mathcal{O}) \cap H_0^1(\mathcal{O})$$

(such a solution exists [8]) then $\bar{u} \in W^{2,p}(\mathcal{O})$ and $\|\bar{u}\|_{W^{2,p}(\mathcal{O})} \leqslant C\|f\|_{L^p(\mathcal{O})}$ where C does not depend on f. In particular $\|\bar{u}\|_{C^1(\bar{\mathcal{O}})} \leqslant C'\|f\|_{L^p(\mathcal{O})}$). But u-$\bar{u}$ is the solution of (1) with $f = 0$, and ψ replaced by $\psi - \bar{u}$. Thus we just need to prove (2) in the case where $f = 0$.

Step 2 An easy computation shows that u also satisfies

(1′) $$a(u, v-u) \geqslant 0 \qquad \forall v \in H_0^1(\mathcal{O}) \qquad v \leqslant -\psi^- \text{ a.e.}$$
$$u \in H_0^1(\mathcal{O}) \qquad u \leqslant -\psi^- \text{ a.e.}$$

where $-\psi^-$ is the non-positive part of ψ. But $-\psi^- \in W_0^{1,\infty}(\mathcal{O})$ and $\|\nabla(-\psi^-)\|_{L^\infty(\mathcal{O})} \leqslant \|\nabla\psi\|_{L^\infty(\mathcal{O})}$ (see Stampacchia [11]).

Now from the following lemma, which we temporarily admit, we see that we just need to prove (2) in the case where $f = 0$ and $\psi \in W^{2,p}(\mathcal{O}) \cap W_0^{1,p}(\mathcal{O})$ (for any $1 < p < +\infty$).

Lemma Let $\psi \in W_0^{1,\infty}(\mathcal{O})$, then the solution ψ_ε of

$$(7) \qquad -\Delta\psi_\varepsilon + \frac{1}{\varepsilon}\psi_\varepsilon = \frac{1}{\varepsilon}\psi; \quad \psi_\varepsilon \in W^{2,p}(\mathcal{O}), \forall p < +\infty; \quad \psi_\varepsilon|_\Gamma = 0$$

satisfies

$$\|\nabla\psi_\varepsilon\|_{L^\infty(\mathcal{O})} \leqslant \|\nabla\psi\|_{L^\infty(\mathcal{O})}$$

and

$$\psi_\varepsilon \underset{C(\overline{\mathcal{O}})}{\longrightarrow} \psi$$

Remark 3 Actually from the well known Schauder estimates, $\psi_\varepsilon \in C^{2,\alpha}(\overline{\mathcal{O}})$ $\forall \alpha < 1$.

Step 3 Estimate of ∇u on the boundary. We introduce the penalized problem associated with (1). Let ψ_ε be the solution of

$$(8) \; Au_\varepsilon + \frac{1}{\varepsilon}(u_\varepsilon - \psi)^+ = 0 \text{ a.e. in } \mathcal{O}; \quad u_\varepsilon \in W^{2,p}(\mathcal{O}), \quad \forall p < +\infty; \quad u_\varepsilon|_\Gamma = 0$$

We know that $u_\varepsilon \underset{\varepsilon \to 0_+}{\rightharpoonup} u$ in $W^{2,p}(\mathcal{O})$ weakly ($\forall p < +\infty$) (see [8]). We remark also that $u_\varepsilon \in W^{3,p}(\mathcal{O})$ $\forall p < +\infty$. We are going to prove L^∞ estimates for ∇u_ε (uniform in ε of course). First we work on Γ and we have to assume that $\mathcal{O}$ has, at every point y in Γ, a uniform exterior sphere, i.e.

$$(9) \qquad \begin{aligned} &\exists \rho > 0, \forall y \in \Gamma, \exists \hat{y} \text{ in } \mathbb{R}^N - \overline{\mathcal{O}} \text{ such that} \\ &\{z / |z - \hat{y}| \leqslant \rho\} \cap \overline{\mathcal{O}} = \{y\} \end{aligned}$$

(This assumption is just a regularity assumption on Γ, and can be dropped by an approximation process).

Let us fix $y \in \Gamma$, and let $\hat{y}$ be as in (9). We introduce

$$w(x) = \exp(-k\rho^2) - \exp(-k|x - \hat{y}|^2)$$

A simple calculation gives: if k is large enough (depending on $A, \overline{\mathcal{O}}$ but not on y),

$$Aw \geqslant \alpha > 0 \;\; \forall x \in \overline{\mathcal{O}} \qquad w|_\Gamma \geqslant 0 \qquad w(y) = 0$$

Moreover

$$|\nabla w| = k|x - \hat{y}| \exp(-k|x - \hat{y}|^2) \geq k\rho \exp(-k\rho^2)$$

(if $k\rho$ is large enough) for any x in $\overline{\mathcal{O}}$. Let us denote by

$$\beta = |\nabla w(y)| = k\rho \exp(-k\rho^2) = \left|\frac{\partial w}{\partial \upsilon}(y)\right|$$

where υ is the unit exterior normal).

We consider

$$\widetilde{w} = \|\nabla \psi\|_{L^\infty(\mathcal{O})}\, \beta^{-1} w$$

we have $A\widetilde{w} \geq 0$ a.e., $\widetilde{w}|_\Gamma \geq 0$, $\widetilde{w}(y) = 0$ and $|\nabla \widetilde{w}(x)| \geq \|\nabla \psi\|_{L^\infty(\mathcal{O})}$ $\forall\ x \in \overline{\mathcal{O}}$.

First we prove that $-\widetilde{w} \leq \psi \leq \widetilde{w}$. We consider $\widetilde{w}_\eta = \widetilde{w} + \eta\beta^{-1} w$ and $\eta > 0$; we want to prove that, for all $\eta > 0$, $\psi \leq \widetilde{w}_\eta$. Indeed if $\widetilde{w}_\eta - \psi$ has a negative minimum in $\overline{\mathcal{O}}$, at this point, we should have $|\nabla \widetilde{w}_\eta| = |\nabla \psi|$ but $|\nabla \widetilde{w}_\eta| > \max_{\overline{\mathcal{O}}} |\nabla \psi|$ and we have a contradiction. Thus $|\psi| \leq \widetilde{w}$.

But now

$$A(+\widetilde{w}) + \frac{1}{\varepsilon}(+\widetilde{w} - \psi)^+ \geq \frac{1}{\varepsilon}(\widetilde{w} - \psi)^+ \geq 0 \qquad \widetilde{w}|_\Gamma \geq 0$$

$$A(-\widetilde{w}) + \frac{1}{\varepsilon}(-\widetilde{w} - \psi)^+ \leq -A\widetilde{w} \leq 0 \qquad -\widetilde{w}|_\Gamma \leq 0$$

thus by the maximum principle $|u_\varepsilon| \leq \widetilde{w}$. This implies

$$\left|\frac{\partial u_\varepsilon}{\partial \upsilon}(y)\right| = |\nabla u_\varepsilon(y)| \leq |\nabla \widetilde{w}(y)| = \|\nabla \psi\|_{L^\infty(\mathcal{O})}$$

and we have proved that

$$\sup_\Gamma |\nabla u_\varepsilon(x)| \leq \|\nabla \psi\|_{L^\infty(\mathcal{O})}$$

Step 4 Final estimate. To conclude, we use a generalization of a method due to Beinstein that we have used for solving quasi-linear equations [9]. We introduce

$$w_\varepsilon(x) = |\nabla u_\varepsilon(x)|^2 + \lambda(+C + u_\varepsilon)^2$$

where $\lambda > 0$, $C = \|\psi\|_{L^\infty}$. Remark that $|u_\varepsilon| \leq C$.

Now in view of step *3*, we may assume that $w_\varepsilon(x)$ reaches its maximum at some point x_0 in $\{|\nabla u_\varepsilon(x)|^2 > \|\nabla \psi\|^2_{L^\infty(\mathcal{O})}\}$. But an easy calculation similar to the one made in [9] shows that, if λ is large enough, $\exists\, \alpha > 0$, for x in a neighbourhood of x_0

$$Aw_\varepsilon(x) \leqslant -\lambda\alpha \mid u_\varepsilon(x)\mid^2 + \left[\sum_k 2\frac{\partial u_\varepsilon}{\partial x_k}\left(\frac{\partial \psi}{\partial x_k} - \frac{\partial u_\varepsilon}{\partial x_k}\right)\frac{1}{\varepsilon}\right]_{(u_\varepsilon \geqslant \psi)}(x)$$

$$+ 2\lambda(+C + u_\varepsilon)[Au_\varepsilon - cu_\varepsilon]$$

but

$$\sum \frac{\partial u}{\partial x_k}\left(\frac{\partial \psi}{\partial x_k} - \frac{\partial u}{\partial x_k}\right)(x) \leqslant \frac{1}{2}\mid\nabla\psi(x)\mid^2 - \frac{1}{2}\mid\nabla u(x)\mid^2 \leqslant 0$$

for x in a neighbourhood of x_0. On the other hand $C + u_\varepsilon \geqslant 0$ and $Au_\varepsilon \leqslant f$, thus we have if $|x - x_0| \leqslant \eta(\eta > 0)$ $\exists \alpha > 0$,

$$Aw_\varepsilon(x) \leqslant -\lambda\alpha \mid u_\varepsilon(x)\mid^2 + 2\lambda(C + u_\varepsilon)[f - cu_\varepsilon]$$

Now using the Bony maximum principle [1], we can complete the proof of Theorem 1.

Step 5 Proof of lemma. First, we remark that the method used in step 3 gives immediately

$$|\nabla\psi_\varepsilon(x)| \leqslant \|\nabla\psi\|_{L^\infty(\mathcal{O})} \qquad \forall x \in \Gamma$$

Now if we differentiate (7) we have

$$-\Delta\left(\frac{\partial\psi_\varepsilon}{\partial\xi}\right) + \frac{1}{\varepsilon}\frac{\partial\psi_\varepsilon}{\partial\xi} = \frac{1}{2}\frac{\partial\psi}{\partial\xi}$$

for all unit vectors ξ and by maximum principle

$$\forall x \in \overline{\mathcal{O}} \quad \left|\frac{\partial\psi_\varepsilon}{\partial\xi}(x)\right| \leqslant \max_{y\in\Gamma}\left(\left\|\frac{\partial\psi}{\partial\xi}\right\|_{L^\infty(\mathcal{O})}, \left|\frac{\partial\psi_\varepsilon}{\partial\xi}(y)\right|\right)$$

Thus

$$|\nabla\psi_\varepsilon(x)| = \sup_\xi\left|\frac{\partial\psi_\varepsilon}{\partial\xi}(x)\right| \leqslant \max_{y\in\Gamma}\left(\|\nabla\psi\|_{L^\infty(\mathcal{O})}, |\nabla\psi_\varepsilon(y)|\right)$$

$$\leqslant \|\nabla\psi\|_{L^\infty(\mathcal{O})}$$

This completes the proof of the lemma.

Remark 4 In the case where a_{ij}, b, c do not depend on x then there exists $C_p > 0$ (independent of f) such that

$$\|\nabla u\|_{L^\infty(\mathcal{O})} \leqslant C_p \|f\|_{L^p(\mathcal{O})} + \|\nabla\psi\|_{L^\infty(\mathcal{O})}$$

3. AN APPLICATION TO THE ELASTIC-PLASTIC PROBLEM

In this section we apply the methods developed above to derive new results about the equivalence between the solution of the so-called elastic-plastic problem

$$(10)\quad \begin{cases} (\nabla u, \nabla(v-u)) \geqslant (f, v-u)\ \forall v \in H_0^1(\mathcal{O}) & \| v\|_{L^\infty(\mathcal{O})} \leqslant 1 \\ u \in H_0^1(\mathcal{O}) & \| u\|_{L^\infty(\mathcal{O})} \leqslant 1 \end{cases}$$

and the solution of a special obstacle problem

$$(1'')\quad \begin{cases} (\nabla u, \nabla(v-u)) \geqslant (f, v-u)\ \forall v \in H_0^1(\mathcal{O}) & v(x) \leqslant \delta(x) \text{ a.e.} \\ u \in H_0^1(\mathcal{O}) & u(x) \leqslant \delta(x) \text{ a.e.} \end{cases}$$

where $\delta(x) = \text{dist}(x, \Gamma)$ $(\delta \in W_0^{1,\infty}(\mathcal{O}))$.

Equivalence of these solutions was remarked on by Brezis and Sibony [2], where they consider the special case of f equal to a positive constant. In the following, we generalize this condition and we introduce another problem equivalent to (10)-(1″). To simplify notation, we shall use the following definition.

Definition We say that for given f, there is equivalence of variational inequalities if the solutions of (10) and (1″) are identical and if moreover they satisfy:

$$(11)\qquad -\Delta u \leqslant f \text{ a.e.} \qquad |\nabla u| \leqslant 1 \text{ a.e.} \qquad (-\Delta u - f)(|\nabla u| - 1) = 0 \text{ a.e.}$$

(We remark that, in view of the results of Brezis and Stampacchia [3], if $f \in L^p(\mathcal{O})$ $(1 < p < +\infty)$, the solution u of (10) satisfies $u \in W^{2,p}(\mathcal{O})(1 < p < +\infty)$).

Theorem 2
Let f be in $W^{1,\infty}(\mathcal{O})$, $f \geqslant 0$, satisfying

$$(12)\qquad \|f\|_{W^{1,\infty}(\mathcal{O})} \leqslant \frac{1}{C_1}$$

or

$$(12')\qquad \|f\|_{W^{1,\infty}(\mathcal{O})} > \frac{1}{C_1} \text{ and a.e. } \frac{f^2}{N} \geqslant |\nabla f| \left(C_1 \|f\|_{W^{1,\infty}}\right)$$

where C_1 depends only on $\|\delta\|_{L^\infty(\mathcal{O})}$. Then for such a function f, there is equivalence of variational inequalities (V.I.).

Remark 5 (1) If f = constant, (12) or (12′) are always satisfied: (2) If f satisfies (12′) then, $\forall\ t \geqslant 0$, tf satisfies (12) or (12′). (3) If f satisfies (12) or (12′) then, $\forall\ t \leqslant 1$, tf satisfies (12) or (12′).

Remark 6 A closer examination of the proof of Theorem 2 leads to more precise conditions than (12′) which we do not state for the sake of simplicity.

Remark 7 A very simple argument gives that if $N = 1$ and if $f \geqslant 0$ a.e. there is equivalence of V.I.

Remark 8 We shall see (in a work to appear) a stochastic interpretation of equivalence of V.I.

Proof of Theorem 2: First we remark that if we prove that for f satisfying the condition (12) or (12′) we have

$$(13) \qquad \forall\ \psi \in W^{1,\infty}(\mathcal{O}) \qquad \psi = 0 \text{ on } \Gamma \qquad \|\nabla\psi\|_{L^\infty(\mathcal{O})} \leqslant 1$$

the solution u of (1) satisfies

$$\|\nabla u\|_{L^\infty(\mathcal{O})} \leqslant 1$$

then Theorem 2 is easily proved. But because of the lemma we just need to prove (13) for ψ smooth.

Thus let u be the solution of (1) with f satisfying the condition (12) or (12′) and let $\psi \in W^{2,p}(\mathcal{O})$ ($\forall\ p < +\infty$) such that $\psi \geqslant 0$ in $\mathcal{O}$, $|\nabla\psi| \leqslant 1$ in $\mathcal{O}$ and $\psi|_\Gamma = 0$.

We are going to prove first that $\forall\ \alpha > 0 \quad \exists C(\alpha)$

$$(14) \qquad \|\nabla u\|_{L^\infty(\mathcal{O})} \leqslant \max\left(\|\nabla\psi\|_{L^\infty(\mathcal{O})},\ \alpha\|\psi\|_{L^\infty(\mathcal{O})} + C(\alpha)\|f\|_{W^{1,\infty}(\mathcal{O})}\right)$$

then choosing $\alpha\|\delta\|_{L^\infty(\mathcal{O})} = 1/2$, $C_1 = 2C(\alpha)$ (for example), we shall conclude.

(1) To prove (14), we introduce again u_ε solution of (8) and we remark that

$$\max_\Gamma |\nabla u_\varepsilon| \to \max_\Gamma |\nabla u| = \max_\Gamma \left|\frac{\partial u}{\partial \nu}\right| \qquad (\text{since } u^\varepsilon \to u \text{ in } W^{2,p}(\mathcal{O}) \text{ weakly})$$

and

$$0 \leqslant u \leqslant \psi \quad \text{implies} \quad \max_\Gamma |\nabla u| \leqslant \max_\Gamma |\nabla\psi|$$

Now we introduce $w_\varepsilon(x) = |\nabla u_\varepsilon(x)|^2 + \lambda |u_\varepsilon(x)|^2$ where $\lambda > 0$, and we consider x_0 a point of $\overline{\mathcal{O}}$ where w_ε attains its maximum: it is clear that the only case which remains, in order to prove (14), is the case where $x_0 \in \mathcal{O} \cap \{|\nabla u_\varepsilon(x)|^2 > |\nabla\psi(x)|^2\}$. But at this point we have by the maximum principle proved in [1]:

$$0 \leqslant -2 \sum_k \frac{\partial u_\varepsilon}{\partial x_k} \frac{\partial f}{\partial x_k} - 2\lambda |\nabla u_\varepsilon|^2 + 2\lambda u_\varepsilon f$$

thus

$$\forall x \in \overline{\mathcal{O}} \ \ \forall \alpha < \lambda \quad |\nabla u_\varepsilon(x)|^2 \leqslant \max_{x \in \overline{\mathcal{O}}} \left(\max_\Gamma |\nabla u_\varepsilon|, \frac{1}{4\alpha(\lambda-\alpha)} |\nabla f|^2 + \frac{2\lambda}{\lambda-\alpha} + |u_\varepsilon|\,|f| + \lambda\,|u_\varepsilon(x)|^2 \right)$$

hence taking $\varepsilon \to 0$, $\alpha = \lambda/2$ and λ small enough we deduce (14).

(2) Then if f satisfies (12)

$$\|\nabla u\|_{L^\infty(\mathcal{O})} \leqslant \max \left(\|\nabla \psi\|_{L^\infty(\mathcal{O})}, 1 \right) = 1$$

On the other hand if f satisfies (12′)

$$\|\nabla u\|_{L^\infty(\mathcal{O})} \leqslant \frac{1}{2} + \frac{C_1}{2} \|f\|_{W^{1,\infty}(\mathcal{O})} \leqslant C_1 \|f\|_{W^{1,\infty}(\mathcal{O})}$$

Now let us consider the open set $\hat{\mathcal{O}} = \{|\nabla u|^2 > 1\}$, which we assume to be non-empty. On $\hat{\mathcal{O}}$ we have $-\Delta u = f$ a.e. Indeed if on some subset B of $\hat{\mathcal{O}}$ of positive measure we had $-\Delta u < f$, we should have $u = \psi$ a.e. in B, thus $|\nabla u|^2 = |\nabla \psi|^2 \leqslant 1$ a.e. in B (see [11]) which contradicts the definition of $\hat{\mathcal{O}}$. Now locally in $\hat{\mathcal{O}}$ $u \in W^{3,p}_{loc}(\hat{\mathcal{O}})$ $\forall p < +\infty$ and we have

$$\begin{aligned} -\Delta |\nabla u|^2 &= -2 \sum_{i,j} \left(\frac{\partial^2 u}{\partial x_i \partial x_j} \right)^2 + 2 \sum_i \frac{\partial u}{\partial x_i} \frac{\partial f}{\partial x_i} \leqslant 2 \sum_i \left(\frac{\partial^2 u}{\partial x_i \partial x_i} \right)^2 + 2|\nabla u|\,|\nabla f| \\ &\leqslant -\frac{2}{N} (\Delta u)^2 + 2|\nabla u|\,|\nabla f| \leqslant -2\frac{f^2}{N} + 2C_1 \|f\|_{W^{1,\infty}(\mathcal{O})} |\nabla f| \leqslant 0 \text{ in } \hat{\mathcal{O}} \end{aligned}$$

which proves the contradiction. This completes the proof of Theorem 2.

Corollary Let f be in $W^{1,\infty}(\mathcal{O})$, $f \geqslant 0$, satisfying

$$\frac{f^2}{N} \geqslant |\nabla f| \left(1 + M \|\nabla f\|_{L^\infty(\mathcal{O})} \right) \text{ a.e.} \tag{15}$$

or

$$|\nabla f| \leqslant M^{-1} \text{ a.e. and } \frac{f^2}{N} \geqslant |\nabla f| \text{ a.e.} \tag{15'}$$

where

$$M = \max_{\mathcal{O}} \varphi$$

and φ is defined by $-\Delta\varphi = 1$ in $\mathcal{O}$, $\varphi|_\Gamma = 0$; then there is equivalence of V.I. for such a function f.

Proof: First remark that on $\hat{\mathcal{O}}$ $-\Delta u = f$ and this implies

$$\|\nabla u\|_{L^\infty(\mathcal{O})} \leqslant 1 + M \|\nabla f\|_{L^\infty(\mathcal{O})}$$

$$\left(\text{as } |\nabla u| \,|_{\partial\hat{\mathcal{O}}} \leqslant 1, \text{ and } \hat{\mathcal{O}} \subset \mathcal{O}\right)$$

Next, we remark that in $\hat{\mathcal{O}}$ we have

$$-\Delta |\nabla u|^2 \leqslant -2 \frac{f^2}{N} + 2 |\nabla u| |\nabla f|$$

then if f satisfies (15), we have $-\Delta |\nabla u|^2 \leqslant 0$ and we conclude.

If f satisfies (15′), we notice that we have in

$$-\Delta |\nabla u|^2 \leqslant |\nabla u| |\nabla f| - 2 |\nabla f| = 2(|\nabla u| - 1) |\nabla f|$$

thus

$$\|\nabla u\|^2_{L^\infty(\mathcal{O})} \leqslant 1 + 2M \left(\|\nabla u\|_{L^\infty(\mathcal{O})} - 1\right) \|\nabla f\|_{L^\infty(\mathcal{O})} \leqslant 1 + 2 \left(\|\nabla u\|_{L^\infty(\mathcal{O})} - 1\right)$$

This completes the proof.

Remark 9 We can prove similar results when the operator $-\Delta$ is replaced by more general operators, but we shall not state such results.

REFERENCES

[1] J.M. Bony. 'Principe du maximum dans le espaces de Sobolev'. *C.R. Acad. Sci., Paris*, **265** (1967) 333-6.

[2] H. Brezis, and M. Sibony. 'Equivalence de deux inéquations variationnelles et applications'. *Arch. Rat. Mech. Anal.*, **41** (1971) 254-65.

[3] H. Brezis, and G. Stampacchia. 'Sur la régularité de la solution d'inéquations elliptiques'. *Bull. Soc. Math. Fr.*, **96** (1968), 153-80.

[4] M. Chipot. 'Sur la régularité lipschitzienne de la solution d'inéquations elliptiques'. To appear.

[5] M. Giaquinta, and G. Modica. 'Regolarità Lipschitziana per la soluzione di alcuni problemi di minimo con vincolo'. *Ann. di Mat.*, **106** (1975) 95.

[6] H.Lewy and G. Stampacchia. 'On existence and smoothness of solutions of some non-coercive variational inequalities.' *Arch. Rat.Mech. Anal.* **41** (1971), p. 241-253.

[7] P.L. Lions. 'Problèmes elliptiques non coercifs'. *Bull. Unione Mat. Ital.* To appear.

[8] P.L. Lions. 'Problèmes elliptiques du 2ème ordre non sous forme de divergence'. *Proc. Roy. Soc. Edim.* To appear.

[9] P.L. Lions. 'Résolution d'équations quasilinéaires'. *Arch. Rat. Mech. Anal.* To appear. (see also *Thèse de 3ème cycle*, Paris VI, 1978).

[10] G. Stampacchia. 'Formes bilinéaires coercives sur les ensembles convexes'. *C.R. Acad. Sci., Paris*, **258** (1964), 4413-5.
[11] G. Stampacchia. 'Le problème de Dirichlet pour les équations du second ordre à coefficients discontinus'. *Ann. Inst. Fourier,* **15** (1965), 189-257.

Chapter 17

The Complementarity Problem for Maximal Monotone Multifunctions (*)

L. McLinden

1. INTRODUCTION

This chapter surveys some of the results we have obtained recently for the non-linear complementarity problem. We treat the class of problems in which the non-linear function is maximal monotone, possibly multivalued in fact and not everywhere defined. This kind of multifunction arises frequently, for example in the contexts of optimization problems and, more generally, variational inequalities. The class of maximal monotone multifunctions includes, in particular, the generalized gradient multifunctions arising from both closed proper convex functions and (with a certain twist) also closed proper saddle functions. It includes, in addition, many non-linear multifunctions not of gradient type, and hence provides quite a broad setting.

Within this framework we impose only one assumption, namely that there exists at least one strictly positive vector whose image under the multifunction contains a strictly positive vector. This condition is analogous to the concept of strict feasibility in the optimization literature. From this blanket hypothesis follow many results, some of which we now summarize.

The solution set is non-empty, compact, and convex in a strong sense. A certain parametrized class of 'approximate complementarity problems' possesses unique solutions. By varying the parameters, uniquely determined natural trajectories of solutions to the approximating problems can thus be specified. These trajectories depend continuously on the parameters, and if the original problem is non-degenerate they converge to some particular solution, even though there may be many solutions. This limit solution depends only on a certain direction, which itself is a pre-selected parameter associated with the trajectory. The limit solution is characterized as that unique sol-

(*) - Research supported in part by the National Science Foundation, under grant number MCS 75-08025 A01 at the Univ. of Illinois at Urbana-Champaign, and by the Center for Operations Research and Econometrics, at the Université Catholique de Louvain.

ution which, from among all the solutions, provides the strong Pareto optimum with respect to the direction parameter. The limiting behaviour just described still holds, moreover, when the prescribed direction parameter is achieved by the trajectory only asymptotically. This last fact, combined with the earlier continuity fact and the relatively more tractable form of the approximating problems, suggests that numerical approaches based on this theory may not be unreasonable.

The original problem is also shown to have certain stability and generic properties. For example, for all sufficiently small perturbations, where the perturbations correspond geometrically to general translations of the graph of the multifunction, the resulting perturbed complementarity problem also has a non-empty compact convex set of solutions which, moreover, depends upper semicontinuously on 'the perturbations' except for a negligible (that is, zero Lebesgue measure) subset of these small perturbations, the perturbed problems have unique solutions. In particular, the original problem can generically be expected to have a unique solution.

The results here are all stated for $\mathbb{R}^n$. Many of them, however, admit infinite dimensional extensions. These and additional related results, as well as all proofs, will appear elswhere.

2. PROBLEM STATEMENT AND EXAMPLES

We suppose some *multifunction* $T : \mathbb{R}^n \to \mathbb{R}^n$ is given, that is, a function which assigns to each $x \in \mathbb{R}^n$ a subset $T(x) \subset \mathbb{R}^n$. The possibilities that $T(x)$ may be empty or contain more than one point are expressly permitted. It will be convenient to let $G(T)$ denote the *graph* of a multifunction T:

$$G(T) = \{(x, y) \in \mathbb{R}^n \times \mathbb{R}^n : y \in T(x)\}$$

By the *complementarity problem* corresponding to T we mean the problem of finding a pair (x, y) such that

$$(C) \qquad y \in T(x), \quad x \geqslant 0, \quad y \geqslant 0, \quad \langle x, y \rangle = 0$$

(Here $x \geqslant 0$ means that $x = (x_1, ..., x_n)$ satisfies $x_k \geqslant 0$ for each $k = 1, ..., n$, and $\langle x, y \rangle = \Sigma_1^n x_k y_k$.) Most authors present the problem as that of finding only the x component of the pair just described, and when T is actually a function that clearly suffices. In treating general multifunctions, however, it is both convenient and natural to place both components of the pair on an equal footing.

A multifunction T is called *monotone* if it satisfies the condition

$$\langle x' - x, y' - y \rangle \geqslant 0 \quad \text{whenever} \quad (x, y), (x', y') \in G(T)$$

A monotone multifunction is called *maximal monotone* if its graph is not properly contained in the graph of any other monotone multifunction between the same spaces. Throughout this paper we shall be concerned with T's which are maximal monotone. For this class of multifunctions, we have found that a single hypothesis suffices to ensure that problem (C) possesses a rather wide variety of strong properties. In addition, for many of the results the condition is also necessary in the case of arbitrary maximal monotone T. This key hypothesis is that

$$(*) \qquad \text{there exists some } (x, y) \text{ such that } x > 0, \quad y > 0, \quad y \in T(x)$$

(Here $x > 0$ means that every coordinate of x is strictly positive.) At this point we do not yet impose $(*)$ as a blanket assumption, since we want first to indicate a variety of T's which are maximal monotone and also to give several illustrations of optimization and variational inequality situations which give rise to problem (C) for such T's.

Example 1
Let T be the affine function given by $T(x) = Mx + q$, where q is an n-vector and M is an $n \times n$ real matrix whose symmetric part is positive semidefinite.

Example 2
Let $T(x) = \partial f(x)$, where ∂f is the subdifferential of any closed proper convex function f from $\mathbb{R}^n$ to $(-\infty, +\infty]$. Recall that

$$\partial f(x) = \{y \in \mathbb{R}^n : f(x') \geqslant f(x) + \langle x' - x, y \rangle, \forall x' \in \mathbb{R}^n\}$$

Example 3
Let T be a 'twisted subdifferential' of a saddle function. That is, put

$$T(x, y) = \{(v, -u) : (v, u) \in \partial H(x, y)\}$$

for any closed proper convex-concave function H from $\mathbb{R}^n = \mathbb{R}^p \times \mathbb{R}^q$ to $[-\infty, +\infty]$.

Example 4
Let T be defined on $\mathbb{R}^2$ by

$$T(x) = \{Mx\} + \partial \psi_B(x),$$

where M is the 2×2 real matrix sending (x_1, x_2) to $(-x_2, x_1)$ and

$$\psi_B(x) = \begin{cases} 0 & \text{if } x \in B \\ +\infty & \text{if } x \in \mathbb{R}^2 \backslash B \end{cases}$$

for closed (Euclidean) unit ball B. This yields

$$T(x) = \begin{cases} \{Mx\} & \text{if } x \in \text{int } B \\ \{Mx + \lambda x : \lambda \geqslant 0\} & \text{if } x \in B \backslash \text{int } B \\ \emptyset & \text{if } x \in \mathbb{R}^2 \backslash B \end{cases}$$

(It is interesting to make a sketch depicting the image sets $T(x)$ for selected non-negative vectors in B.)

The T described in each of Examples 1-4 is a maximal monotone multifunction. This paper intersects the topic of linear complementarity theory via specializing the results described below to the T in Example 1. Since the T's in Examples 2-3 are defined in terms of subdifferentials, problem (C) for these T's is associated naturally with certain extremum problems (see Examples 5-8 below). For a comprehensive treatment of Examples 2-3, see [2]. Example 4, due to Rockafellar, is an especially simple instance of a maximal monotone multifunction not covered by Examples 1-3.

This section concludes with six more examples, to illustrate some of the situations which fit the mould of problem (C) for maximal monotone T.

Example 5

Consider $T = \partial f$ as in Example 2. It is known (e.g. [2]) that problem (C), which here assumes the form

$$(1) \qquad y \in \partial f(x), \quad x \geqslant 0, \quad y \geqslant 0, \quad \langle x, y \rangle = 0,$$

describes the optimality conditions associated with (and indeed linking) the dual convex minimization problems

$$(2) \qquad \min_{x \geqslant 0} f(x), \qquad \min_{y \geqslant 0} f^*(y)$$

Here f^* denotes the Fenchel transform of f, given by

$$f^*(y) = \sup_x \{\langle x, y \rangle - f(x)\}$$

In terms of it one can characterize ∂f as follows:

$$y \in \partial f(x) \quad \text{if and only if} \quad f(x) + f^*(y) = \langle x, y \rangle$$

Under the assumption that (either) one of the problems in (2) has a Slater point and bounded level sets (e.g. for the first problem in (2) this means that there exists some $x > 0$ such that $f(x) \in \mathbb{R}$ and there exists some $\alpha \in \mathbb{R}$ such that $\{x \in \mathbb{R}^n : x \geqslant 0,\ f(x) \leqslant \alpha\}$ is bounded), then both the problems in (2) have solutions and the product of their solution sets equals the solution set to problem (C) for $T = \partial f$. It can be shown that the assumption just mentioned is equivalent to condition (*) for $T = \partial f$.

Example 6

Let $f_0, f_1, \ldots, f_m$ be given closed proper convex functions on $\mathbb{R}^n$, and for simplicity assume that

$$\{x : f_k(x) < \infty\} = \{x : f_0(x) < \infty\} \qquad \text{for } k = 1, \ldots, m,$$

where this common set is closed. The ordinary convex programming problem is

$$\text{(3)} \qquad \min \{f_0(x) : f_1(x) \leqslant 0, \ldots, f_m(x) \leqslant 0\}$$

It is considered mild to assume that this problem has a Slater point and bounded level sets (i.e. there exists some $\bar{x}$ such that $f_1(\bar{x}) < 0, \ldots, f_m(\bar{x}) < 0$, and there exists some $\alpha \in \mathbb{R}$ such that $\{x : f_0(x) \leqslant \alpha, f_1(x) \leqslant 0, \ldots, f_m(x) \leqslant 0\}$ is bounded). Under this assumption it is known [2] that the associated Kuhn-Tucker optimality conditions

$$\text{(4)} \qquad \left\{ \begin{array}{l} f_k(x) \leqslant 0 \qquad y_k \geqslant 0 \qquad y_k f_k(x) = 0 \qquad \text{for } k = 1, \ldots, m \\ \text{and} \\ 0 \in f_0(x) + \sum y_k \partial f_k(x) \end{array} \right.$$

(where the summation extends only over those indices for which $y_k > 0$) are solvable and exactly characterize the pairs $(x, y) \in \mathbb{R}^n \times \mathbb{R}^m$ such that x solves (3) and y is an optimal Lagrange multiplier for (3).

We contend that this situation can be treated as an elaborately structured (but, in a certain sense to be mentioned, seemingly trivial) instance of Example 5. The idea is to take as the function in Example 5 the perturbation function associated with (3). This is the function p on $\mathbb{R}^m$ given by

$$p(u) = \inf \{f_0(x) : f_1(x) + u_1 \leqslant 0, \ldots, f_m(x) + u_m \leqslant 0\}$$

(Note that the u_k here are placed on the lower side of the inequalities, in contrast to general practice). At first glance, doing this may seem rather pointless, since the coordinatewise non-decreasing nature of p (as defined here) means that the first problem in (2), namely $\min_{u \geqslant 0} p(u)$ trivially has $u = 0$ as a solution. It can be shown, though, that the assumptions on (3) mentioned above imply not only that p is closed proper convex, but also that the infimum defining p is achieved whenever it is not $+\infty$, that condition (*) is fulfilled for $T = \partial p$, and that solving (1) for ∂p is equivalent to solving (4). (If (x, y) solves (4), then (u, y) solves (1), where $u = -(f_1(x), \ldots, f_m(x))$. Conversely, if (u, y) solves (1) then (x, y) solves (4), where x is any vector yielding the infimum in $p(u)$.) Finally, the dual problem of (2), $\min_{y \geqslant 0} p^*(y)$, is (up to a minus sign) precisely the ordinary dual problem usually associated with (3). All of these facts still hold if the space $\mathbb{R}^n$ on which the functions $f_0, f_1, \ldots, f_m$ are given is replaced by a general infinite-dimensional space. The reason for this is that the complementarity conditions involved in (4) are operating only in the image space $\mathbb{R}^m$ of the vector constraint function $x \to (f_1(x), \ldots, f_m(x))$.

The upshot of the preceding observations is that all of the finite dimensional results in sections 3-4 below concerning the complementarity problem (C) can be applied to convex programs having finitely many inequality constraints and set in general spaces. In particular, the 'approximate complementarity problems (C^z) treated in section 3, when applied to the multifunction $T = \partial p$, furnish a gradient-level solution scheme which corresponds directly to the logarithmic penalty method. In this way new properties of that well known method are established and its validity is extended to infinite dimensional problems. Also, the results in section 4, when interpreted for $T = \partial p$, provide apparently new information concerning convex programs of the type (3).

Example 7

Let T be defined as in Example 3 for some closed proper convex-concave function H. It is known [1] that problem (C) for such a T describes the optimality conditions associated with each of the following four extremum problems:

$$\operatorname*{minimax}_{x \geqslant 0,\, y \geqslant 0} H(x, y), \qquad \operatorname*{maximin}_{v \geqslant 0,\, u \geqslant 0} K(v, u),$$

$$\min_{x \geqslant 0,\, u \geqslant 0} F(x, u), \qquad \max_{v \geqslant 0,\, y \geqslant 0} G(v, y)$$

The functions F, G, and K are closed proper convex, concave, and concave-convex, respectively. Each is related to H (and to the others, as well) via one or at most two partial Fenchel transformations. For example,

$$F(x,u) = \sup_{y} \{H(x,y) + \langle u,y\rangle\},$$

$$G(v,y) = \inf_{x,u} \{F(x,u) - \langle x,v\rangle - \langle u,y\rangle\},$$

$$K(v,u) = \inf_{x} \{F(x,u) - \langle x,v\rangle\}.$$

If one makes the assumption that the convex minimization problem (or alternatively, the concave maximization problem) admits a Slater point and has bounded level sets, then all four of these extremum problems have solutions and each of their solution pairs arises from a solution to problem (*C*) for the given T. The assumption just mentioned is equivalent to condition (*) for this T.

The situation in Example 7 is actually equivalent, in a certain sense, to that of Example 5. Indeed, it can be shown that the T in Example 7 satisfies

$$\begin{aligned} T(x,y) &= \{(v,u) : (v,-u) \in \partial H(x,y)\} \\ &= \{(v,u) : (v,y) \in \partial F(x,u)\}, \end{aligned}$$

so that T corresponds to a 'partial inverse' of ∂F for a certain closed proper convex F. Since the complementarity requirements (i.e. the non-negativity of variables combined with orthogonality in problem (*C*)) are preserved under such a 'partial inversion', it follows that problem (*C*) for this T is equivalent to problem (*C*) for ∂F. Conversely, whenever one has a problem of the form (*C*) for a subdifferential multifunction ∂F, to each of the ways one can decompose the domain space into a product of two spaces there corresponds, via this partial inversion process, the four-way extremum problem correspondence described above.

Example 8
Here we consider the non-negativity constrained variant of problem (3):

$$(5) \qquad \min\{f_0(x) : x \geqslant 0,\ f_1(x) \leqslant 0, \ldots, f_m(x) \leqslant 0\}$$

For simplicity, let all the f_k's be differentiable convex on all of $\mathbb{R}^n$. Under the assumption that there exists some $\bar{x} > 0$ such that $f_1(\bar{x}) < 0, \ldots, f_m(\bar{x}) < 0$ and there exists some $\alpha \in \mathbb{R}$ such that the set

$$\{x : f_0(x) \leqslant \alpha,\ x \geqslant 0,\ f_1(x) \leqslant 0, \ldots, f_m(x) \leqslant 0\}$$

is bounded, the Kuhn-Tucker conditions for (5) are solvable and exactly characterize the pairs $(x,y) \in \mathbb{R}^n \times \mathbb{R}^m$ such that x solves (5) and y is an optimal Lagrange multiplier for (5). The Kuhn-Tucker conditions for this problem can be written in the form

$$(6) \qquad (x,y) \geqslant (0,0), \quad (v,u) \geqslant (0,0), \quad \langle (x,y),(v,u) \rangle = 0\,,$$

$$(7) \qquad \begin{aligned} v &= -\left(\nabla f_0(x) + \sum_{k=1}^{m} y_k \, \nabla f_k(x) \right), \\ u &= -\left(f_1(x), \ldots, f_m(x) \right) \end{aligned}$$

Conditions (6) describe problem (*C*) for the multifunction T on $\mathbb{R}^n \times \mathbb{R}^m$ given by

$$(8) \qquad T(x,y) = \begin{cases} \{(v,u)\} & \text{if } (x,y) \geqslant (0,0) \\ \emptyset & \text{otherwise}, \end{cases}$$

where the v and u here are specified by (7). It can be shown that under the conditions stated this T is maximal monotone and satisfies condition (*). This example is essentially Example 7 applied to the Lagrangian saddle function H associated with problem (5).

Example 9

This [3] involves generalizing the preceding example by replacing the gradient term $\nabla f_0(x)$ in (6)-(8) by $A_0(x)$, for a general maximal monotone multifunction A_0 on $\mathbb{R}^n$. If A_0 is not single-valued, then of course (7)-(8) must be rephrased accordingly. If the point $\bar{x}$ assumed in Example 8 satisfies $A_0(\bar{x}) \neq \emptyset$, then it can be shown that the resulting T is maximal monotone. Condition (*) for this T is that there exist vectors x, y, w such that

$$x > 0\,, \quad y > 0\,, \quad w \in A_0(x)\,,$$

$$f_k(x) < 0 \qquad \text{for } k = 1, \ldots, m\,,$$

$$w + \sum_{k=1}^{m} y_k \, \nabla f_k(x) < 0$$

The extremum problem interpretation (5) is, of course, no longer applicable in general to this situation, but the resulting complementarity problem for this multifunction can be viewed as a basic, finite dimensional approximation to a variational inequality in Hilbert space (involving a maximal monotone A of which A_0 is an approximation). See [4], where a strong case is made for viewing this example as being of fundamental interest. The results of section 3 below constitute a 'direct' non linear approximation approach to this problem. The alternative in general would be to go on to 'linearize' this (presumably non linear) complementarity problem, in order to bring to bear the much more developed linear complementarity methodology.

Example 10
Suppose we wish to find the fixed points of a given multifunction F having both domain and range contained in the non negative orthant of $\mathbb{R}^n$. This problem is equivalent to solving problem (C) for the multifunction $T : \mathbb{R}^n \to \mathbb{R}^n$ induced by F via $T = I - F$ (that is, $T(x) = \{x\} - F(x)$ for $x \geqslant 0$ and $T(x) = \emptyset$ otherwise). The results of the present paper apply, then, to all F's for which the induced T is maximal monotone. (This includes all multifunctions $F : \mathbb{R}^n \to \mathbb{R}^n$ such that (i) $x \geqslant 0$ and $y \geqslant 0$ whenever $(x, y) \in G(F)$, and (ii) the multifunction $-F$ is maximal monotone). Condition (*) for the induced T corresponds to the existence of some $(x, y) \in G(F)$ satisfying $x > y \geqslant 0$.

3. EXISTENCE AND PROPERTIES OF TRAJECTORIES OF APPROXIMATE SOLUTIONS

From now on, we assume that T is some given maximal monotone multifunction on $\mathbb{R}^n$, and that it satisfies the condition

$$(*) \qquad G(T) \cap (P \times P) \neq \emptyset,$$

where $P = \{x \in \mathbb{R}^n : x > 0\}$. It is helpful to introduce another multifunction S on $\mathbb{R}^n$ in order to describe the complementarity constraints of problem (C):

$$G(S) = \{(x, y) \in \mathbb{R}^n \times \mathbb{R}^n : x \geqslant 0, y \geqslant 0, \langle x, y \rangle = 0\}$$

Problem (C) can now be reformulated as that of finding a pair (x, y) such that

$$(C) \qquad (x, y) \in G(T) \cap G(S)$$

The solution set to (C) will be denoted by Ω. (Thus, $\Omega = G(T) \cap G(S) \subset \mathbb{R}^n \times \mathbb{R}^n$).

Theorem 1
The set Ω of solutions to (C) is non empty, compact, and convex.

Note that the convexity assertion here is stronger in general than asserting that the two obvious projections of Ω are convex, because Ω need not fill up all of the product of the two projections. We remark that Ω can also be shown non empty under various conditions other than (*); this will be detailed elesewhere.

A major part of the difficulty in complementarity problems generally can be traced to the 'non-differentiable' (or alternatively, 'Boolean') character of S. The idea naturally arises, then, of attempting to modify the problem (C) somehow, in order to mitigate the difficulties. The remainder of section 3 centres around an approach directed toward this objective.

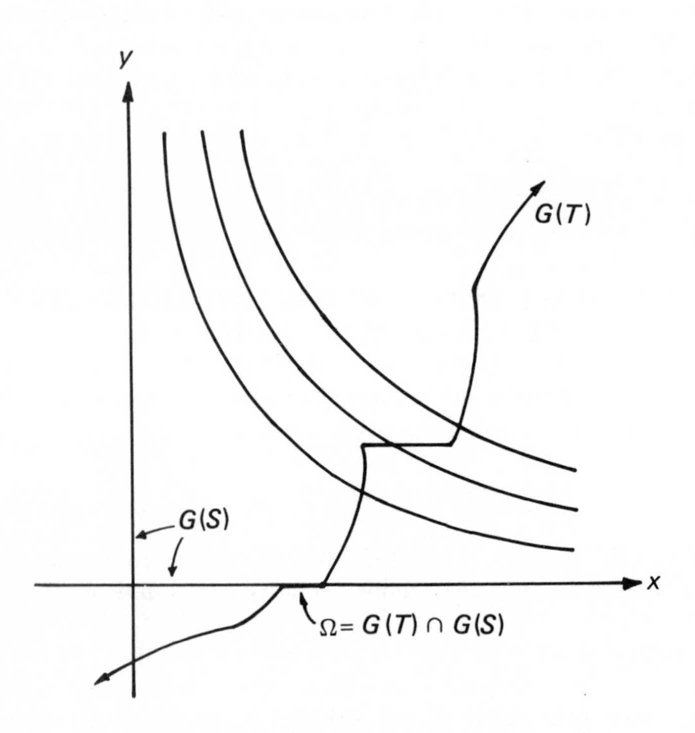

Figure 1 The general idea, portrayed for $n = 1$.

It is helpful to introduce the underlying idea geometrically, in the very special, but suggestive, simple case of dimension $n = 1$. We refer to Figure 1, in which $G(T)$ is shown as a 'complete non decreasing curve', reflecting its maximal monotone character, and $G(S)$ is the 'right-angle' curve formed by the two non negative axes. The remaining curves are various hyperbolas in the first quadrant. Each of these appears to meet $G(T)$ at a unique point, and as the hyperbolas become progressively 'steeper', their points of intersection with $G(T)$ appear to converge towards $G(T) \cap G(S) = \Omega$. Since maximal monotone multifunctions in dimension $n = 1$ are so exceedingly nice, it is *a priori* highly problematical that this idea, however appealing it may seem intuitively, would carry over somehow to the general case $n > 1$.

Yet it does, as we now proceed to outline.

The first task is to find some parametrized family of relatively nice multifunctions that, for general n, can play the role of the 'successively steeper hyperbolas' to approximate S. This is accomplished by defining, for each $z \in Q = \{x \in \mathbb{R}^n : x \geqslant 0\}$, a multifunction S_z on $\mathbb{R}^n$ via setting

$$G(S_z) = \{(x, y) \in \mathbb{R}^n \times \mathbb{R}^n : x \geqslant 0, y \geqslant 0, x_k y_k = z_k, \forall\ k\}$$

Next, for each such multifunction define the corresponding *approximate complementarity problem* to be the problem of finding an (x, y) such that

$$(C^z) \qquad (x, y) \in G(T) \cap G(S_z)$$

Notice that the original problem corresponds to the parameter choice $z = 0$. When T is actually a function, for example, for each $z \in P$ problem (C^z) amounts to solving the non linear monotone equation system

$$T(x_1, \ldots, x_n) = \left(\frac{z_1}{x_1}, \ldots, \frac{z_n}{x_n}\right)$$

for some $x = (x_1, \ldots, x_n)$ strictly positive.

Since $G(S_z) \subset P \times P$ for $z \in P$, it is clear that a *necessary* condition for (C^z) to have a solution for such z is that T satisfy the condition (*). We find it surprising that this condition is also *sufficient* for existence in general.

Theorem 2

For each $z \in P$, problem (C^z) has a unique solution denoted by (x^z, y^z).

If the approximation idea sketched above is to work, these unique solutions ought to depend in a regular manner on strictly positive z.

Theorem 3

The function $z \to (x^z, y^z)$ is continuous on P.

We digress briefly now to list two new geometric properties of $G(T)$ which follow from Theorems 2 and 3.

Theorem 4

The set $\Gamma = G(T) \cap (P \times P)$ is maximal monotone with respect to $P \times P$. It is also homeomorphic to P via the function $(x, y) \to z$ defined by $z_k = x_k y_k$, $\forall\ k$.

Returning now to the approximation theme, if the problems (C^z) are to be of use in connection with (C), then their solutions must approach the set Ω in some sense as $z \to 0$ through values in P. The next few results bear on this issue. For their statements it is helpful to introduce the concept of support.

Let the *support* of a *vector* $w \in Q$ be defined as the subset $K(w) = \{k : w_k > 0\}$ of the indices $\{1,\ldots, n\}$. Now define the *support* of a non empty convex set $W \subset Q$ to be that subset of $\{1,\ldots, n\}$ given by the expression

$$K(W) = \max \{K(w) : w \in W\}$$

The convexity of W implies that this formula is well defined, i.e. specifies uniquely a maximal subset of $\{1,\ldots n\}$. One has $K(W)=\emptyset$ if and only if W consists of only the zero vector. We want to apply this notion to the set Ω. By Theorem 1, Ω is a non empty convex subset of $Q \times Q$, and hence its projections

$$X = \{x \in \mathbb{R}^n : \exists\ y, (x, y) \in \Omega\}$$

$$Y = \{y \in \mathbb{R}^n : \exists\ x, (x, y) \in \Omega\}$$

onto the domain and range spaces of T are non empty convex subsets of Q. Write

$$I = K(X), \quad J = K(Y)$$

The complementarity requirements embodied in S, together with the convexity of Ω, imply that

$$I \cap J = \emptyset, \tag{9}$$

and clearly one also has that

$$I \cup J \subset \{1, \ldots, n\} \tag{10}$$

Since Ω has also a well defined support, how does that relate to I and J? Because $\Omega \subset \mathbb{R}^n \times \mathbb{R}^n$ rather than $\mathbb{R}^n$, a little care must be taken. For intuitive purposes one can think of $K(\Omega)$ as $I \cup J$, but it is more appropriate to regard $K(\Omega)$ as corresponding to the ordered pair (I, J). In any event, due to (9)-(10) one has

$$0 \leq \text{card } K(\Omega) = \text{card } I + \text{card } J \leq n$$

The next result contains the surprising fact that all 'limits' of sequences of approximate solutions (x^z, y^z), as z approaches the zero vector from some fixed 'direction', even asymptotically, are elements of Ω having maximal support.

Theorem 5
Let $\{z^m\}_{m=1}^{\infty} \subset P$ be any sequence converging to the zero vector in such a way that, for some vector a satisfying

$$a \in P, \quad a_1 + \ldots + a_n = 1 \tag{11}$$

one has

$$\lim_{m \to \infty} a_k^m = a_k, \quad k = 1, \ldots, n,$$

where

$$a_k^m = z_k^m \left(\sum_{\ell=1}^{n} z_\ell^m \right)^{-1}, \quad \forall\, m \quad \forall\, k$$

Then the sequence

$$\left\{ \left(x^{z^m}, y^{z^m} \right) \right\}_{m=1}^{\infty} \tag{12}$$

of solutions to the approximate complementarity problems (C^{z^m}) has at least one accumulation point. Moreover, every such point, for instance (x^0, y^0), has the properties that

$$(x^0, y^0) \in \Omega, \quad K(x^0) = I, \quad K(y^0) = J \tag{13}$$

and also satisfies the estimate

$$\sum_I ak\left(\frac{x_k}{x_k^0}\right) + \sum_J ak\left(\frac{y_k}{y_k^0}\right) \leqslant 1 \tag{14}$$

for each $(x, y) \in \Omega$.

Corollary 1 If Ω is a singleton, then the sequence (12) converges to that pair.

What can be said about uniqueness of accumulation points to (12) in general? According to the next corollary, for 'reasonable' problems (C), the sequence (12) must converge. The statement invokes the following technical lemma and a definition.To handle cases in which I and/or J is empty in problem (15) below, we adopt the convention $\prod_{\phi} = 1$.

Lemma For any $a \in P$ (not necessarily satisfying $a_1 + \ldots + a_n = 1$), the problem

$$\max \left\{ \prod_I (x_k)^{a_k} \cdot \prod_J (y_k)^{a_k} : (x, y) \in \Omega, K(x) = I, K(y) = J \right\} \tag{15}$$

has a unique solution, which will be denoted $(\widetilde{x}, \widetilde{y})$.

We shall say that (C), or equivalently Ω, is *nondegenerate* if $I \cup J = \{1, \ldots, n\}$. *Degenerate* problems, namely those for which $I \cup J$ is properly contained in $\{1, \ldots, n\}$ (see (10) above), are those which are dimensionally deficient in a certain sense.

Corollary 2 If (C) is nondegenerate, then the sequence (12) converges, in fact to the pair $(\widetilde{x}, \widetilde{y})$ solving problem (15).

It may be noted that in order to use this convergence criterion in practice, it is not necessary to know *a priori* whether (C) is nondegenerate. Indeed, by conclusion (13) of Theorem 5 one is guaranteed that any (every) accumulation point of (12) will single out the special index sets I and J and one can tell simply by inspection whether (C) is degenerate or not.

Corollary 2 is a special case (corresponding to $\beta = 1$) of the following result, which applies even to degenerate problems. It is a consequence of the estimate (14). The assumption $I \cup J \neq \emptyset$ here is harmless, since the case of $I \cup J = \emptyset$ corresponds to $\Omega = \{(0,0)\}$, which is already covered by Corollary 1.

Corollary 3 Assume that $I \cup J \neq \phi$. If β denotes the number $\sum_{I \cup J} a_k$ and μ denotes the optimal value in problem (15), then each accumulation point (x^0, y^0) of the sequence (12) satisfies

$$\beta^\beta \cdot \mu \leqslant \prod_I (x_k^0)^{a_k} \cdot \prod_J (y_k^0)^{a_k} \leqslant \mu$$

The special pair $(\widetilde{x}, \widetilde{y})$ in Ω solving the problem (15) depends only on the 'direction' vector a in Theorem 5 describing the asymptotic direction of approach of $\{z^m\}_{m=1}^\infty$ to 0. It is the strong Pareto optimum element of Ω with respect to the direction a. Figure 2 below depicts a representative situation. The set Q^I is the smallest (closed) face of Q containing X and is given by

$$Q^I = \{x \in \mathbb{R}^n : x \in Q, K(x) \subset I\}$$

Similarly for Q^J and Y. The curves in Figure 2 are intended to indicate level surfaces of the criterion function involved in problem (15).

Before continuing with further properties of the trajectories of approximate solutions, we digress once more to mention a particularly strong result concerning non degeneracy. For this result only, we depart from our usual blanket assumption (*).

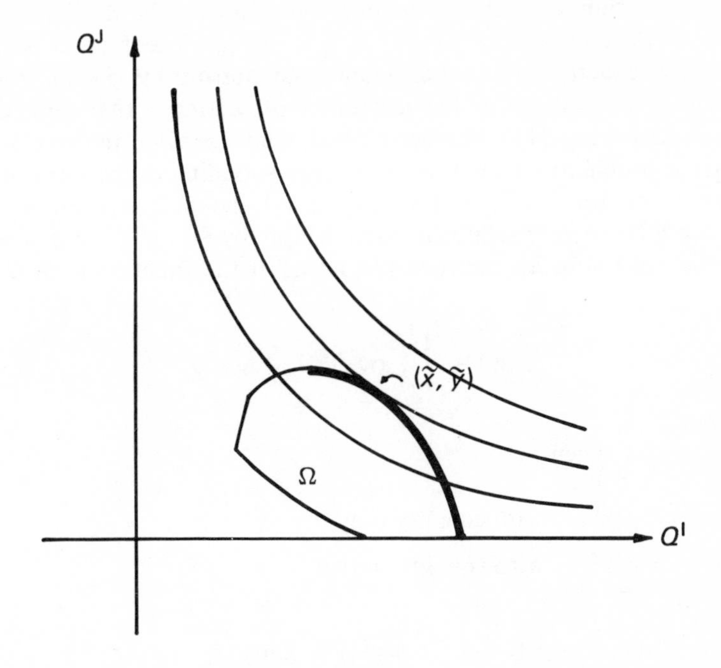

Figure 2 The solution set Ω, with the strong Pareto optimum solution $(\tilde{x},\tilde{y})$. The darkened portion of the boundary of Ω is the Pareto frontier, accessible by varying the direction of approach parameter a.

Theorem 6
Let $T = \partial f$ for any proper polyedral convex function f on $\mathbb{R}^n$. If (C) is feasible, that is, if

$$G(T) \cap (Q \times Q) \neq \emptyset ,$$

then (C) is non degenerate.

The remaining results of this section concern the manner in which the two components of (x^z, y^z) approach the projections X and Y of Ω. Some strong monotonicity properties are revealed, which can serve as precise criteria for gauging improvement monotonically in the overall course of the approximation process (i.e. as z is sent to 0), this despite the fact that the general situation being treated is on the operator-theoretic level with no criterion function evident. We assume the (numerically idealized) situation in which the convergence of $\{z^m\}_{m=1}^{\infty}$ to 0 in Theorem 5 is from the direction a exactly, rather than merely asymptotically (i.e. $a_k^m = a_k$, $\forall m$, $\forall k$). Thus, we assume for the remainder of section 3 that some direction vector a satisfying (11) has been fixed, and consider the one-parameter subclass of problems of the form (C^{ζ}) corresponding in the obvious way to vectors $z = \zeta a$ for $0 \leqslant \zeta < \infty$. For each $\zeta > 0$ the unique solution of (C^{z}) assured by Theorem 2 will be denoted simply by (x^{ζ}, y^{ζ}). The monotonicity properties to follow are expressed in terms of the function ρ on Q defined by

$$\rho(w) = \prod_{k=1}^{n} (w_k)^{a_k} \quad \forall\, w \in Q$$

Theorem 7
Let $0 < \zeta' \leqslant \zeta$. Then

$$0 < \rho(x^{\zeta'}) \leqslant \rho(x^{\zeta}) \quad \text{with equality only if} \quad x^{\zeta'} = x^{\zeta}$$

$$0 < \rho(y^{\zeta'}) \leqslant \rho(y^{\zeta}) \quad \text{with equality only if} \quad y^{\zeta'} = y^{\zeta}$$

and

$$(\rho(x^{\zeta'}), \rho(y^{\zeta'})) = (\rho(x^{\zeta}), \rho(y^{\zeta})) \quad \text{only if} \quad \zeta' = \zeta$$

Corollary 4 Let $\zeta > 0$. Then

$$x^{\zeta} \in X \quad \text{implies} \quad x^{\zeta'} = x^{\zeta} \quad \text{for all} \quad 0 < \zeta' \leqslant \zeta$$

and

$$y^{\zeta} \in Y \quad \text{implies} \quad y^{\zeta'} = y^{\zeta} \quad \text{for all} \quad 0 < \zeta' \leqslant \zeta$$

Corollary 5 For each $0 < \zeta < \infty$, let $\xi(\zeta) = \rho(x^{\zeta})$ and $\eta(\zeta) = \rho(y^{\zeta})$. Then the set

$$\Gamma_a = \{(\xi(\zeta), \eta(\zeta)) : 0 < \zeta < \infty\}$$

is a maximal monotone subset of $(0, \infty) \times (0, \infty)$, and the mapping

$$h_a(\zeta) = (\xi(\zeta), \eta(\zeta))$$

is a homeomorphism between $(0, \infty)$ and Γ_a, with inverse given by

$$h_a^{-1}(\xi, \eta) = \alpha^{-1} \xi\eta, \qquad \alpha = \rho(a)$$

Corollary 5 guarantees that, as $\zeta \downarrow 0$, the trajectory (x^{ζ}, y^{ζ}) improves strictly in either the x or the y component (or both). Figure 3 depicts the typical situation for the set Γ_a. We say typical because, according to the next result, if Γ_a did not approach (0, 0) as $\zeta \downarrow 0$ then problem (*C*) would essentially not be a real complementarity problem.

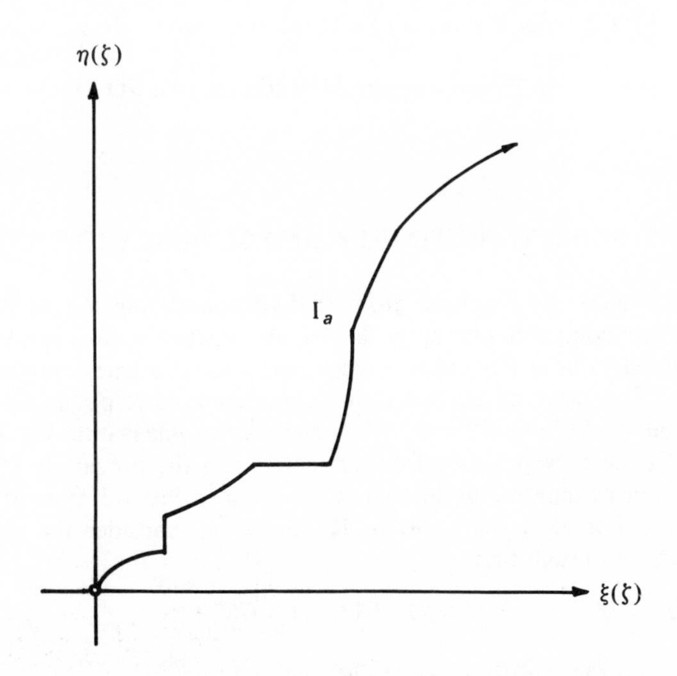

Figure 3 The typical situation for Γ_a.

Theorem 8

One and only one of the following is the case:

(*a*) $\lim_{\zeta \downarrow 0} x^\zeta = \tilde{x} \in P$ and $\lim_{\zeta \downarrow 0} y^\zeta = \tilde{y} = 0$;

(*b*) $\lim_{\zeta \downarrow 0} x^\zeta = \tilde{x} = 0$ and $\lim_{\zeta \downarrow 0} y^\zeta = \tilde{y} \in P$;

(*c*) $\lim_{\zeta \downarrow 0} \rho(x^\zeta) = 0$ and $\lim_{\zeta \downarrow 0} \rho(y^\zeta) = 0$

(Recall that $(\tilde{x}, \tilde{y})$ denotes the unique solution of problem (15).)

Case (*a*) says that a solution to (*C*) for T can be found by solving the 'unconstrained' problem

$$0 \in T(x), \quad x > 0,$$

while case (b) says that a solution to (C) for T can be found by solving the 'unconstrained' problem

$$0 \in T^{-1}(y), \quad y > 0,$$

where T^{-1} is the (maximal monotone) multifunction on $\mathbb{R}^n$ given by

$$G(T^{-1}) = \{(y, x) \in \mathbb{R}^n \times \mathbb{R}^n : (x, y) \in G(T)\}$$

4. STABILITY, CONTINUITY, AND GENERIC PROPERTIES

Now we change the emphasis from that of considering (C) in relation to approximate complementarity problems, and instead focus mainly on questions related to how (C) behaves when perturbed as a true complementarity problem. The class of perturbations considered corresponds to arbitrary translations of $G(T)$ in $\mathbb{R}^n \times \mathbb{R}^n$. However, since some results can be proved with both the new perturbation parameters and the previously studied approximation parameter, we introduce a combined 'hybrid' type of problem, as follows. For each $(u, v) \in \mathbb{R}^n \times \mathbb{R}^n$ and $z \in Q$, consider the problem of finding an (x, y) such that

$$(C^z(u, v)) \qquad (x, y) \in G(T_{u,v}) \cap G(S_z),$$

where $T_{u,v}$ is the multifunction on $\mathbb{R}^n$ defined by

$$G(T_{u,v}) = G(T) - \{(u, v)\},$$

that is

$$T_{u,v}(x) = T(x + u) - \{v\} \qquad \forall\, x \in \mathbb{R}^n$$

For $z = 0$, problem $(C^0(u, v))$ is thus to find an (x, y) such that

$$y \in T(x + u) - \{v\}, \qquad x \geqslant 0, \qquad y \geqslant 0, \qquad \langle x, y \rangle = 0,$$

or equivalently, to find an (x, y) such that

$$y \in T(x), \quad x - u \geqslant 0, \quad y - v \geqslant 0, \quad \langle x - u, y - v \rangle = 0$$

Of course, problem $(C^0(0, 0))$ is exactly the original problem (C). Dependence on the parameters (u, v) and z will be indicated in an obvious fashion. When the parameter z is absent, it is to be understood that $z = 0$. This situation, as just seen, corresponds to a 'true' complementarity problem, as opposed to approximate.

We are interested primarily in those perturbation parameters (u, v) chosen from the set

$$W = G(T) - (P \times P)$$

Note that this open set contains $(0, 0)$, by virtue of the blanket assumption $(*)$.

Theorem 9

For each $z \in P$ and $(u, v) \in W$ problem $(C^z(u,v))$ has a unique solution, denoted by $(x^z(u, v), y^z(u, v))$, and the function

$$(z, u, v) \to (x^z(u, v), y^z(u, v))$$

is continuous from $P \times W$ into $P \times P$

Theorem 10

(i) For each $(u, v) \in \mathbb{R}^n \times \mathbb{R}^n$ one has the estimate

$$\langle x - x^0, v\rangle + \langle u, v\rangle + \langle u, y - y^0\rangle \geqq \langle x, y^0\rangle + \langle x^0, y\rangle \geqq 0$$

whenever $(x, y) \in \Omega(u, v)$ and $(x^0, y^0) \in \Omega$.

(ii) The sets $\Omega(u, v)$ are uniformly bounded for all (u, v) in some neighbourhood of $(0, 0)$.

(iii) For each $(u, v) \in W$ the set $\Omega(u, v)$ is nonempty, compact and convex, and the multifunction

$$(u, v) \to \Omega(u, v)$$

is upper semicontinuous from W into $Q \times Q$.

(iv) For almost all $(u, v) \in W$ in the sense of Lebesgue measure, the sets $\Omega(u, v)$ are singletons. Restricted to such (u, v)'s, the multifunction in Theorem 10 (iii) is thus a continuous function.

(v) If Ω is a singleton, then $K(x) \supset I$ and $K(y) \supset J$ whenever $(x, y) \in \Omega(u, v)$, for all (u, v) ranging over some neighbourhood of $(0, 0)$. Thus, if Ω is nondegenerate and a singleton, then $\Omega(u, v)$ is also nondegenerate, and has the same support as Ω, for all (u, v) in some neighbourhood of $(0, 0)$.

We conclude by giving variants of parts (iv) and (v) of Theorem 10. We use the notation

$$U = \{u : (u, 0) \in W\}, \qquad Y(u) = \{y : \exists\, x, (x, y) \in \Omega(u, 0)\},$$
$$V = \{v : (0, v) \in W\}, \qquad X(v) = \{x : \exists\, y, (x, y) \in \Omega(0, v)\}$$

Theorem 11

(i) For almost all $v \in V$ the sets $X(v)$ are singletons. For almost all $u \in U$ the sets $Y(u)$ are singletons.

(ii) If X is a singleton, then $K(x) \supset I$ whenever $x \in X(v)$, for all v ranging over some neighbourhood of 0. If Y is a singleton, then $K(y) \supset J$ whenever $y \in Y(u)$, for all u ranging over some neighbourhood of 0. (The sets X, Y, I, J here are as defined preceding Theorem 5.)

REFERENCES

[1] R. T. Rockafellar. 'A general correspondence between dual minimax problems and convex programs'. *Pacific J. Math.*, 25 (1968), 597-611.

[2] R. T. Rockafellar. '*Convex Analysis*'. Princeton University Press, Princeton, N. J. (1970).

[3] R. T. Rockafellar. 'Monotone operators and augmented Lagrangian methods in non-linear programming'. In *Nonlinear Programming*, vol. 3, eds. O.L. Mangasarian, R.R. Meyer and S.M. Robinson, Academic Press, New York (1978), 1-25.

[4] R. T. Rockafellar. 'Lagrange multipliers and variational inequalities'. This volume, chap. XX.

Chapter 18

On Some Non-Linear Quasi-Variational Inequalities and Implicit Complementarity Problems in Stochastic Control Theory

U. Mosco

1. INTRODUCTION

In their dynamic programming approach to the problems of stochastic impulse and continuous optimal control, Bensoussan and Lions have introduced a class of elliptic and parabolic non-linear *quasi-variational inequalities* (in short, QVI).

In the stationary case, these QVI involve an 'implicit' obstacle in the state variables $x \in \sigma$, of type

$$M(u)(x) = 1 + \inf_{\substack{\xi \geq 0 \\ x+\xi \in \bar{\sigma}}} u(x+\xi)$$

σ being some open subset of $\mathbb{R}^N$, and a second-order p.d.o. (partial differential operator) of the form

$$A(u) = Lu + G(u)$$

where Lu is a second-order linear elliptic operator and $G(u)$ is a first-order non-linear p.d.o. of the form

$$G(u) = -\min_{d \in U} [g_0(x, d) + Du \cdot g_1(x, d)]$$

Here $U \subset \mathbb{R}^Q$, $Q \geq 1$, is the set of *admissible controls* and $g_0 : \sigma \times U \to \mathbb{R}$, $g_1 : \sigma \times U \to \mathbb{R}^N$ are given, and $Du = (\partial u/\partial x_1, \ldots, \partial u/\partial x_N)$.

We refer to a forthcoming book [3] for a complete illustration of the theory that has been developed so far by relying on both analytic and probabilistic methods [(1)].

(1) For some recent theoretical contributions, we also refer to the forthcoming *Proceedings of the Colloquium on 'Recent methods of nonlinear analysis and applications'*, Rome, May 1978.

The specific problem we shall discuss in the present paper is that of the existence of *strong solutions* to the above-mentioned QVI's.

We shall only consider the *stationary case* and *Dirichlet boundary conditions*. By strong solution we mean a solution u of the QVI which belongs to the Sobolev space $W^{2,q}(\sigma)$ for some (or all) $q \geq 2$.

We shall also confine ourselves to the *non-linear monotone* case, as described in section 1 below.

In [6] this problem was first dealt with in the *linear* case of pure impulse control (i.e. $G \equiv 0$), by assuming all coefficients of the operator L to be constant in σ. The main tool used there were the so-called Lewy-Stampacchia dual estimates for weak solutions of variational inequalities [9, 13, 14].

In [11, 12] this method was adapted to the non-linear monotone case, by only assuming that the *higher-order* coefficients a_{ij} of the operator L are constant in σ. By relying on non-linear Lewy-Stampacchia estimates, [10] a constructive proof of the existence of a strong solution can indeed be given. These results will be summarized in section 4 below.

More recently, in dealing with the linear case and variable a_{ij}, Friedman and Caffarelli [4] observed that the implicit obstacle $M(u)$ is less irregular than expected, due to the fact that it inherits some regularity from the smoothness properties that the weak solution u enjoys in the continuation set $\{u < M(u)\}$. Then [4] they use this property of $M(u)$, together with some local regularity results for solutions of variational inequalities, to prove that the solution u of the linear QVI belongs *locally* to the spaces $W^{2,q}$. Their proof uses a local representation of $M(u)$ which seems, however, to require some further refinements.

In section 2 and 3 of the present paper, by exploiting the basic remark about $M(u)$ just mentioned, we show how to apply one dual estimate method in order to prove a *global* regularity result for the non-linear QVI and variable a_{ij}.

2. STRONG REGULARITY OF WEAK SOLUTIONS

Let σ be a bounded open subset of $\mathbb{R}^N$, $N \geq 1$. We shall denote the boundary of σ by Γ and we shall assume that Γ is sufficiently smooth; for example, that Γ is of class C^2.

The QVI we consider in this section can be formulated as follows:

$$(1) \quad \begin{cases} u \in H_0^1(\sigma) \cap L^\infty(\sigma) \\ u \leq M(u) \quad \text{a.e. (almost everywhere) in } \sigma \\ a(u, u-v) + (G(u), u-v) \leq (f, u-v) \\ \forall v \in H_0^1(\sigma) \text{ such that } v \leq M(u) \quad \text{a.e. in } \sigma \end{cases}$$

The form $a(u, v)$ is given by

(2) $$a(u,v)=\int_\sigma \left(\sum_{i,j=1}^{N} a_{ij} \frac{\partial u}{\partial x_i} \frac{\partial v}{\partial x_j} + \sum_{j=1}^{N} a_j \frac{\partial u}{\partial x_j} v + a_0 uv \right) \, dx$$

where the coefficients are supposed to satisfy the following conditions

(3) $$a_{ij}, a_j, a_0 \in L^\infty(\sigma) \qquad a_0 \geqslant c > 0, i, j = 1, \dots, N$$

(4) $$\sum_{i,j=1}^{N} a_{ij}\, \xi_i\, \xi_j \geqslant \beta \sum_{i=1}^{N} \xi_i^2 \qquad \text{a.e. in } \sigma, \quad \forall \xi \in \mathbb{R}^N$$

for some constants $c > 0$ and $\beta > 0$.

The operator $G(u)$ is a first-order non-linear p.d.o. of the form

(5) $$G(u) = -H(x, Du)$$

where $Du = (\partial u/\partial x_1, \dots, \partial u/\partial x_N)$ and $H(x, p)$ is a measurable function of its numerical arguments $x \in \sigma$ and $p \in \mathbb{R}^N$ that satisfies the following conditions

(6) $$|H(x,p)| \leqslant h(x) + c_0 |p| \qquad x \in \sigma, \; p \in \mathbb{R}^N$$

(7) $$|H(x,p') - H(x,p'')| \leqslant c_0 \, |p' - p''| \qquad x \in \sigma, \; p', p'' \in \mathbb{R}^N$$

for some constant $c_0 \geqslant 0$ and some function $h \in L^\infty(\sigma)$.

The function f is also given and satisfies

(8) $$f \in L^\infty(\sigma)$$

By $(\cdot\,,\;\cdot)$ we denote the inner product in $L^2(\sigma)$.

The implicit obstacle $M(u)$ appearing in (1) is of the form

(9) $$M(u)(x) = k + \inf_{\substack{\xi \geqslant 0 \\ x + \xi \in \overline{\sigma}}} [c(\xi) + u(x + \xi)]$$

where

(10) $$k > 0$$

is some given constant and the function $c(\xi)$ is a non-negative continuous function on $\mathbb{R}_+^N$ which is assumed to be subadditive and non-decreasing in ξ.

The space $H_0^1(\sigma)$ is the usual Sobolev space of all square-integrable functions on σ whose distribution first-order derivatives also belong to $L^2(\sigma)$ and whose trace on Γ vanishes.

The main problem we are concerned with in this section consists in finding a solution u of problem (1) that satisfies, in addition, the following regularity conditions

$$u \in C(\overline{\sigma}) \tag{11}$$

$$Lu \in L^2(\sigma) \tag{12}$$

where Lu is the operator associated with the form $a(u, v)$, namely

$$Lu = -\sum_{i,j=1}^{N} \frac{\partial}{\partial x_i}\left(a_{ij}\frac{\partial u}{\partial x_j}\right) + \sum_{j=1}^{N} a_j \frac{\partial u}{\partial x_j} + a_0 u \tag{13}$$

Let us first remark that the *existence* of a solution u of problem (1) can be shown by applying to the case at hand the monotonicity methods introduced by Bensoussan and Lions, provided the following assumptions are satisfied in addition to those listed above:

$$\inf_{\sigma} f - \sup_{\sigma} h \geqslant -K \inf_{\sigma} a_0 \tag{14}$$

$$\inf_{\sigma} a_0 \geqslant (2\beta)^{-1}\left(\sum_{j=1}^{N} \|a_j\|_\infty^2 + 2c_0^2\right) \tag{15}$$

where $\|\cdot\|_\infty = \|\cdot\|_{L^\infty(\sigma)}$ and k, C_0, β are the constants appearing in (9), (6), (4), respectively.

The *continuity* of u up to the boundary Γ, that is, property (11), can then be proved either by probabilistic methods as in [1] or by applying a recent analytic result [5]. In order to apply the last result we need to assume, in addition, that $H(x, p)$ is concave in p for every x and that the strict inequality holds in (14).

Let us also note that under these additional hypotheses the (continuous) solution u of (1) is *unique,* as follows by adopting the Laetsch's argument to the case at hand [8, 5]. For further comments on the hypotheses (14) and (15) we refer the reader to [12].

In order to show that the solution u has the stronger regularity (12), we need some further smoothness of the *data.* We then assume that the following hypotheses are satisfied

(16) Γ is of class $C^{2,\alpha}$

(17) $a_{ij} \in C^{1,\alpha}(\overline{\sigma}) \quad a_j, a_0 \in C^{0,\alpha}(\overline{\sigma}) \quad i,j = 1, \ldots, N$

(18) $f \in C^{0,\alpha}(\overline{\sigma})$

(19) $H(\cdot, p) \in C^{0,\alpha}(\overline{\sigma})$ uniformly in p

for some $\alpha \in]0, 1[$.

Moreover, we introduce the function

(20) $$\gamma(x) = \inf_{\substack{\xi \geq 0 \\ x+\xi \in \Gamma}} c(\xi) \qquad x \in \overline{\sigma}$$

and we assume that

(21) $$\gamma \in C^2(\sigma)$$

Let us remark, however, that condition (21), which is essentially a smoothness assumption on Γ, could be weakened to require that γ belongs to the Sobolev space $H^1(\sigma)$ and that $L(\gamma)$ is bounded from below by some constant, as a distribution in σ.

We now prove the following theorem.

Theorem 1

Under all above assumptions, the (unique) continuous solution u of problem (1) satisfies the additional regularity condition

(22) $$u \in W^{2,p}(\sigma) \qquad \text{for all } p \geq 2$$

and, moreover,

(23) $$Lu \in L^\infty(\sigma)$$

Corollary We have, in particular, $u \in C^{1,\alpha}(\overline{\sigma})$ for all $\alpha \in]0, 1[$.

The proof of Theorem 1 is an immediate consequence of the usual Lewy-Stampacchia dual estimates for solutions of variational inequalities, once the following proposition has been established.

Proposition 1

Let M be the operator defined by (9). Let u be any function that satisfies the following condition:

$$u \in H_0^1(\sigma) \cap C(\sigma) \cap W^{2,\infty}(D) \tag{24}$$

for every open subset D of σ whose closure $\overline{D}$ is contained in the set

$$C = \{x \in \overline{\sigma} : u(x) < M(u)(x)\} \tag{25}$$

Then, $M(u)$ has the following properties

$$M(u) \in \text{Lip}\,(\overline{\sigma}) \tag{26}$$

$$A(M(u)) \geqslant -c \tag{27}$$

in σ in the distribution sense, where c is some positive constant.

We denote by A the operator

$$A(u) = Lu + G(u)$$

which is defined on the space $H^1(\sigma)$ to the dual $H^{-1}(\sigma)$ of $H_0^1(\sigma)$, by the identity

$$\langle A(u), v\rangle = a(u, v) + (G(u), v)$$

$u \in H^1(\sigma)$, $v \in H_0^1(\sigma)$. Let us remark that, as a consequence of our previous assumption – in particular, of assumption (15) – A is a strictly monotone coercive operator.

Proof of Theorem 1: Let us first remark that u is a generalized solution, in the sense of [7], of the equation

$$A(u) = f$$

in the open set $C \cap \sigma$ and, moreover, $u = 0$ on $C \cap \Gamma$. Therefore, it follows from the smoothness assumptions verified by the *data*, that $u \in W^{2,\infty}(D)$ for every D as in Proposition 1 (see theorem 12.1 in [7]).

Therefore, by Proposition 1, the obstacle $M(u)$ in (1) satisfies properties (26) and (27). This enables us to apply the non-linear Lewy-Stampacchia estimate, as given in [10], theorem 4.1, and we obtain

$$f \geqslant A(u) \geqslant f \wedge A(M(u)) \tag{28}$$

in the distribution sense in σ, where

$$f \wedge A(M(u)) = f - (f - A(M(u)))^{+}$$

By taking (27) again into account, it follows from (28) that

$$A(u) \in L^{\infty}(\sigma)$$

and hence Theorem 1 has been proved.

3. PROOF OF PROPOSITION 1

For every $x \in \overline{\sigma}$ we denote by $\Sigma(x)$ the set of all $y \in \overline{\sigma}$ of the form $y = x + \eta$, $\eta \geqslant 0$, such that

$$c(\eta) + u(x + \eta) = \min_{\substack{\xi \geqslant 0 \\ x+\xi \in \overline{\sigma}}} [c(\xi) + u(x + \xi)] \tag{29}$$

Lemma 1 (See [4]). For every $x \in \overline{\sigma}$, we have

$$\Sigma(x) \subset C \tag{30}$$

where C is the set (25).

Proof: Let $y \in \Sigma(x)$. Then

$$k + c(\eta) + u(y) = M(u)(x) \leqslant c(\eta) + M(u)(y) \tag{31}$$

the last inequality being a consequence of the subadditivity of $c(\cdot)$. Since $k > 0$, (31) implies $u(y) < M(u)(y)$, i.e. $y \in C$. This completes the proof.

For every $x \in \mathbb{R}^N$ and $\rho > 0$, we denote by $B_\rho(x)$ the open ball of radius ρ centred at x and we also write $B_\rho(x) = x + I_\rho$, where $I_\rho = \{y \in \mathbb{R}^N : |y| < \rho\}$.

We denote by $\tilde{u}$ the function defined for every $x \in \mathbb{R}^N$ by setting $\tilde{u}(x) = u(x)$ if $x \in \overline{\sigma}$ and $\tilde{u}(x) = 0$ for all $x \in \mathbb{R}^N - \overline{\sigma}$.

Lemma 2 For every $x_0 \in \overline{\sigma}$, there exist a ball $B_{\rho_0}(x_0)$, $\rho_0 > 0$, a subset $T_{x_0} \subset \mathbb{R}^N_+$ and an open subset $V_{x_0} \subset \mathbb{R}^N$, such that:

$$M(u)(x) = k + \inf_{\xi \in T_{x_0}} [c(\xi) + \tilde{u}(x + \xi)] \quad \text{for all } x \in B_{\rho_0}(x_0) \cap \sigma \tag{32}$$

$$B_{\rho_0}(x_0) + T_{x_0} \subset V_{x_0} \tag{33}$$

$$\overline{V_{x_0} \cap \sigma} \subset C \tag{34}$$

where C is the set (25).

Proof: Let us set, for every x

$$\Pi_x = \Sigma(x) - x \tag{35}$$

By Lemma 1, $\Sigma(x_0)$ is a compact subset of C. We can then choose $\delta_0 > 0$ so that the open subset of $\mathbb{R}^N$

$$V_{x_0} = \Sigma(x_0) + I_{\delta_0} \tag{36}$$

satisfies condition (34). Let now δ_1 be such that $0 < \delta_1 < \delta_0$. As it is easily verified, the multivalued mapping $x \to \Sigma(x)$ is upper semicontinuous. Therefore, there exists $\rho_1 > 0$ such that

$$\Sigma(x) \subset \Sigma(x_0) + I_{\delta_1} \tag{37}$$

for all $x \in B_{\rho_1}(x_0) \cap \sigma$ hence also *a fortiori* for all $x \in B_{\rho_0}(x_0) \cap \sigma$, where $\rho_0 > 0$ has been chosen so as to satisfy $0 < \rho_0 < \rho_1$, and

$$2\rho_0 + \delta_1 < \delta_0 \tag{38}$$

It follows from (37) that

$$\Pi_x \subset \Pi_{x_0} + I_{\rho_0 + \delta_1}$$

Hence if we define

$$T_{x_0} = \bigcup_{x \in B_{\rho_0}(x_0) \cap \sigma} \Pi_x \tag{39}$$

we have

$$T_{x_0} \subset \Pi_{x_0} + I_{\rho_0 + \delta_1}$$

This implies

$$B_{\rho_0}(x_0) + T_{x_0} \subset x_0 + \Pi_{x_0} + I_{2\rho_0 + \delta_1} \subset \Sigma(x_0) + I_{\delta_0}$$

the last inclusion being a consequence of (38). Thus (33) is satisfied.

Finally, (32) is an immediate consequence of (39), since $\Pi_x \subset T_{x_0}$ for every $x \in B_{\rho_0}(x_0) \cap \sigma$ and $c(\cdot)$ is non-decreasing. This completes the proof.

Lemma 3 We have $M(u) \in \mathrm{Lip}(\overline{\sigma})$.

Proof: It suffices to prove that for every $x_0 \in \sigma$ there exists $\rho_0 > 0$ such that

(40) $$M(u) \in \mathrm{Lip}(\overline{B_{\rho_0}(x_0) \cap \sigma})$$

We choose $B_{\rho_0}(x_0)$, T_{x_0} and V_{x_0} as in Lemma 2. Since u, as a consequence of our assumption (24), satisfies a Lipschitz condition on

(41) $$D = \overline{V_{x_0} \cap \sigma}$$

we have that the functions

(42) $$v_\xi(x) = c(\xi) + \tilde{u}(x + \xi) \qquad \xi \in T_{x_0}$$

in consequence of (33), (34), are *equilipschitzian* on $\overline{B_{\rho_0}(x_0) \cap \sigma}$.
Thus (40) is an immediate consequence of the representation (32) of $M(u)$. This completes the proof.

It follows, in particular, from Lemma 2, that $M(u) \in H^1(\sigma)$. Therefore, $A(M(u))$ is well defined as a distribution in σ and it is an element of the dual $H^{-1}(\sigma)$ of $H_0^1(\sigma)$.

Lemma 4 Let $(v_\alpha)_\alpha$ be a family of functions on σ that satisfy the following conditions

(43) $$v_\alpha \in H^1(\sigma) \qquad \text{for every } \alpha$$

(44) $$\bigwedge_\alpha v_\alpha \in H^1(\sigma)$$

(45) $$A(v_\alpha) \geqslant g \qquad \text{for every } \alpha \text{ as a distribution in } \sigma$$

where $g \in L^2(\sigma)$ is some given function. Then,

(46) $$A(\bigwedge_\alpha v_\alpha) \geqslant g \qquad \text{as distribution in } \sigma$$

Proof: The proof can be given along the lines of the proof of theorem 3.2 in [10].

We are now in a position to complete the proof of Proposition 1 by proving the following lemma.

Lemma 5 We have $A(M(u)) \geqslant -c$ in the distribution sense in σ, for some constant $c > 0$.

Proof: Again it suffices to prove that for every $x_0 \in \overline{\sigma}$ there exists $\rho_0 > 0$, such that

(47) $$A(M(u)) \geqslant -c_0 \qquad \text{as distribution in } B_{\rho_0}(x_0) \cap \sigma$$

where $c_0 > 0$ is some constant, possibly depending on x_0.

As in the proof of Lemma 3, we choose $B_{\rho_0}(x_0)$, T_{x_0} and V_{x_0} to satisfy all properties (32), (33), (34).

Let us remark that, as a consequence of (32), we can also write

$$M(u)(x) = k + \inf_{\xi \in T_{x_0}} v_\xi \wedge \gamma(x) \tag{48}$$

for all $x \in B_{\rho_0}(x_0) \cap \sigma$, where $\gamma(x)$ is the function given by (20) and v_ξ is given by (42).

We now define, for every $\xi \in T_{x_0}$ and every $\varepsilon > 0$,

$$v_{\xi,\varepsilon}(x) = v_\xi(x) \wedge [\gamma(x) - \varepsilon] \tag{49}$$

as a function on $B_{\rho_0}(x_0) \cap \sigma$.

For every fixed $\varepsilon > 0$, we have

$$B_{\rho_0}(x_0) \cap \sigma = \sigma_1^{\xi,\varepsilon} \cup \sigma_2^{\xi,\varepsilon} \tag{50}$$

where the open subsets $\sigma_1^{\xi,\varepsilon}$, $\sigma_2^{\xi,\varepsilon}$ are defined as follows

$$\sigma_1^{\xi,\varepsilon} = \{x \in B_{\rho_0}(x_0) \cap \sigma : v_\xi(x) < \gamma(x) - \tfrac{1}{2}\varepsilon\}$$

$$\sigma_2^{\xi,\varepsilon} = \{x \in B_{\rho_0}(x_0) \cap \sigma : v_\xi(x) > \gamma(x) - \varepsilon\}$$

Let us estimate the distribution $A(w_{\xi,\varepsilon})$ on $\sigma_1^{\xi,\varepsilon}$. We have

$$\sigma_1^{\xi,\varepsilon} + \xi \subset V_{x_0} \cap \sigma$$

hence, by (24), all functions v_ξ, together with their first-order and second-order partial derivatives, are *bounded* on $\sigma_1^{\xi,\varepsilon}$ by constants that do not depend on $\xi \in T_{x_0}$ and on $\varepsilon > 0$. Therefore, $A(v_\xi)$ is bounded on $\sigma_1^{\xi,\varepsilon}$ by a constant that does not depend on ξ and ε. By applying Lemma 4 to the open set $\sigma_1^{\xi,\varepsilon}$ and the two functions $v_\xi(x)$ and $\gamma(x) - \varepsilon$ considered on $\sigma_1^{\xi,\varepsilon}$, we then find, by taking (21) into account, that

$$A(w_{\xi,\varepsilon}) \geqslant (-c) \wedge [A(\gamma) - a_0\varepsilon] \geqslant (-c) \wedge (-c - a_0\varepsilon) \tag{51}$$

as distribution in $\sigma_1^{\xi,\varepsilon}$, for some constants $c > 0$ independent of $\xi \in T_{x_0}$ and $\varepsilon > 0$.

We now estimate the distribution $A(w_{\xi,\varepsilon})$ on $\sigma_2^{\xi,\varepsilon}$. On this set we clearly have

$$w_{\xi,\varepsilon}(x) \equiv \gamma(x) - \varepsilon$$

therefore

(52) $$A(w_{\xi,\varepsilon}) \geqslant A(\gamma) - a_0\,\varepsilon \geqslant -c - a_0\,\varepsilon$$

as distribution in $\sigma_2^{\xi,\varepsilon}$, with c independent of $\xi \in T_{x_0}$ and ε.

It follows from the joint estimates (51) and (52), by taking (50) into account, that

(53) $$A(w_{\xi,\,\varepsilon}) \geqslant -c_0 - a_0\,\varepsilon \qquad \text{on } B_{\rho_0}(x_0) \cap \sigma$$

for c_0 independent of $\xi \in T_{x_0}$ and $\varepsilon \gtrsim 0$.

By letting $\varepsilon \to 0$, we obtain from (53)

(54) $$A[v_\xi \wedge \gamma] \geqslant -c_0 \qquad \text{on } B_{\rho_0}(x_0) \cap \sigma$$

with c_0 independent of $\xi \in T_{x_0}$.

The estimate (47) is then a consequence of Lemma 4, applied to the open set $B_{\rho_0}(x_0) \cap \sigma$ and to the family of functions

$$v_\xi \wedge \gamma \qquad \xi \in T_{x_0}$$

$M(w)$ being given by (48). This completes the proof of proposition 1.

4. ITERATIVE ALGORITHMS

In this section we make the assumption that

(55) the coefficients $a_{ij}, i, j = 1, \ldots, N$, are constant in σ

On the other hand, however, we no more require that the smoothness assumptions (16), . . . , (19) should be satisfied.

We now consider any function u^0 that verifies the conditions

(56) $$u^0 \in H^1(\sigma) \cap L^\infty(\sigma) \qquad Lu^0 \in L^\infty(\sigma)$$

(57) $$u^0 \geqslant -k - \inf_{\substack{\xi > 0 \\ x+\xi \in \bar{\sigma}}} c(\xi)$$

and we define the function u^n, $n = 1, 2, \ldots$, iteratively as the solution of the following *complementarity problem*

(58)
$$\begin{aligned}
&u^n \in H_0^1(\sigma) \cap W^{2,p}(\sigma) \qquad Lu^n \in L^\infty(\sigma)\\
&u^n(x) \leqslant M(u^{n-1})(x) \qquad \forall x \in \overline{\sigma}\\
&Lu^n + G(u^n) \leqslant f \quad \text{a.e. in } \sigma\\
&(u^n - M(u^{n-1}))\,(Lu^n + G(u^n) - f) = 0 \qquad \text{a.e. in } \sigma
\end{aligned}$$

It can be shown that the sequence $(u^n)_{n=1,2}$ converges to the solution u of the implicit complementarity problem

(59)
$$\begin{aligned}
&u \in H_0^1(\sigma) \cap W^{2,p}(\sigma) \qquad Lu \in L^\infty(\sigma)\\
&u(x) \leqslant M(u)(x) \qquad \forall x \in \overline{\sigma}\\
&Lu + G(u) \leqslant f \quad \text{a.e. in } \sigma\\
&(u - M(u))(Lu + G(u) - g) = 0 \qquad \text{a.e. in } \sigma
\end{aligned}$$

which can be seen to be a strong formulation of the quasi-variational inequality considered in section 2. Let us also note that u^n converges to u, in particular, in the topology of the space $C^{1,\alpha}$ for all $\alpha \in]0, 1[$.

More precisely, the following theorem holds.

Theorem 2

(See [11, 12].) The solution u of problem (59) exists and is unique. The solution u^n of problem (58), defined as $n = 1, 2, \ldots,$ exists and is unique for every n. The function u^n converges to u as $n \to \infty$ in the weak topology of the space $W^{2,p}(\sigma)$ for every $p \geqslant 2$, the functions Lu^n being uniformly bounded in σ.

Moreover, if $u^1 \leqslant u^0$ (resp., $u^1 \geqslant u^0$), then the whole sequence $(u^n)_{n=1,2\ldots}$ is non-increasing (resp., non-decreasing).

For example, if we take

$$u^0 = \overline{u}$$

where $\overline{u}$ is the solution of the Dirichlet problem

$$\overline{u} \in H_0^1(\sigma) \qquad A(\overline{u}) = f \quad \text{in } \sigma$$

then u^n converges to u monotonically from above.

On the other hand, if we take for example

$$u^0 = -k - \inf_{\substack{\xi > 0\\ x+\xi \in \overline{\sigma}}} c(\xi)$$

then u^n converges to u monotonically from below.

Further, we assume that the strict inequality holds in (14).

The proof of Theorem 2 is based on the use of *non-linear* dual estimate for solutions of variational inequalities [10] in order to prove that $Lu^n \in L^\infty(\sigma)$ for every n and then on the use of suitable interpolation estimates and *linear* dual estimates in order to prove that the functions Lu^n are bounded *uniformly* in σ.

We refer to the above-mentioned references [11] and [12] for further details.

REFERENCES

[1] A. Bensoussan, and J. L. Lions. 'Contrôle impulsionnel et contrôle continu . Méthode des I. Q. V. non linéaires'. *C. R. Acad. Sci., Paris,* A **278** (1974).

[2] A. Bensoussan, and J. L. Lions. 'Optimal impulse and continuous control: Method of non linear quasi-variational inequalities'. *Trudy matématičeskovo Institut Imeni Steklova.* To appear.

[3] A. Bensoussan, and J. L. Lions. *'Temps optimal, contrôle impusionnel et Applications economiques'.* To appear.

[4] A. Friedman, and L. Caffarelli. 'Regularity of the solution of the quasi-variational inequality for the impulse control problem'. *Comm. on P D F* 3(8), 1978.

[5] B. Hanouzet, and J. L. Joly. 'Convergence uniforme des itérés définissant la solution d'une inéquation quasi variationnelle abstraite'. *C. R. Acad. Sci., Paris,* A **286** (1978).

[6] J. L. Joly, U. Mosco, and G. Troianiello. 'On the regular solution of a quasi variational inequality connected to a problem of stochastic impulse control'. *J. Math. Anal. & Appl.,* **61** (1977).

[7] O.A. Ladyzhênskaja, and N.N. Ural'tseva. 'Linear and quasi linear elliptic Equations'. Academic Press, New York (1976).

[8] T. Laetsch. 'A uniqueness theorem for elliptic quasi-variational inequalities'. *J. Funct. Anal.,* **18** (1975).

[9] H. Lewy, and G. Stampacchia . 'On the smoothness of superharmonics which solve a minimum problem'. *J. Anal. Math.,* **23** (1970).

[10] U. Mosco. 'Implicit variational problems and quasi-variational inequalities'. In *Non-linear Operators and Calculus of Variations,* Brux., ed. Waelbrocck, *Lecture Notes in Mathematics,* No. 543, Springer Berlin (1975).

[11] U. Mosco. 'Regularité forte de la fonction d'Hamilton-Jacobi du contrôle stochastique impulsionnel et continu'. *C. R. Acad. Sci., Paris,* **286** (1978).

[12] U. Mosco. 'Non linear quasi-variational inequalities and stochastic impulse control theory'. *Lect. Equadiff IV Praha,* 1977. (To appear in *Springer Notes.*)

[13] U. Mosco, and G. Troianello.'On the smoothness of solutions of Unilateral Dirichlet Problems'. *Boll. Unione Mat. Ital.,* **8** (1973).

[14] G. M. Troianello. 'On the regularity of solutions of unilateral variational problems'. *Rend. Accad. Sc. Fis. Mat., Napoli,* **41** (1975).

Chapter 19

On Two Variational Inequalities Arising from a Periodic Viscoelastic Unilateral Problem

M. Raous

1. INTRODUCTION

We study the periodic problem connected with the behaviour of a viscoelastic cracked solid submitted to periodic loads. The unilateral conditions on the edges of the crack lead to a complementarity problem coupled with a differential equation for time. We solve the problem in the case of Maxwell behaviour equations. In the linear case, the Maxwell law is among the most difficult because of the presence of a secular term of displacements and strains.

We choose a model where coefficients are functions of time (often, via the temperature, but sometimes also directly (aging concrete [1], for example), and of space (non-homogeneous material). We show how this problem leads to two coupled variational inequalities: the first one is a stationary variational inequality for the secular term; the second one is a variational inequality coupled with a differential equation for the periodic parts of the solutions.

Here, we solve these two problems with suitable numerical methods: a finite element method for space discretization, a Runge-Kutta method on time, an overrelaxation method with projection to solve the inequalities. A convenient combination of these discritization methods is realized and a particular solving method proposed. The associated algorithms are given and applied to an example: we give numerical results and discuss them from a mechanical point of view.

2. THE MECHANICAL PROBLEM

This work takes place in the study of the adaptation of a cracked solid under cyclic thermal fatigue: we try to show from a theoretical point of view that a thermic crack is not brittle. It will be possible with a suitable choice of damage law (generally empirical) to calculate, from the stress solution of the problem with a given length of crack, the progression velocity of the crack.

Here, we calculate the stress and displacement fields for the periodic problem with a given crack configuration. The study we present here is related to the oligocyclic fatigue in cracked turbine blades. The period T will be, for example, the period of the plane flight.

Since the hypothesis of contact without friction is unrealistic in physics, the crack is supposed to lie in a plane of symmetry, which allows us to formulate conditions of contact mathematically identical to those of a contact without friction. The periodic problem leads to a weakly coupled system of variational inequalities equivalent to two complementary systems.

Let us consider a plane medium for wich the classical hypotheses of *infinitesimal displacement* and *plane stresses* are assumed. For these reasons,

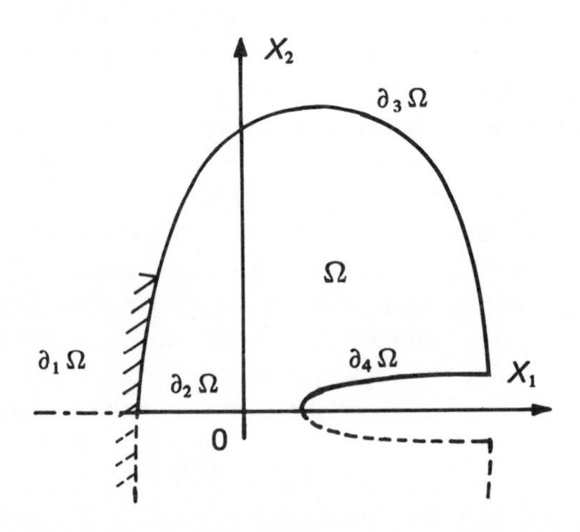

we suppose the problem to be symmetrical with respect to the axis $0x_1$, so that we need only consider the part located in the positive x_2 half-plane. We denote by Ω the open set limiting the medium, $\partial\Omega$ being its boundary.

All the fields considered, given or unknown, will be functions of time t and of position $x = (x_1, x_2)$: for example, the condition

$$\forall\ (t, x) \in [0, T] \times \partial_1 \Omega \qquad u(t, x) = u^0 (t, x)$$

will be written $u = u^0$ on $\partial_1 \Omega$.

The unknown fields of displacements, stresses and strains will be respectively denoted u, s, and e.

A prescribed strain field e^0 is given on Ω. Here, it is a dilation field:

$$e^0 = -\chi\theta \tag{1}$$

where θ is a temperature field and χ a field of thermal dilation symmetrical tensor. We set

$$e = Du + e^0 \tag{2}$$

that means

$$e_{ij} = \frac{1}{2}\left(\frac{\partial u_i}{\partial x_j} + \frac{\partial u_j}{\partial x_i}\right) - \chi_{ij}\theta$$

with tensor notations.

As for now, we shall write the constitutive law under the symbolic form

$$s = \mathscr{K}(e) \tag{3}$$

which represents a functional relation between the stress and strain fields will be specified in section 3.

Equilibrium equations are written

$$s_{ij,j} = -\phi_{1i} \tag{4}$$

where ϕ_1 denotes a given field of forces per unit area on $[0, T] \times \Omega$.

The boundary is divided into four disjoint open arcs. We define

$$u = 0 \qquad \text{on } \partial_1\Omega \tag{5}$$

The symmetry hypothesis is written in the form

$$u_2 = 0 \qquad s_{12} = 0 \qquad \text{on } \partial_2\Omega \tag{6}$$

On $\partial_3\Omega$, we give a field ϕ_2 of forces per unit length:

$$s_{ij}n_j = \phi_{2i} \tag{7}$$

The arc $\partial_4\Omega$ is the upper edge of the crack: it is submitted to contact forces from the lower edge (they are vertical due to the symmetry). These forces are represented by a field of forces $(0, g)$ per unit length on $[0, T] \times \partial_4 \Omega$. Then, the equilibrium equations on $\partial_4 \Omega$ are written:

$$s_{12} = 0 \qquad -s_{22} = g \qquad \text{on } \partial_4 \Omega \tag{8}$$

Let us note that g is unknown.

Now, we write the following unilateral constraints that will lead to the complementarity problem:

$$u_N \geqslant 0 \tag{9}$$

$$g \geqslant 0 \qquad \text{on } \partial_4 \Omega \tag{10}$$

$$u_N \cdot g = 0 \tag{11}$$

where u_N is the second component of the trace of u under $\partial_4 \Omega$.

The condition (9) means that the edges of the crack can be parted but cannot interpenetrate each other; g is a compression force.

Finally, the condition (11) means that the contact force g vanishes when the edges are parted.

3. THE CONSTITUTIVE LAW

We choose for (3) a Maxwell behaviour law that allows one to take into account the flow phenomenom. We write it with the strain parameter ξ (see [4]):

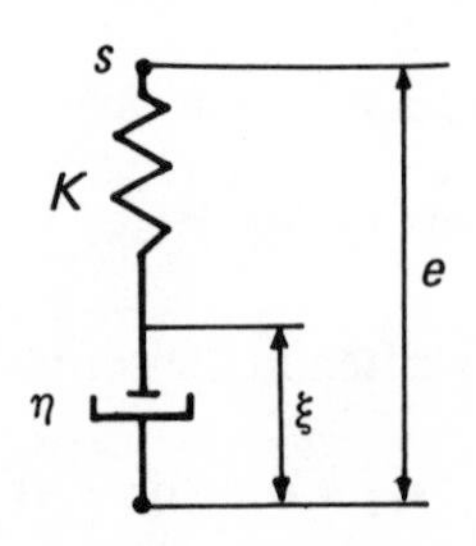

$$(12) \qquad \begin{aligned} s(t,x) &= \eta(t,x)\,\dot{\xi}(t,x) \\ s(t,x) &= K(t,x)\,[e(t,x) - \xi(t,x)] \end{aligned}$$

η and K are mappings defined on $[0, T] \times \Omega$ with values in the space of 3×3 symmetrical matrices: K is the stiffness matrix, and η is the viscosity matrix.

We shall look for a periodic stress solution. We see immediately from equation (12) that $\dot{\xi}$ will be periodic, but in general with a mean value different from zero. Therefore, the strains ξ and e take the form:

$$\xi(t,x) = \alpha(x)t + r(t,x) \tag{13}$$

$$e(t,x) = \alpha(x)t + p(t,x) \tag{14}$$

where $\alpha(x)$ is a strain secular term.

Consequently, the displacement will have also the form:

$$u(t,x) = \beta(x)t + q(t,x) \tag{15}$$

where $\beta(x)$ is a displacement secular term. These secular terms $\alpha(x)$ and $\beta(x)$ mean that, under constant loads, the strains and the displacements are linear with time (flow phenomenom).

The constitutive law (12) shows that the strains will be defined with an arbitrary additive constant on time. Then, the solutions will not be unique. Consequently, we may choose $r(t,x), p(t,x)$ and $q(t,x)$ as periodic functions of time with mean value zero.

4. FUNCTIONAL SCHEME

The mathematical approach of that Signorini problem (see unilateral conditions (9), (10), (11)) are treated in [2, 3] by duality and virtual works techniques. We recall here the functional scheme.

The situation is summed up in the following:

$$\begin{array}{lllll}
u_N \in L(C) & L(C) \subset \Gamma & (\cdot,\ \cdot) & G & g \in G \\
 & L \uparrow & & \downarrow {}^tL & \\
u \in C & C \quad \subset U & \langle\langle \cdot,\ \cdot \rangle\rangle & \Phi \subset U' & \phi \in \Phi \\
 & D \downarrow & & \uparrow {}^tD & \\
e, \xi \in E & \quad E & \langle \cdot,\ \cdot \rangle & S & s \in S
\end{array}$$

with $U = H^1_{\partial_1}(\Omega)^2$ space of displacement defined by:

$$U = \{u = (u_1, u_2) : u \in H^1(\Omega); u = 0 \text{ on } \partial_1\Omega; u_2 = 0 \text{ on } \partial_2\Omega\}$$

C is the cone defined as the set of displacements which have the second component of their trace on $\partial_4\Omega$ positive or zero:

$$C = \{u \in U : Lu \geqslant 0\}$$

The mapping L associates to a displacement u the second component of its trace on $\partial_4 \Omega$. We set $\Gamma = L(U)$. U' is the topological dual of U. A classical subset of U' is a set isomorphic to $\Phi = L^2(\Omega)^2 \times L^2(\partial_3 \Omega)^2$. Φ is the load space.

$E = L^2(\Omega)^3$ is the space of symmetrical strain tensors. $S = L^2(\Omega)^3$ is the space of symmetrical stress tensors. D is the strain operator defined by relation (2): $e = Du + e^0$. ${}^tD$ transposed of D for the duality, is the equilibrium mapping. It is defined by the relations (4), (7), and (8):

$${}^tDs = \phi + g$$

means

$$- \operatorname{div}\, s = \phi_1 \qquad \text{on } \Omega$$

$$s \cdot \mathbf{n} = \phi_2 \qquad \text{on } \partial_3 \Omega$$

$$s \cdot \mathbf{n} = g \qquad \text{on } \partial_4 \Omega$$

(**n** is the normal vector to the considered arc).

The duality bilinear forms are:

$$\langle\langle u, \phi \rangle\rangle = \int_\Omega u(x)\phi_1(x)\,dx + \int_{\partial_2 \Omega} \gamma u(x)\,\phi_2(x)\,dx'$$

$$\langle e, s \rangle = \int_\Omega e(x)\,s(x)\,dx$$

(γu denotes the trace of u on the part $\partial_2 \Omega$).

From a mechanical point of view, $\langle\langle u, \phi \rangle\rangle$ is the work of the loads ϕ in the displacement u, and $-\langle e, s \rangle$ is the work of the stress s in the strain e.

5. THE EQUATIONS

Now, we set the following asymptotic problem because of the presence of the secular terms.

5.1. Problem P1

Let the given data satisfy: $\phi(t)$ is T periodic and $e^0(t)$ is T periodic. Find the time t_0, the stress $s(t)$ (T periodic), the displacement $u(t) = \beta \cdot t + q(t)$

with $q(t)$ T periodic, the strain $e(t) = \alpha t + p(t)$ with $p(t)$ T periodic, the strain $\xi(t) = \alpha t + r(t)$ with $r(t)$ T periodic, and the force $g(t)$, such that

$$\left.\begin{aligned} e &= Du + e^0 \\ {}^tDs &= \phi + {}^tLg \\ s &= \eta\dot{\xi} \\ s &= K(e - \xi) \\ u_N &= Lu \end{aligned}\right\} \quad \forall\, t \in [n + 1)],\ n \in N$$

$$\left.\begin{aligned} u_N &\geqslant 0 \\ g &\geqslant 0 \\ u_N \cdot g &= 0 \end{aligned}\right\} \quad \forall\, t > t_0$$

Comment Indeed, at the points $x \in \partial_4 \Omega$ of the crack where β is not zero, the unilateral conditions are satisfied only after a time t_0.

This is a periodic problem of Signorini type coupled with a differential equation. We shall transform this problem into two coupled complementarity problems: the first one on the secular terms, and the other one on the periodic parts of the solution.

5.2. Parting the unilateral conditions

Theorem 1

The problem P1 is equivalent to problem P2.

Problem P2

Let $\phi\,(t)$ and $e^0\,(t)$ be given as in P1. Find $s(t)$, $u(t)$, $e(t)$, $\xi(t)$, $g(t)$ as in P1 such that:

$$\left.\begin{aligned} e &= Du + e^0 \\ {}^tDs &= \phi + {}^tLg \\ s &= \eta\xi \\ s &= K(e - \xi) \\ \beta_N &= L\beta \\ q_N &= Lq \\ \beta_N &\geqslant 0 \qquad\qquad q_N \geqslant 0 \\ g &\geqslant 0 \quad \text{and} \quad g \geqslant 0 \\ \beta_N \cdot g &= 0 \qquad\qquad q_N \cdot g = 0 \end{aligned}\right\} \quad \forall\, t \in [0, T]$$

where

$$\bar{g} = \frac{1}{T}\int_0^T g(t)\,dt$$

From now on, we shall always denote the mean value of a T-periodic function f:

$$\bar{f} = \frac{1}{T}\int_0^T f(t)\,dt$$

5.3. Parting the problem P2 into two coupled problems

We introduce the restrictive hypothesis that the viscosity does not depend on the time t.

Note In fact, we can solve the problems where the viscosity may be written (see [3]):

$$\eta(t, x) = \eta_0(t)\,\eta(x)$$

where $\eta_0(t)$ is a smooth, scalar, positive, T-periodic function.

Then, using the first equations of the problem P1, we can write:

$$s = \eta\dot{\xi} \quad \text{with} \quad \dot{\xi} = D\beta + \dot{r}$$

and then

$$\bar{s} = \eta D\beta$$

It follows that the problem P2 leads to two coupled complementarity problems; we give here, immediately, the variational form of these problems.

Problem P3

Let $\bar{\phi}$ be given. Find $\beta \in C$ such that

(i) $$\forall\, v \in C \quad \langle\langle v - \beta, {}^tD\eta\; D\beta - \bar{\phi}\rangle\rangle \geqslant 0$$

This is a classical elliptic problem of Signorini which is stationary. The operator ${}^tD\eta D$ is similar to the elasticity operator and has also all the good properties (coercivity, symmetry, . . .). There will be no difficulties to solve it. The solution gives us the secular term of displacement β.

It can be shown that, in the case where only the stiffness matrix K is a function of time, the secular terms are zero if the mean values of the loads are zero.

Problem P4
Let $\phi(t)$ and $e^0(t)$ be given as in P1. Find $q(t) \in C$, and $r(t)$ as in P1, such that

(j) $\quad \forall\, v \in C \quad \langle\langle v - \dot{q}(t), {}^tDK(t)\, Dq(t) + {}^tDK(t)\,[e^0(t) - r(t)] - \phi(t)\rangle\rangle \geqslant 0$

(jj) $\quad \eta\dot{r}(t) = -K(t)\,r(t) + K(t)\,[Dq(t) + e^0(t)] - \eta D\beta$

with $s(t) = \eta D\beta + \eta\dot{r}$ T-periodic.

The connection between the variational inequation (j) and the differential equation (jj) is realized by the variables $r(t)$ and $q(t)$.

6. DISCRETIZATION AND NUMERICAL TREATMENT

6.1. Problem P5

This problem is a classical elliptic problem connected with a stationary variational inequality. We use a finite element method to discretize this problem. We then replace the initial problem with the following approximate problem:

Problem P5
Let $\overline{\phi}$ be given. Find

$$\beta^h = \sum_i \beta_i^h \omega_i^h$$

such that:

$$\forall\, v^h \in C^h \quad \langle\langle v^h - \beta^h, {}^tD\eta\, D\beta^h - \overline{\phi}\rangle\rangle \geqslant 0$$

C^h is a convex subset of U^h which is conveniently assumed to realize an approximation of the given cone C. U^h is the finite-dimensional space depending on a parameter h connected with the mesh.

Let ω_i^h be a basis in U^h $(i = 1,\ldots,n)$. The *approximate problem* P5 leads to a *discrete problem* written in the n-dimensional vector space $\mathbb{R}^n$ (n being the dimension of U^h). We thus exibit an $n \times n$ matrix N^h given by:

$$N_{ij}^h = \langle D\omega_i, \eta D\omega_j\rangle$$

It is completely similar to the classical elasticity matrix because the viscosity matrix η has the same properties as the stiffness matrix K.

We solve that first classical problem by an overrelaxation method with projection on the cone C^h at each step [6, 7]. We are only confronted with the usual numerical difficulties inherent to this problem.

6.2. Problem P6

This problem is more difficult because of the coupling between the variational inequality and the differential equation. Moreover, it is a periodic problem.

A result about the asymptotic stability of the Cauchy problem solution, i.e. of the associated problem with initial condition (instead of the periodic condition) can be found in [2, 3]. Consequently, we have an *explicit result* about the asymptotic convergence of the Cauchy solution towards the periodic solution. That explicit result (see [5]) enables us to estimate *a priori* the length of time of the Cauchy solution to be calculated to have a 'sufficiently' periodic solution. For the examples treated, this time was always less than $10 \times T$ (where T is the period).

Remark 1 For the cases where that convergence would be very slow, we shall consider a direct treatment with a Fourier decomposition on time. The formulation leads to a very large system which is sparse and banded by block. We have treated a one-dimensional example with this method for a case without constraints. The numerical difficulty comes from the very large dimension of that matrix: we shall prefer to use the first method each time it will be reasonable for the computation time.

Remark 2 On the other hand, we have tried to use the Poincaré mapping (or mapping of translation) which is classical for the ordinary periodic differential equation in a Hilbert space. For this, we introduce an integral operator of which the discretization is alas not numerically 'reasonable'. So, that trial fails. However, that formulation enables us to exhibit the asymptotic stability of the Cauchy solution and also the stability condition (because of the use of an explicit integration method) and to give the error estimation (in the case without constraint, for instance).

So, we shall solve the Cauchy problem associated with problem P4, where the condition (find $s = \eta D\beta + \eta\dot{r}$ periodic) is replaced by an initial condition ($r(0) = r_0$ given). And then, we calculate the Cauchy solution until $\|s(nT) - s((n+1)T)\|_s < \varepsilon$.

The approximate problem. On space, we use a *finite element* method: we write the problem P4 in the finite-dimensional space U^h defined above. To integrate the coupled differential equation, we use a *Runge-Kutta* method.

6.2.1 Semi-discretization on time

After several comparative trials with an equivalent number of operations, on several Runge-Kutta methods (order 1: Euler; order 2: Heunn; order 4) and also on a semi-implicit method (prediction-correction), we choose the

Heunn method with a fitting discretization on time. This method is preferable to the order-4 Runge-Kutta method because of the variable discretization on time. Indeed, with an equivalent number of operations, the discretization is more adjusted in this case.

We set

$$f(r, q, t) = -\eta^{-1} K(t)\, r(t) + \eta^{-1} K(t)\, q(t) + \eta^{-1} K(t)\, e^0(t) - D\beta$$

The Heunn relation written on the equation (jj) leads to:

$$(11) \qquad r_{k+1} = r_k + (1 - \rho)\, \Delta t_k f_k + \rho \Delta t_k f_{k'}$$

with $\rho = 3/4$, $f_k = f(r_k, q_k, t_k)$

$$f_{k'} = f(r_k + \frac{\Delta t_k}{2p} f_k, q_{k'}, t_{k'}) \quad \text{and} \quad t_{k'} = t_k + \frac{\Delta t_k}{2p}$$

$q_{k'}$, is a solution of the equation (j) with $t = t_{k'}$, and

$$r_{k'} = r_k + \frac{\Delta t_k}{2p} f_k$$

Let M be the number of samples of the discretization on time ($k = 1,\ldots,M$)

6.2.2. *Spatial discretization*

A finite element scheme applied to the equation (j) leads to relation (2) similar to the problem P5 (with regard to the notations)

$$\forall t_k \quad k = 1, \ldots, M$$

$$q_k^h = \sum_{i=1}^{n} q_{ki}^h\, \omega_i^h$$

$$(1) \qquad \forall v^h \in C^h \quad \langle\langle v^h - q_k^h, A_k q_k^h - B_k r_k^h + F_k^h \rangle\rangle \geqslant 0$$

where A_k is the elasticity operator ${}^tDK_kD$, B_k is the operator ${}^tDK_k$, F_k^h is the term connected with the loads ϕ_k and the prescribed strain e_k^0.

We associate to the problem P4 the following approximate problem P6.

6.2.3. *Problem P6*

Let $\phi(t, x)$ and $e^0(t, x)$ be given T periodic. Find $q_k^h(x) \in C^h$ with

$$q_k^h(x) = \sum_{i=1}^{n} q_{ki}^h\, \omega_i^h(x)$$

with $k = 1,\ldots, M$ such that:

(1) $$\forall v^h \in C^h \qquad \langle\langle v^h - q_k^h, A_k q_k^h - B_k r_k^h + F_k^h \rangle\rangle \geqslant 0$$

(11) $$r_{k+1}^h = r_k^h + (1-\rho)\Delta t_k f_k + \rho \Delta t_k f_{k'}$$

where r_1^h is given, remembering that h is the symbol for space discretization and k is the symbol for discretization on time.

The problem P6 is the *approximate problem* associated with problem P4: we do not give here the explicit form of the *discrete problem* in the n-dimensional vector space $\mathbb{R}^n$ that we solve in fact, so as not to complicate this paper with heavy notations. The exhibited $n \times n$ matrix associated with the operator A is the classical elasticity matrix:

$$\widetilde{A}_{ij} = \langle D\omega_i, KD\omega_j \rangle$$

Therefore, the *algorithm* is:

- we choose an initial condition r_1^h;
- we solve the variational inequality problem (1) by an overrelaxation method with projection on the cone C^h at each step;
- as r_k^h and q_k^h are now known, we use the explicit relation (11) to calculate r_{k+1}^h (in fact, we must solve the variational inequality (1) m times, if m is the order of the used Runge-Kutta method: here $m = 2$, we must solve the inequality (1) on time again to calculate $q_{k'}$, and then $f_{k'}$);
- and so on, until the periodicity of the stress is 'sufficient' i.e.

$$\|s_{(a+1)M}^h - s_{aM}^h\|_\infty < \varepsilon$$

Technical remark We have to solve several times the variational inequality (1); the matrix A remains constant if K does not depend of time, or changes little between two successive times t_k and t_{k+1}. Thus, at each step we have a good initial condition for the overrelaxation method with the result of the preceding calculation. We use iterative methods for this reason.

7. EXAMPLE

After several tests, the application of the method to a first simplified example gives the following results.

It is an example where the secular term is zero. Indeed, we have solved several problems on secular terms (solution $\beta > 0$, or $\beta = 0,\ldots$): there is no specific difficulty. So, we present here an example where $\beta = 0$.

We study a square domain with a 75 nodes mesh (see Figure 1) with the plane stress hypothesis (bidimensional problem). This calculation is a preliminary approach to an example that should allow a comparison with an experimental study on a rectangular symmetric piece of IN 100 (turbine blade) at 1100°C; this experience will be performed by ONERA.

Here, the temperature is then constant and in consequence all the parameters are constants too. Period $T = 600$ s. Relaxation time $\tau = 60$ s. Young's modulus $E = 13000$ kg/mm². Poisson's coefficient $\upsilon = 0.2$. On AB, boundary conditions $u_2 = 0$ and, for the node B, $u_1 = 0$ $u_2 = 0$. On BC, the unilateral conditions. On ED, we give the following loads

$$\phi_{2x_1}(t, x) = 0$$

$$\phi_{2x_2}(t, x) = F_0 \sin \frac{2\pi}{T} t \quad \text{with} \quad F_0 = 10 \text{ kg/mm}^2$$

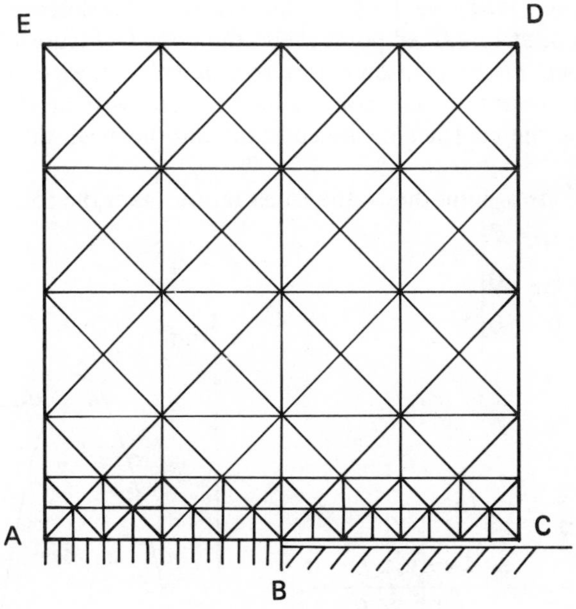

Figure 1 The mesh.

on Ω, $\phi_1(t, x) = 0$ (no volumic loads), $e^0(t, x) = 0$ (constant temperature).

Short computational aspect The dimension of the matrix $\widetilde{A}$ is 150: the band-width is 34. Determination of the *relaxation coefficient:* we find after a dichotomic analysis of the interval $]0, 2[$, $\widetilde{\omega} = 1.919$. We give in table 1, the number of necessary iterations to obtain, in a case without constraints, the same residue as that with the use of a direct pivoting method. We note the great sensitivity of the method with regard to the relaxation coefficient.

Table 1 Optimal relaxtion coefficients.

$\tilde{\omega}$	1.7	1.8	1.9	1.91	1.919	1.93	1.95
Number of iterations	496	341	167	144	115	123	155

We calculate the solution of the Cauchy problem on 3 periods: the stress s is then periodic with a precision of 3%. On time, we take $M=41$. The overall computation time (including the exploitation of the results) is of 8′ on the UNIVAC 1110. We use 15k-words of the central memory and store the solutions in a file of 30K-words.

The solution First, we look at the stress and displacement variations along a period, at one fixed point (here, the node C (Figure 1)), to show the unilateral aspect of the solution *before reaching the asymptotic state.* (In this case, for the asymptotic solution, we have contact with no stresses along the crack because the coefficients are constant and the mean value of the load is zero).

With plane stress hypothesis, the stress tensor takes the form:

$$s = \begin{vmatrix} s_{11} & s_{12} & 0 \\ s_{12} & s_{22} & 0 \\ 0 & 0 & 0 \end{vmatrix}$$

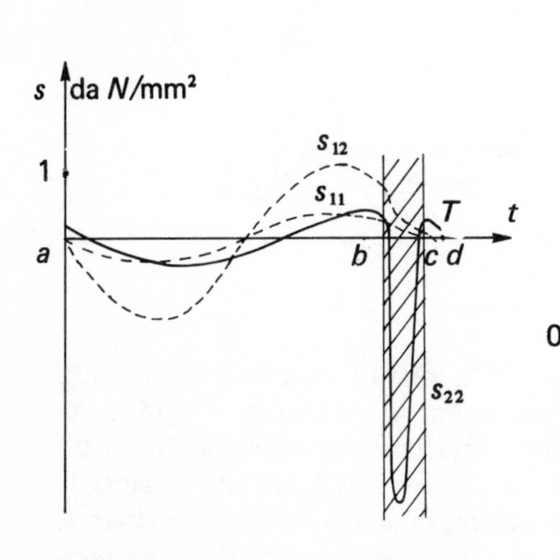

Figure 2 Stress $s=(s_{11}, s_{22}, s_{12})$ at the node C along a period.

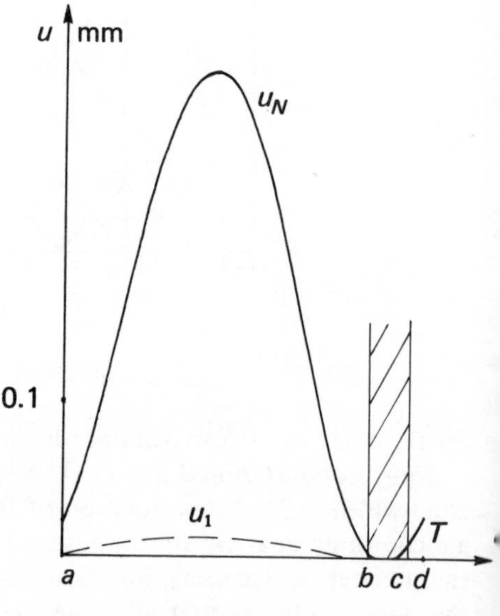

Figure 3 Displacement U_N of the node C along a period.

Figures 2 and 3 show the unilateral aspect of the solution on the edges of the crack: when $t \in [a, b]$ or $t \in [c, d]$, the crack is *open* and we have $u_N > 0$ and $g = 0$; when $t \in [b, c]$, the crack is *closed* and we have $u_N = 0$ and $g > 0$: the sudden variation of s_{22} shows the compression effect on the edge of the crack.

On the other hand, we can look at the displacement and the stress field in the whole domain at one fixed time.

On Figure 4, we show the stress level curves (3 components) for $t = T/4$ (maximum stress).

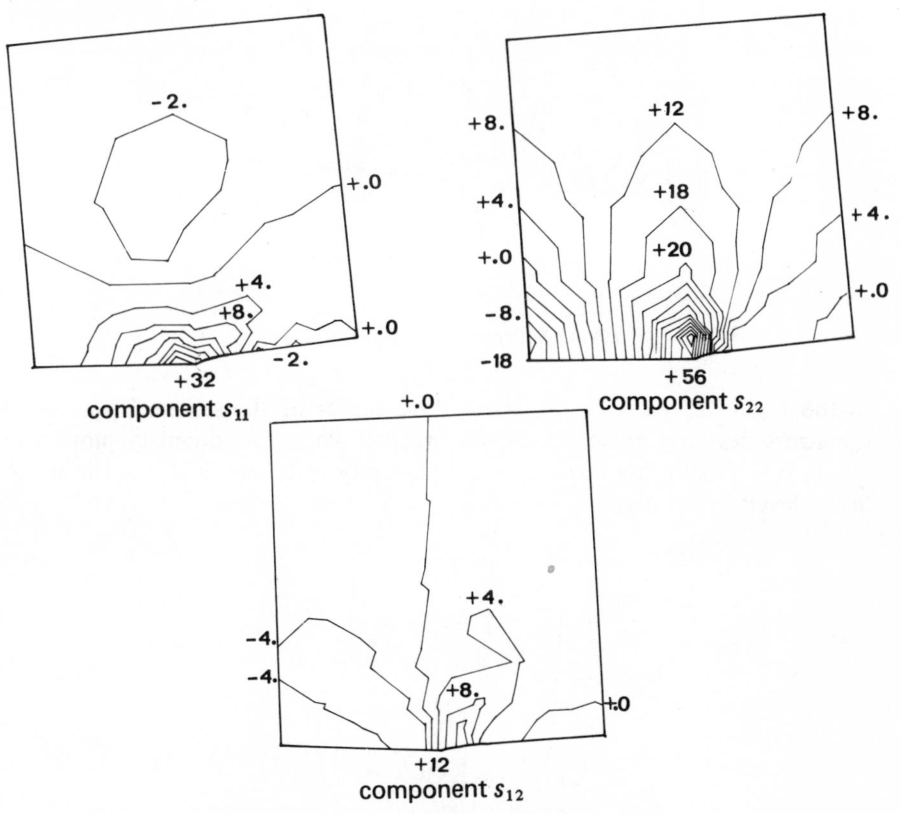

Figure 4 Stress tensor components for $t = T/4$

On figures 4,5 and 6, the displacements are multiplied by a coefficient 10 to see them clearly.

Figure 5 gives a more synthetic aspect of the stress state of the plate with the principal stresses (again for $t = T/4$).

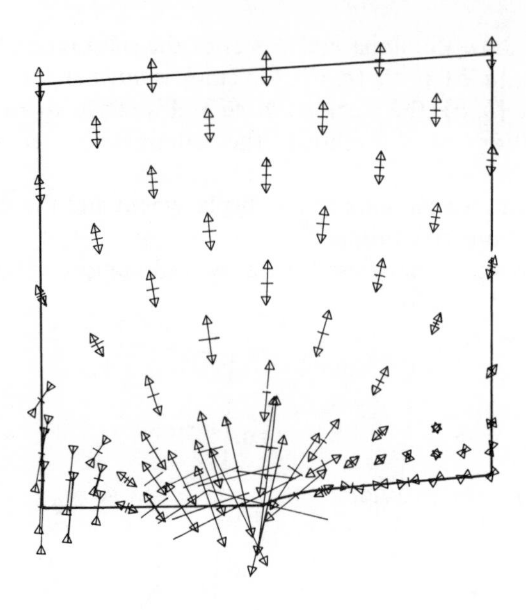

Figure 5 Principal stresses for $t = T/4$

In the Figure 6, we give the stress level curves of the second invariant of the stress deviator tensor in all the cracked plate. The quantity plotted in Figure 6 is usually used as a classical plasticity test; here, it shows the solicitation levels in the plate.

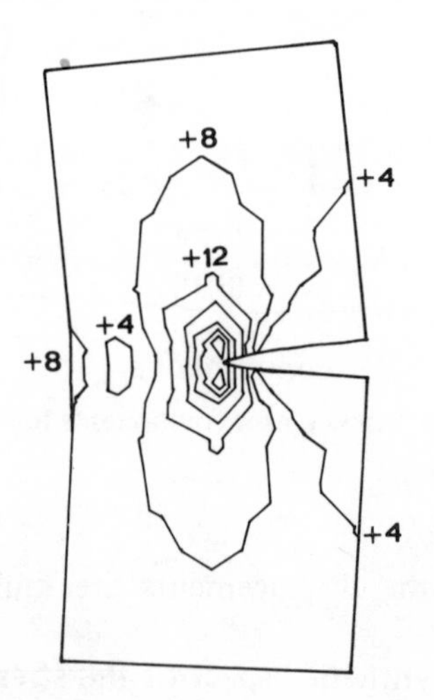

Figure 6 Second invariant of the stress deviator tensor in all the cracked plate for $t = T/4$

These presentations exhibit the high stress gradient at the top of the crack. For a better approach to the stress on that part, we shall use special elements (special functions defined on that zone; see [8]).

8. OTHER ASPECTS AND REFERENCES

The functional analysis aspect of the problem can be found in [2,3]. In this paper, existence and 'uniqueness' theorems for the Cauchy Signorini problem and for the periodic Signorini problem are given; a result on asymptotic stability of the Cauchy problem is also established.

Moreover, we have established several numerical analysis results with Geymonat [5] in the *case where* $C=U$ (i.e. the variational inequalities become only *variational equalities*): that is the case of the problem without crack. In this paper, we first give results of existence and asymptotic stability for the problems by another means than in [3]. But the main result given in [5] is to obtain the numerical stability condition under an explicit form.

Furthermore, the condition on the choice of Δt is independent of the space discretization: $\Delta t_k < 2\tau$ (where τ is the smallest value of the relaxation time). This interesting result occurs because we have a bounded operator (order zero). We also give for this case, the error estimates for space and time discretization.

REFERENCES

[1] E.P. Bazant, and S.T. Wu. 'Rate type creep law for aging concrete based on Maxwell chain'. *Rilem Marériaux et Constructions,* **7** (1974) 45-60.

[2] R. Bouc, G. Geymonat, M. Jean, and B. Nayroles. Hilbertian unilateral problems in viscoelasticity'. *Lecture notes in mathematics,* n. 503, Springer Verlag, Berlin (1976).

[3] R. Bouc, G. Geymonat, M. Jean, and B. Nayroles. 'Cauchy and periodic unilateral problems for aging linear viscoelastic materials'. *J. Math. Anal. Appl.,* **61**, 1977.

[4] P. Germain. 'Course de mécanique des milieux continus', Masson, Paris (1973).

[5] G. Geymonat, and M. Raous. 'Eléments finis en viscoélasticité périodique' *Lecture notes in Mathématics,* n. 606, Springer Verlag, Berlin (1977). *(International meeting on mathematical aspect of finite element method,* Dec. 10-12, 1975.*)*

[6] R. Glowinski, J.L. Lions, and R. Trémolieres. *'Analyse numérique des inéquations variationelles'.* Dunod, Paris (1976).

[7] U. Mosco. 'An introduction to the approximate solutions of variational inequalities'. Course CIME *'Constructive aspect of functional anlysis',* 1974 Cremonese, Roma (1973).

[8] G. Strang, and J. Fix. *'Analysis of the finite element method'*. Prentice Hall, New York (1973).

[9] G. Duvaut, and J.L. Lions. *'Les inéquations en mécanique et en physique'*. Dunod, Paris (1972).

[10] I.C. Cormeau, and O.C. Zienkiewicz. 'Visco-plasticity – plasticity and creep in elastic solids – a unified numerical solution approach'. *Int. J. Numer. Meth. Eng.*, 8, (1974) 821-45.

Chapter 20

Lagrange Multipliers and Variational Inequalities

R. T. Rockafellar

1. INTRODUCTION

Variational inequalities have been used to characterize the solutions to many problems involving partial differential equations with unilateral constraints. The complementarity problem in mathematical programming concerns a special type of variational inequality in finite dimensions that has been the focus of important algorithmic developments. General variational inequalities can be reduced to this special type through discretization and the introduction of Lagrange multipliers, and this provides an approach to computation.

Other approaches are suggested by analogies with convex programming. Many variational inequalities actually express the condition for the minimum of a convex functional relative to a certain convex set which, in the course of 'discretization', is represented by a finite system of convex (or linear) inequalities. A broader class of variational inequalities, covering perhaps the majority of applications, is obtained by replacing the gradient mapping associated with the convex minimand by a mapping that is 'monotone' in the general sense due to Minty. To the extent that algorithms for convex programming can be formulated entirely in terms of the gradient of the minimand, rather than the minimand itself (including such numerical considerations as stopping criteria), one can get computational procedures for variational inequalities that may offer advantages in some cases over complementarity.

For example, penalty methods have been used in the solution of variational inequalities, since they reduce a constrained problem to a sequence of 'unconstrained' problems to which classical numerical techniques can be applied. Nowadays in mathematical programming, penalty methods in pure form are in disrepute because of their inherent numerical instabilities. They have been supplanted by methods that are based on augmented Lagrangian functions and include varying Lagrange multiplier values (dual variables) as well as penalty parameters.

Some efforts have been made to apply such penalty-duality methods to variational inequalities, but only, it seems, in the case of equality constraints, and even then in terms of what is analogous to exact minimization in each unconstrained problem. For greater effectiveness, it is important to have procedures that are capable of handling inequality constraints and can be shown to converge under more practical criteria (tolerance levels for certain accessible quantities).

The purpose of this chapter, besides explaining some of the background to such matters, is to draw attention to a new penalty-duality method that has been designed with these requirements in mind. It converts a variational inequality for a monotone mapping (and a convex set defined by a finite system of differentiable convex constraint functions) into a sequence of unconstrained subproblems, in each of which one calculates an 'approximate' root of a non-linear equation for a certain strongly monotone mapping. The overall convergence rate is generically linear, with a ratio that approaches zero as the penalty parameter is increased and therefore yields superlinear convergence if the penalty parameter goes to infinity. From the close parallel with earlier methods, which are known to be highly effective in non-linear programming but do not carry over so easily in their formulation to general variational inequalities, one can hope for very good results. However, the verdict must await more testing. It is also clear that for applications to variational inequalities it would be helpful to incorporate additional features and flexibility beyond what has seemed possible within the present theoretical framework.

2. THE GENERAL PROBLEM

Let K be a non-empty closed convex set in a real Hilbert space V (finite or infinite-dimensional), and let $A : V \to V$ be a mapping (single-valued) that may be linear or non-linear. The *variational inequality problem* for K, A and an element $a \in V$ is to determine an element u satisfying the conditions

$$
\begin{aligned}
&u \in K \\
&\langle A(u) - a, v - u \rangle \geqslant 0 \qquad \text{for all } v \in K
\end{aligned}
\tag{1}
$$

Here $\langle \cdot, \cdot \rangle$ denotes the inner product in V.

If K is all of V, (1) reduces to the equation $A(u) = a$. More generally it expresses a normality condition very familiar to everyone who has studied optimization theory. The normal cone to K at a point u consists by definition of all the normal vectors to half-spaces that support K at u:

$$(2) \qquad N_K(u) = \{w : \langle w, v - u\rangle \leqslant 0 \quad \text{for all} \quad v \in K\}$$

This is a closed convex cone containing the origin, and in terms of it we write (1) in the form

$$(3) \qquad u \in K \quad \text{and} \quad a - A(u) \in N_K(u)$$

If there is a differentiable function F on V whose gradient mapping satisfies

$$(4) \qquad A(v) - a = \nabla F(v) \quad \forall\, v \in V$$

the variational inequality expresses the fundamental first-order necessary condition for u to be a local solution to the optimization problem

$$(5) \qquad \text{minimize } F(v) \quad \text{over all } v \in K$$

The condition is sufficient for global optimality when F is convex. Note that when A is linear, the existence of F satisfying (4) is equivalent to A being symmetric, in which event one has

$$(6) \qquad F(v) = \tfrac{1}{2}\langle A(v), v\rangle + \langle a, v\rangle + \text{constant}$$

An important concept in this context is that of monotonicity in the sense of Minty. The mapping A is *monotone* if

$$(7) \qquad \langle A(v) - A(\bar{v}), v - \bar{v}\rangle \geqslant 0 \quad \forall v, \forall \bar{v}$$

and *strongly monotone* (with modulus $\alpha > 0$) if

$$(8) \qquad \langle A(v) - A(\bar{v}), v - \bar{v}\rangle \geqslant \alpha\, \|v - \bar{v}\|^2 \quad \forall v, \forall \bar{v}$$

If A is linear, the expression on the left reduces to

$$\langle A(v - \bar{v}), v - \bar{v}\rangle = \langle A_s(v - \bar{v}), v - \bar{v}\rangle$$

where $A_s = \frac{1}{2}(A + A^*)$ is the symmetric part of A. Monotonicity then means that A is positive semidefinite, while strong monotonicity corresponds to positive definiteness (these terms being employed whether or not A is itself symmetric).

If F is a functional such that (4) holds (with A not necessarily linear), the monotonicity condition (7) can be written as

$$\langle \nabla F(\bar{v} + z) - \nabla F(\bar{v}), z\rangle \geqslant 0 \quad \forall z, \forall \bar{v}$$

It is not difficult to show that this is true if and only if F is *convex*. Thus the optimization problems (5), where a diffentiable convex function is minimized over a convex set, correspond to a special class of variational inequalities where A is monotone. However, not every variational inequality with A monotone can be interpreted in this way, since, for example, if A is linear, but not symmetric, a gradient representation (4) is impossible. It will be seen below that variational inequalities, where the mapping is monotone and not the gradient of any functional, can nevertheless give the optimality conditions for some minimization problems of convex type, when Lagrange multipliers are brought into the picture. No one knows whether by some extension of these ideas, all 'monotone' variational inequalities can be interpreted as arising from 'optimization'. What is clear, though, is that monotone mappings not of gradient type do arise in a number of ways, in particular in certain physical problems involving friction.

In applications to partial differential equations, V is usually a space of real-valued functions, such as a Sobolev space defined on a region Ω in $\mathbb{R}^N$. The mapping A represents a differential operator, and K incorporates various boundary conditions. Functional analytic considerations are then crucial, for instance in making sure that V, K, and A are well chosen for the problem one has in mind. The theory of distributions is used in studying the existence of weak and strong solutions, the regularity of solutions and other aspects. These aspects are important in the discussion of schemes of discretization, but for someone in mathematical programming who is interested mainly in solving problems that are already discretized, they are not essential.

Incidentally, if variational inequalities are viewed in the form (3), there is a very natural generalization to the case where K is non-convex: interpret $N_K(u)$ as the normal cone in the sense of Clarke [1] (see also Rockafellar [2, 3]). Clarke's work with the calculus of variations (for example, [4, 5]) and mathematical programming [6] indicates that first-order necessary conditions for many non-convex problems of optimization can be written as variational inequalities in this sense. As far as the theory of variational inequalities is concerned, such generalizations have not yet been explored. Most of the existing results are based on the consequences of convexity and the monotonicity idea.

3. DISCRETIZATION

For the purpose of computation, a variational inequality (1) is often 'approximated' by one involving a *finite* constraint system in a *finite*-dimensional space. Thus K is replaced by a set K_0 that lies in a finite-dimensional subspace V_0 of V and has the form

$$K_0 = \{v \in V_0 : F_1(v) \leq 0, \ldots, F_m(v) \leq 0\} \tag{9}$$

where each F_i is a finite *convex* function on V_0. (Equality constraints for affine functions are also permissible but will be kept out of the discussion temporarily for notational simplicity.) The variational inequality is reduced accordingly to

$$\begin{aligned} & u \in K_0 \\ & \langle A(u) - a, v - u \rangle \geq 0 \qquad \text{for all} \quad v \in K_0 \end{aligned} \tag{10}$$

This condition really depends only on the orthogonal projections of $A(u)$ and a on V_0:

$$A_0(u) = \text{proj}\, A(u) \qquad a_0 = \text{proj}\, a \tag{11}$$

We therefore get another variational inequality which is entirely in V_0, namely the one for K_0, the mapping $A_0 : V_0 \to V_0$, and the element $a_0 \in V_0$. In terms of normal cones, we have discretized condition (3) to obtain

$$u \in K_0 \quad \text{and} \quad a_0 - A_0(u) \in N_{K_0}(u) \tag{12}$$

In the gradient case (4), the projection step (11) in this procedure amounts to restricting the functional F to V_0. Denoting the restriction by F_0, one has

$$A_0(v) - a_0 = \nabla F_0(v) \qquad \forall v \in V_0 \tag{13}$$

where the gradient is taken in the sense of the space V_0. The corresponding optimization problem (5) is thereby discretized to

$$\begin{aligned} &\text{minimize} \qquad F_0(v) \qquad \text{over all} \quad v \in V_0 \\ &\text{satisfying} \qquad F_1(v) \leq 0, \ldots, F_m(v) \leq 0 \end{aligned} \tag{14}$$

It is important to recall that when A is monotone, the functional F is convex, and hence F_0 too is convex. The discretized problem then falls into the classical pattern of *convex programming.*

Optimization problems and their extensions typically involve not only primal variables, but also dual variables that lend themselves to interpretation as Lagrange multipliers associated with various constraints. This is why a process of discretization (in the sense used in this context — mathematical programmers are accustomed to speaking of discretization only in the extreme case where all 'continuous' structure is abandoned) must aim at a finite system of constraint functions, as well as a primal space that is finite-dimensional.

Dual variables in the case of (12) arise in representing the normal cone $N_{K_0}(u)$. Let us assume that $F_1, \ldots, F_m$ are not only convex but differentiable and that

$$\exists\, v \in V_0 \quad \text{with} \quad F_1(v) < 0, \ldots, F_m(v) < 0 \tag{15}$$

Then as is well known, $N_{K_0}(u)$ for a point $u \in K_0$ is the convex cone (containing 0) generated by the gradients $\nabla F_i(u)$ of the *active* constraints at u (i. e. those having $F_i(u) = 0$ rather than $F_i(u) < 0$). In other words,

$$N_{K_0}(u) = \left\{ \sum_{i=1}^{m} y_i \nabla F_i(u) : y_i \geqslant 0, y_i F_i(u) = 0 \right\} \tag{16}$$

The variational inequality (12) then can be rewritten as a condition in both $u \in V_0$ and $y = (y_1, \ldots, y_m) \in \mathbb{R}^m$:

$$\begin{aligned} &F_i(u) \leqslant 0 \quad y_i \geqslant 0 \quad \text{and} \quad y_i F_i(u) = 0 \qquad \text{for } i = 1, \ldots, m \\ &A_0(u) - a_0 + \sum_{i=1}^{m} y_i \nabla F_i(u) = 0 \end{aligned} \tag{17}$$

In the gradient case (13) these are known as the Kuhn-Tucker conditions for problem (14), and they play an enormous role in computational techniques. The conclusion we wish to emphasize is that *monotone* variational inequalities, after discretization, correspond to an extended form of the classical optimality conditions for convex programming, in the sense of involving a monotone term $A_0(u) - a_0$ that does not have to be the gradient of anything. One can attempt to generalize methods for solving (17) in the case of

$$A_0(u) - a_0 = \nabla F_0(u)$$

to this broader framework. Since monotone mappings have many powerful properties that seem to place them within the realm of convex analysis, it is natural therefore to regard the solution of monotone variational inequalities as a kind of 'extended convex programming'.

When equality constraints are of interest (the functions in question being affine), one can, of course, express an equation $F_i(u) = 0$ by a pair of inequalities $F_i(u) \leq 0$ and $- F_i(u) \leq 0$, at least for the sake of theoretical uniformity. A familiar refinement of assumption (15) can then be invoked in passing to the representation (17): strict inequality can be relaxed to weak inequality for each constraint function that is affine.

4. COMPLEMENTARITY

Many situations lead to the following model. Given a mapping $M : \mathbb{R}^N \to \mathbb{R}^n$ and vector $q \in \mathbb{R}^N$, determine $z \in \mathbb{R}^N$ such that

$$(18) \qquad z \geq 0 \qquad M(z)+q \geq 0 \qquad z \cdot [M(z)+q] = 0$$

This is the *complementarity problem* for M and q. The notation $z \geq 0$ means that z belongs to the non-negative orthant $\mathbb{R}^N_+$. Thus in terms of $z = (z_1, ..., z_N)$ and $M(z)+q = w = (w_1, ..., w_N)$, the condition says that

$$(19) \qquad \begin{array}{l} \text{for each} \quad j = 1, ..., N, \quad \text{one has} \quad z_j \geq 0, \;\; w_j \geq 0 \\ \text{and either} \quad z_j = 0 \;\; \text{or} \;\; w_j = 0 \end{array}$$

Another version of the condition can be stated in terms of the normal cones associated with $\mathbb{R}^N_+$. Observing that

$$N_{\mathbb{R}^N_+}(z) = \{-w : w \geq 0, \; z \cdot w = 0\} \quad \text{for} \quad z \geq 0$$

we see that (18) is the same as

$$(20) \qquad z \in \mathbb{R}^N_+ \quad \text{and} \quad -q - M(z) \in N_{\mathbb{R}^N_+}(z)$$

The complementarity problem is thus simply the variational inequality problem for $K = \mathbb{R}^N_+$. Its simple form has lent itself to the development of a number of algorithms, mainly for linear mappings M (see other chapters in this volume). Again in this context the monotonicity of the mapping plays an important role.

Not so obvious is the fact that *every* monotone variational inequality in the discretized form above can be reformulated as a monotone complementarity problem. Thus the special case is not so special after all.

To demonstrate the truth of this assertion, let us first look at a somewhat simpler, but very common case where $K_0 \subset \mathbb{R}^n_+$ (the space V_0 being identified now notationally with $\mathbb{R}^n$). Specifically, let us suppose that in the constraints in (9) one has $m = n + p$ and for $v = (v_1, ..., v_n)$

$$F_{p+j}(v) = -v_j \quad \text{for} \quad j = 1, ..., n$$

Then in terms of

$$\hat{G}(v) = (F_1(v), ..., F_p(v)) \tag{21}$$

and $\hat{y} = (y_1, ..., y_p)$, $\tilde{y} = (y_{p+1}, ..., y_m)$, one can write (17) as

$$\begin{aligned} &\hat{y} \geq 0 \quad \hat{G}(u) \leq 0 \quad \hat{y} \cdot \hat{G}(u) = 0 \\ &u \geq 0 \quad \tilde{y} \geq 0 \quad \tilde{y} \cdot u = 0 \\ &A_0(u) - a_0 + \sum_{i=1}^{p} \hat{y}_i \nabla F_i(u) = -\tilde{y} \end{aligned} \tag{22}$$

Setting $q = (-a_0, 0)$ and

$$M(u, \hat{y}) = \left(A_0(u) + \sum_{i=1}^{p} \hat{y}_i \nabla F_i(u), \ -\hat{G}(u)\right) \tag{23}$$

we get (22) into the complementarity form

$$(u, \hat{y}) \geq 0 \quad M(u, \hat{y}) + q \geq 0 \quad (u, \hat{y}) \cdot [M(u, \hat{y}) + q] = 0 \tag{24}$$

For the general case of (17) the reformulation as a complementarity problem is similar but requires some elementary tricks well known in optimization theory. The equation must be written as a pair of inequalities, and the vector u must be expressed as the difference of vectors u^+ and u^- that are constrained to be non-negative. Thus one sets $q = (-a_0, a_0, 0)$ and

$$M(u^+, u^-, y) = (H(u^+ - u^-, y), -H(u^+ - u^-, y), -G(u^+ - u^-)) \tag{25}$$

where

$$G(u) = (F_1(u), ..., F_m(u)) \tag{26}$$

$$H(u, y) = A_0(u) + \sum_{i=1}^{m} y_i \nabla F_i(u) \tag{27}$$

and this puts (17) in the form

$$z \geq 0 \quad M(z) + q \geq 0 \quad z \cdot [M(z) + q] = 0 \quad \text{for} \quad z = (u^+, u^-, y) \tag{28}$$

Proposition 1
If the mapping A_0 in the expanded variational inequality (17) is monotone and continuous (and each F_i is convex), then the mapping M in the corresponding complementarity problem (28) (or (24)) is monotone and continuous relative to the non-negative orthant. However, M cannot be strongly monotone (even if A_0 is strongly monotone), when the constraint system (9) is non-vacuous.

To prove the first assertion, with M given by (25), we observe that the form

$$[M(z)-M(\bar{z})]\cdot[z-\bar{z}] \quad \text{for} \quad z=(u^+,u^-,y),\ \bar{z}=(\bar{u}^+,\bar{u}^-,\bar{y})$$

reduces to

$$(29)\qquad [H(u,y)-H(\bar{u},\bar{y})]\cdot[u-\bar{u}] - [G(u)-G(\bar{u})]\,[y-\bar{y}] =$$

$$= [A_0(u)-A_0(\bar{u})]\cdot[u-\bar{u}] +$$

$$+\sum_{i=1}^{m}\ [(y_i\,\nabla F_i(u)-\bar{y}_i\,\nabla F_i(\bar{u}))\,(u-\bar{u})+(y_i-\bar{y}_i)(F_i(u)-F_i(\bar{u}))]$$

where $u=u^+-u^-$ and $\bar{u}=\bar{u}^+-\bar{u}^-$. If A_0 is monotone, we have

$$[A_0(u)-A_0(\bar{u})]\cdot[u-\bar{u}] \geqslant 0$$

On the other hand, the convexity of each F_i yields

$$(y_i\,\nabla F_i(u)-\bar{y}_i\,\nabla F_i(\bar{u}))(u-\bar{u})+(y_i-\bar{y}_i)(F_i(u)-F_i(\bar{u}))\geqslant 0$$

through combination of the two inequalities

$$\bar{y}_iF_i(u)\geqslant \bar{y}_iF_i(\bar{u})+\bar{y}_i\nabla F_i(\bar{u})\cdot(u-\bar{u})$$

$$y_iF_i(\bar{u})\geqslant y_iF_i(u)+y_i\nabla F_i(u)\cdot(\bar{u}-u)$$

(Non-negativity of y_i and $\bar{y}_i$ is needed here.) The expression in (29) is then non-negative (when $y\geqslant 0$, $\bar{y}\geqslant 0$), and the monotonicity of M with respect to the non-negative orthant, when A_0 is monotone, is established. The continuity of M follows from the continuity of A_0 and the fact that a differentiable convex function F_i is always continuously differentiable. Strong monotonicity is impossible because (19) vanishes when $u=\bar{u}$ but $y\neq\bar{y}$.

Incidentally, it can also be proved that a mapping M of form (25) (or (23)) *cannot* be the gradient of any function, even if A_0 is of gradient type. Thus, through the process of reformulation using Lagrange multipliers, we see that monotone mappings that are not strongly monotone nor of gradient type have an essential role to play in the theory. Even when we start from a problem of optimization, we may be led to a complementarity problem, and hence a particular kind of variational inequality, involving such a mapping.

5. LINEARITY VERSUS NON-LINEARITY

At the present stage of development, most of the techniques for solving the complementarity problem, at least the ones that might take advantage of monotonicity or other such structure rather than just reducing everything to the location of a general fixed point, concern only the case of a linear mapping. The following elementary fact is therefore central in any discussion of solving variational inequalities by way of such techniques.

Proposition 2
If A_0 in the expanded 'discretized' variational inequality (17) is linear and each F_i is affine, then M is linear in the corresponding complementarity problem (28) (or 24)). The converse is also true.

Of course, in discretizing a general variational inequality (1), we obtain A_0 as a restriction and projection of A, so A_0 is linear if A is. On the other hand, the closed convex set K is the intersection of all the closed half-spaces containing it, so in principle we may regard K as defined by a possibly infinite system of linear inequalities. Discretization could be achieved by passing to a finite-dimensional subsystem.

Thus in a certain sense, the infinite-dimensional variational inequality problems that can be approximated by *linear* complementarity problems can be identified simply as the ones in which the mapping A is linear. In the optimization context, they are the ones that correspond to infinite-dimensional quadratic/linear programming.

The linear complementarity approach to computation could be pushed a bit further in considering a scheme in which a non-linear A is linearized iteratively to get subproblems that fall within the guidelines just laid down. However, no such scheme has been shown to be attractive for convex programming, until recently, and then in forms that may not readily extend to general variational inequalities.

Many important variational inequalities do involve linear monotone mappings and constraints which in passing to a finite-dimensional subspace V_0 (based on a triangularization of the domain Ω over which the function space V is defined) do reduce neatly to finite systems of linear inequalities. Lest one take this sort of behaviour for granted, though, it is well to look at a classical case where the constraints cannot be handled in such a simple fashion. This will also illustrate how there may be an advantage in using convex (non-linear) inequalities in the discretization process, although the original variational inequality appears to be just 'linear'.

A good example is the potential problem studied by Stampacchia in the early days of the theory of variational inequalities. In this problem there is a bounded open domain Ω in $\mathbb{R}^N$ and a non-empty closed set $E \subset \Omega$. The Sobolev space $H_0^1(\Omega)$ serves as V, while K consists of the closure of the set of 'test functions' (C^∞ functions) v in $H_0^1(\Omega)$ that satisfy

$$(30) \qquad \begin{aligned} v(x) &\geqslant 0 \quad \text{for all } x \in \Omega \setminus E \\ v(x) &\geqslant 1 \quad \text{for all } x \in E \end{aligned}$$

Note that (30) has the appearance of an infinite system of linear inequalities indexed by E and $\Omega \setminus E$:

$$(31) \qquad \begin{aligned} f_1^x(v) &\leqslant 0 \quad \text{for all } x \in E \\ f_2^x(v) &\leqslant 0 \quad \text{for all } x \in \Omega \setminus E \end{aligned}$$

where

$$(32) \qquad f_1^x(v) = 1 - v(x) \qquad f_2^x(v) = -v(x)$$

The trouble is that the functionals f_1^x and f_2^x are not continuous or even defined everywhere on $V = H_0^1(\Omega)$. Indeed, the elements v of this space are really just equivalence classes of functions that differ only on a set of measure zero. Evaluation at a point x therefore does not make sense unless an equivalence class can be identified with a distinguished member that is continuous, say (or in the case of the 'test functions', infinitely differentiable).

An idea that might at first look tempting in this situation would be to represent the constraint by an abstract inequality in a partially ordered space. Thus if $H_0^1(\Omega)$ is given the natural ordering induced by the cone of functions that are non-negative almost everywhere, one could hope to write (30) as

$$(33) \qquad v \geqslant \chi_E$$

where χ_E is the characteristic function of E. But χ_E does not belong to the space $H_0^1(\Omega)$, so this formulation falls short. In any event, like the preceding formulation, it fails to take into account the fact that K is merely the *closure* of the set of *test functions* satisfying the constraints.

The lack of continuity of the functionals (32) also bodes ill for schemes of discretization where the infinite system (31) is replaced by a finite subsystem. Although the finite-dimensional subspace V_0 that is chosen may be such that these functionals are well defined, there is little to guarantee that the discretized problem really 'approximates' the given one.

One way around the impasse is to forgo the linearity of the constraint representation. For 'test functions' ν in $H_0^1(\Omega)$, define

$$(34) \qquad \begin{aligned} F_1(\nu) &= \max \{ 1 - \nu(x) : x \in E \} \\ F_2(\nu) &= \max \{ -\nu(x) : x \in \mathrm{c}\ell\, [\Omega \backslash E] \} \end{aligned}$$

(see equations (32)). Then define $G : H_0^1(\Omega) \to \mathbb{R} \cup \{+\infty\}$ by

$$G(\nu) = \liminf_{\nu' \to \nu} \; [\max \{ F_1(\nu'), F_2(\nu') \}]$$

where the limit is taken over all 'test function' sequences converging to ν in the norm topology of $H_0^1(\Omega)$. The set we are interested in is, by definition,

$$K = \{ \nu \in H_0^1(\Omega) : G(\nu) \leqslant 0 \}$$

Note that G is a lower semi-continuous *convex* functional. On certain subspaces V_0, in particular those whose elements are all test functions, it will be true that

$$(35) \qquad G(\nu) = \max \{ F_1(\nu), F_2(\nu) \}$$

with F_1 and F_2 still well defined by (34). Then

$$(36) \qquad K \cap V_0 = \{ \nu \in V_0 : F_1(\nu) \leqslant 0, F_2(\nu) \leqslant 0 \}$$

Taking the latter to be K_0, we have a discretization where the constraint system is expressed by a pair of finite *convex* functions. As a means of approximate the original problem, this approach is much more stable.

Obviously the convex functions F_1 and F_2 will not usually be *differentiable* on V_0, but this need not be a serious obstacle. In the theory of convex programming, much attention has recently been given to non-differentiable functions. In part, this is due precisely to their role in reformulating problems that might otherwise involve more constraints than may be handled conveniently at one time. A kind of 'aggregation of dual variables' is involved. If the infinite system (31) were approximated by a large finite subsystem, there would be a correspondingly large number of Lagrange multipliers that would have to be kept track of. In the representation (36) there are only two multipliers. In evaluating F_1 or F_2 or one of their subgradients at a point ν, as may be required by an algorithm, we need only invoke a subroutine for solving the maximization problems in x that are embedded in the formulas (34). Thus, in effect, we can generate, as we go along, the affine functions f_1^x or f_2^x that turn out to be important, rather than having to treat all of them individually.

6. GENERALIZATION TO MULTIFUNCTIONS

If non-differentiable functions are to be treated, the gradient mappings that appear in (17) (and (13)) must be replaced by something more general. For a convex functional $F : V \to \mathbb{R} \cup \{+\infty\}$, it will be recalled that the set of *subgradients* of F at v is

(37) $$\partial F(v) = \{w \in V : F(v') \geq F(v) + \langle w, v' - v\rangle, \forall v' \in V\}$$

The multifunction ∂F associates, in other words, with each $v \in V$ a certain closed convex (possibly empty) set in V. (For the appropriate definition of ∂F when F is not convex, see [1,3].)

The connection between convexity and monotonicity remains strong under this generalization. A multifunction (set-valued mapping) $T : V \rightrightarrows V$ is said to be *monotone* if

(38) $$\langle w - \overline{w}, v - \overline{v}\rangle \geq 0 \qquad \text{for all } v, \overline{v} \text{ in } V \text{ and all } w \in T(v), \overline{w} \in T(\overline{v})$$

It is *maximal* monotone if, in addition, there does not exist any monotone multifunction $T' : V \rightrightarrows V$ with $T'(v) \supset T(v)$ for all v and $T'(v) \neq T(v)$ for at least one v.

Moreau [7] proved in the Hilbert space case that ∂F is maximal monotone if F is a convex function on V that is lower semicontinuous and not identically $+\infty$. (For the generalization to Banach spaces and the precise characterization of the class of maximal monotone multifunctions that are obtained in this fashion, see Rockafellar [8]).

The general variational inequality (1) in the case of a *multifunction* A takes the form

(39) $$u \in K \qquad w \in A(u)$$
$$\langle w - a, v - u\rangle \leq 0 \qquad \text{for all } v \in K$$

or in other words

(40) $$u \in K \quad \text{and} \quad a \in A(u) + N_K(u)$$

The complementarity problem for a multifunction M has (18) replaced by

(41) $$z \geq 0 \quad w \geq 0 \quad z \cdot w = 0 \quad w \in M(z) + q$$

so that the condition is

(42) $$-q \in M(z) + N_{\mathbb{R}_+^N}(z)$$

In the reformulations discussed earlier, (17) becomes

$$(43)\quad \begin{aligned} &F_i(u) \leqslant 0 \quad y_i \geqslant 0 \quad \text{and } y_i F_i(u) = 0 \quad \text{for } i = 1, \ldots, m \\ &a_0 \in A_0(u) + \sum_{i=1}^{m} y_i \, \partial F_i(u) \end{aligned}$$

while in the definition (25) of M one takes

$$(44)\qquad H(u,y) = A_0(u) + \sum_{i=1}^{m} y_i \, \partial F_i(u)$$

instead of (27).

Proposition 3
Suppose A_0 is maximal in (43) (and each F_i is convex), and that (15) holds for an element v such that $A_0(v) \neq \emptyset$. Then the multifunction M in the corresponding complementarity problem (defined by (25) and (44)) is maximal monotone.

Proof: This follows from results in [9] when the sum of two maximal monotone mappings is again maximal monotone – for the argument, see [10, proposition 5].

The assumption about (15) holding with $A_0(v) \neq \emptyset$ can be replaced by the following. There exists $v \in K_0 \cap \text{int } D(A_0)$ such that $F_i(v) < 0$ for all non-affine functions F_i, where $D(A_0)$ is the set of points where A_0 is non-empty-valued. This modification is needed in treating the case of equality constraints.

The connection between Proposition 3 and Proposition 2 is this: a monotone mapping $A : V \to V$ that is weakly continuous relative to each line segment in V is maximal monotone when viewed as a multifunction. Furthermore, if $A : V \rightrightarrows V$ is a maximal monotone multifunction and $V_0 \cap \text{int } D(A) \neq \emptyset$, then the projection $A_0 : V_0 \to V_0$ is maximal monotone. This is a consequence of [9, theorem 1].)

Whether the extended problems in terms of multifunctions can be solved effectively, remains a largely unexplored question. But in view of the trend in convex programming, there is certainly hope in this direction.

7. PENALTY-DUALITY METHODS

Variational inequality problems that occur in applications are very often generalizations of classical boundary-value problems for partial differential equations. Therefore, it is very natural to look for ways of solving them in terms of reducing them to a succession of classical problems for which numerical tchniques are already highly developed. Sometimes this can be accomplished by the introduction of penalties to force the satisfaction of constraints. However, there now appears to be a possibility of using more sophisticated methods based on ideas of recent years in non-linear programming.

To see these ideas in their original and simplest form, let us consider first the case of a non-linear programming problem with *equality* constraints:

$$\text{(45)}\quad \text{minimize } F_0(v) \quad \text{subject to} \quad F_1(v) = 0, \ldots, F_m(v) = 0$$

The *augmented Lagrangian* function for this problem is

$$\text{(46)}\quad L(v,y,r) = F_0(v) + \sum_{i=1}^{m} [\, y_i F_i(v) + \frac{r}{2} F_i(v)^2 \,]$$

where r is a non-negative variable that serves as a penalty parameter.

A fundamental class of penalty-duality algorithms can be described as follows. At iteration k we have a vector $y^k = (y_1^k, \ldots, y_m^k)$ in $\mathbb{R}^m$ and a value $r_k \geqslant 0$. We determine u^{k+1} as an 'approximate' solution to the (unconstrained!) problem

$$\text{(47)}\quad \text{minimize } L(v, y^k, r_k) \quad \text{over all } v$$

Then, by means of some rule that may involve information gleaned during this process, we generate y^{k+1} and r_{k+1} and repeat the step. The aim is to get a sequence $\{u^k\}$ that in some sense yields in the limit an optimal solution to the constrained problem (15).

Taking $y^k \equiv 0$ and $r_k \nearrow \infty$, we obtain the classical quadratic penalty method. Pure duality methods have non-trivial sequences $\{y^k\}$ but $r_k \equiv 0$. In 1968, Hestenes [11] and Powell [12] independently proposed the very easy rule

$$\text{(48)}\quad y_i^{k+1} = y_i^k + r_k F_i(u^{k+1})$$

The sequence $\{r_k\}$ was non-decreasing but did *not* have to tend to infinity in order to ensure that $\{u^k\}$ approached a solution to the constrained problem. (Powell actually allowed a different value of r_k for each F_i.) This was an important breakthrough, since faster convergence was demonstrated than with pure penalty methods, and some of the intrinsic numerical instability in the latter was sidestepped.

In the case of an inequality constraint $F_i(v) \leqslant 0$ in (45), the expression

$$y_i F_i(v) + \frac{r}{2} F_i(v)^2$$

in the augmented Lagrangian should be replaced by

$$(49) \quad \begin{cases} y_i F_i(v) + \frac{r}{2} F_i(v)^2 & \text{if } F_i(v) \geq -\frac{y_i}{r} \\ -\frac{y_i^2}{2r} & \text{if } F_i(v) \leq -\frac{y_i}{r} \end{cases}$$

The corresponding form of rule (48) is

$$(50) \qquad y_i^{k+1} = \max \{0, y_i^k + r_k F_i(u^{k+1})\}$$

For the theory of this case and a discussion of the criteria that can be used in solving (47) 'approximately', see [13, 14, 15, 16]. The theory makes heavy use of properties of maximal monotone multifunctions.

The Hestenes-Powell method has already been applied to certain variational inequalities that correspond to boundary-value problems of (convex) optimization type with equality constraints only; see Glowinski/Marrocco [17] and Mercier [18]. This work supposes u^{k+1} to be an *exact* solution to (47) at each iteration. Of course, solving (47) exactly is equivalent to solving a certain equation in v, namely

$$(51) \qquad \begin{aligned} 0 &= \nabla_v L(v, y^k, r_k) \\ &= \nabla F_0(v) + \sum_{i=1}^{m} y_i(v, y_i^k, r_k) \nabla F_i(v) \end{aligned}$$

where

$$(52) \qquad Y_i(v, y_i, r) = \begin{cases} y_i + rF_i(v) & \text{for equalities} \\ \max \{0, y_i + rF_i(v)\} & \text{for inequalities} \end{cases}$$

This could easily be generalized to variational inequalities not of optimization type by replacing the term $\nabla F_0(v)$ by $A_0(v) - a_0$.

Unfortunately, exact solution at each iteration cannot be obtained except in very special cases. Yet the kind of criterion for 'approximate' minimization in (47) that has been used in proofs of global convergence is

$$L(u^{k+1}, y^k, r_k) \leq \inf_v L(v, y^k, r_k) + \delta_k$$

and this has no analogue for variational inequalities not of optimization type.

8. PROXIMAL METHOD OF MULTIPLIERS

We have proposed [10] a modified version of the Hestenes-Powell algorithm that does carry over to generalized variational inequalities as represented in the form (17). There is no space here to explain the natural motivation; suffice it to say that the theory of maximal monotone multifunctions is deeply involved.

To apply the method one needs to specify a parameter value $s > 0$, and positive sequences $\{r_k\}$ and $\{\varepsilon_k\}$ satisfying

(53) $$r_k \uparrow r_\infty \leq \infty \quad \text{and} \quad \sum_{i=0}^{\infty} \varepsilon_k < \infty$$

One also makes an 'initial guess' (u^0, y^0) of the solution to (17). (In fact u^0 and y^0 can be chosen arbitrarily, for instance both zero). In iteration k, one forms the mapping

(54) $$T_k(v) = \frac{1}{s}(v - u^k) + A_0(v) - a_0 + \sum_{i=1}^{m} Y_i(v, y_i^k, r_k)\nabla F_i(v)$$

and looks for an approximate solution u^{k+1} to the equation $T_k(v) = 0$. Specifically, u^{k+1} is taken to be any vector satisfying

(55) $$\| T_k(u^{k+1}) \| \leq \frac{\varepsilon_k}{r_k} \max \{1, \|(u^{k+1}, y^{k+1}) - (u^k, y^k)\|_s\}$$

where

(56) $$\|(u,y)\|_s = [s^{-2}\|u\|^2 + \|y\|^2]^{1/2}$$

Then y^{k+1} is defined by

(57) $$y_i^{k+1} = Y_i(u^{k+1}, y_i^k, r_k) \quad \text{for} \quad i = 1,\dots,m$$

(Here Y_i is given by (52); this formulation covers any mixture of equality and inequality constraints. In the case of an equality constraint, the corresponding conditions in the first line of (17) are replaced simply by $F_i(u) = 0$.)

In problems of optimization type, the task of solving $T_k(v) = 0$ approximately reduces to that of minimizing $L(v, y^k, r_k) + (1/2s)\|v - u^k\|^2$ approximately.

Theorem

(see [10]) If A_0 is a continuous monotone mapping and each F_i is convex and differentiable (or in the case of an equality constraint, affine), then the mapping T_k in each iteration is strongly monotone with modulus $(1/s)$.

Assuming also that the expanded variational inequality (17) has at least one solution (u, y), it will be true that

(58) $$(u^k, y^k) \to (u^\infty, y^\infty)$$

where (u^∞, y^∞) is some particular solution (even though there may be more than one solution!)

Moreover there is a constant $q_s \in [0, \infty]$ such that

(59) $$\limsup_{k\to\infty} \frac{\|(u^{k+1}, y^{k+1}) - (u^\infty, y^\infty)\|_s}{\|(u^k, y^k) - (u^\infty, y^\infty)\|_s} \leqslant \theta\left(\frac{q_s}{r_\infty}\right)$$

where

$$\theta\left(\frac{q_s}{r_\infty}\right) = \begin{cases} \left(\frac{q_s}{r_\infty}\right)/[1 + \left(\frac{q_s}{r_\infty}\right)^2]^{1/2} < 1 & \text{if } q_s < \infty, r_\infty < \infty \\ 0 & \text{if } q_s < \infty, r_\infty = \infty \\ 1 & \text{if } q_s = \infty \end{cases}$$

If q_s is finite, and more will be said about this in a moment, the last part of the theorem guarantees linear convergence at a rate that can be controlled by how high the penalty parameter values r_k are allowed to go. The case where $r_k \nearrow \infty$ yields superlinear convergence. This is much superior to what would happen with a pure penalty method. In practice, the improved rate of convergence means that a satisfactory termination can be reached before r_k gets so high as to cause numerical instabilities. As $s \nearrow \infty$, q_s decreases to the constant that would appear in the corresponding convergence rate formula for the Hestenes-Powell algorithm, in the case of convex programming in $\mathbb{R}^N$.

If $q_s = \infty$, one still has global convergence (58) from any starting point (u^0, y^0), but it is no longer possible to establish a linear rate. However, it can be shown that this case is generically rare.

Proposition 4

Suppose $F_i(v) = G_i(v) - b_i$ for $i = 1,...,m$. Then for almost every choice of $a_0 \in V_0$ (finite-dimensional) and $b = (b_1,...,b_m) \in \mathbb{R}^m$ such that the variational inequality (17) has a solution,the corresponding constant q_s is finite.

This follows from a theorem of Mignot on the almost-everywhere differentiability of a maximal monotone multifunction on the interior of its effective domain in $\mathbb{R}^n$. For the argument, see [10].

The choice of s in the proximal point algorithm is somewhat problematical. It s is too low, q_s may be high, and the convergence ratio $\theta(q_s/r_\infty)$ will suffer. While this could be compensated for by choosing r_k high, there might be a price paid in numerical stability, After all, the whole point of the augmented Lagrangian approach is to succeed without r_k getting too high. It would be an improvement if instead of a fixed value for s the argorithm could be expressed in terms of a non-decreasing sequence $s_k \nearrow s_\infty \leqslant \infty$. This would allow some control over the phenomenon, just as in the case of the penalty parameter. Better still, in order to take account of second-order information, the factor $1/s$ in the definition (54) of T_k might be generalized to S_k^{-1}, where S_k is a positive definite matrix. The use of separate values of r_k in (57) for each constraint function, as proposed by Powell, could also be restored. Undoubtedly such developments are possible and desirable, but a significant enlargement of the theoretical apparatus (in terms of maximal monotone multifunctions) would be required.

Another direction of generalization, as mentioned earlier, would be to allow A_0 to be a multifunction and F_i to be non-differentiable. For this, the existing theory carries over with $w^{k+1} = T_k(u^{k+1})$ replaced by an element $w^{k+1} \in T_k(u^{k+1})$ in (55) at each interation. But effective numerical techniques would need to be developed for determining such u^{k+1} and w^{k+1}.

REFERENCES

[1] F.H. Clarke. 'Generalized gradients and applications'. *Trans. Am. Math. Soc.*, **205** (1975), 247-62.

[2] R.T. Rockafellar. 'Clarke's tangent cones and the boundaries of closed sets in $\mathbb{R}^n$'. *Non-linear Anal.* **3** (1979), 145-54.

[3] R.T, Rockafellar. 'Generalized derivatives and subgradients of nonconvex functions'. *Can. J. Math.* To appear.

[4] F.H. Clarke. 'The generalized problem of Bolza'. *SIAM J. Control & Optimiz.* **14** (1976), 682-99.

[5] F.H. Clarke. 'Extremal arcs and extended Hamiltonian systems'. *Trans. Am. Math. Soc.*, **231** (1977), 349-67.

[6] F.H. Clarke. 'A new approach to Lagrange multipliers'. *Math. Oper. Res.*, **1** (1976), 165-74.

[7] J.J. Moreau. 'Proximité et dualité dans un espace hilbertien'. *Bull. Soc. Math. Fr.*, **93** (1965), 273-99.

[8] R.T. Rockafellar. 'On the maximal monotonicity of subdifferential mappings'. *Pacific J. Math.*, **33** (1970), 209-16.

[9] R.T. Rockafellar. 'Maximality of sums of nonlinear monotone operators'. *Trans. Am. Math. Soc.*, **149** (1970), 75-88.

[10] R.T. Rockafellar. 'Monotone operators and augmented Lagrangian methods in nonlinear programming'. In *Nonlinerar Programming* vol. 3, eds. O.L. Mangasarian et al., Academic Press, New York (1978), 1-25.

[11] M.R. Hestenes. 'Multiplier and gradient methods'. *J. Optimiz. Theor. & Appl.*, **4** (1969), 303-20.

[12] M.J.D. Powell. 'A method for nonlinear constraints in minimization problems'. In *Optimization* ed. R. Fletcher, Academic Press, New York (1969), 283-98.

[13] R.T. Rockafellar. 'The multiplier method of Hestenes and Powell applied to convex programming'. *J. Optimiz. Theor. & Appl.*, **12** (1973), 555-62.

[14] R.T. Rockafellar. 'Solving a nonlinear programming problem by way of a dual problem'. *Symposia Mathematica* (Istituto Nazionale Alta Matematica) vol. XIX, Academic Press, New York (1976), 135-60.

[15] R.T. Rockafellar. 'Augumented Lagrangians and applications of the proximal point algorithm in convex programming'. *Math. Oper. Res.*, **1** (1976), 97-116.

[16] D.P. Bertsekas. 'Multiplier methods: a survey'. *Proc. IFAC 6th Triennial World Congress*, Part 1B.

[17] R.Glowinski and A. Marocco. 'Sur l'approximation, par éléments finis d'ordre un, et la resolution, par pénalisation-dualité, d'une classe de problems de Dirichlet non linéaires'. *C.R. Acad. Sci., Paris,* **278** (1974), 1649.

[18] B. Mercier. 'Approximation par éléments finis et resolution, par un algorithme de penalisation-dualité, d'un problème d'élasto-plaslicité'. *C.R. Acad. Sci., Paris,* **280** (1975), 287-90.

Chapter 21

A Numerical Approach to the Solution of Some Variational Inequalities

F. Scarpini

PREAMBLE

I think it is right, in a volume dedicated to the memory of Professor Guido Stampacchia, to state in advance some remarks on the numerical treatment of variational inequalities carried out by his pupils. I refer, in particular, to the work done by some teachers of the Mathematical Institute at the University of Rome, while for more ample references, I refer to [25].

Beginning in 1967, U. Mosco, in some papers [25] was interested in approximations of the solutions of variational inequalities and their convergence. In 1969, in Rome, the numerical aspects of the problem were considered, in a concrete way, and some applications on a computer [12, 32, 7] carried out. The problem of projecting an assigned function on the positive cone in a Sobolev space was considered, the convenience of using, in the finite-dimensional case, a complementarity system formulation of the problem was pointed out. In a first paper [12] the above problem was studied in the space of the positive measures so constructing a dual solution method, by carrying out a convenient discretization via the approximation of the Green function for the Dirichlet problem and by constructing, in the end, an algorithm for the solution of the complementarity system so achieved [32].

Later, the finite element method was applied to treat the same problem directly in the primal Sobolev space. As a consequence algorithms for the numerical solution of the complementarity systems with one or two constraints [26, 30, 17] were studied. Only lately have we considered the case of non-linear differential operators of monotone type and the related algorithms that are explained in the present paper [31].

1. INTRODUCTION

We consider the following variational problem:

$$(\mathrm{P}_c) \qquad \begin{aligned} & u \in \mathbb{K}_c \\ & \langle Au - f, v - u\rangle \geqslant 0 \qquad \forall v \in \mathbb{K}_c, \; c = 1, 2 \end{aligned}$$

where Ω is a bounded open subset of $\mathbb{R}^n$ with sufficiently smooth boundary $\partial\Omega = \Gamma$; $\mathscr{H}$ is a real reflexive Banach space, a complete vector lattice for the partial order relation induced by a closed positive cone $P = \{v, \in \mathscr{H} : v \geqslant 0\}$; V is a subspace, a complete sublattice of $\mathscr{H}$; V' is the dual of V; $\langle \cdot \, , \cdot \rangle$ is the duality pairing between V' and V; $\| \cdot \|$ ($\| \cdot \|_*$) is the norm on V (on V'); $[u^0, +\infty]$ is the ordered unbounded interval $\{v : v \in \mathscr{H} / v \geqslant 0\}$; $[u_0, v_0]$ is the ordered interval $\{v : v \in \mathscr{H} / v_0 \geqslant v \geqslant u_0\}$, $v_0 - u_0 \in P$; $\mathbb{K}_1 = [u_0, +\infty] \cap V$ ($\mathbb{K}_2 = [u_0, v_0] \cap V$); $f \in V'$; $A : V \to V'$ a bounded operator which is:
(i) hemicontinuous: $t \to \langle A(u + tv), \omega\rangle$ is continuous $\forall u, v, \omega \in V$;
(ii) strictly monotone: $\langle Av - Au, v - u\rangle > 0 \quad u, v \in V, u \neq v$;
(iii) coercive:

$$\exists z_0 \in V / \ \frac{\langle Av, v - z_0\rangle}{\|v\|} \to +\infty$$

when $\|v\| \to +\infty$.

The problem (P_1) under the hypotheses (i)-(iii) (the problem (P_2) under the hypotheses (i), (ii)) admits only a solution u (see [18]).

We apply the finite affine element method to (P_c). To simplify we shall suppose Ω is a bounded, convex, open subset of $\mathbb{R}^2$. Given $0 < h \leqslant 1$, we first inscribe a convex polygon Ω_h in Ω, whose vertices belong to Γ and whose sides have a length not exceeding h. We then decompose Ω_h with 'a regular triangulation' [35] and use the following notation: x_i are the nodes in Ω, $i \in I = \{1, \ldots, N\}$; $\varphi_i^h(x)$ is the finite affine element such that $\varphi_i^h(x_i) = \delta_{ij}$ (Kronecker δ) $i, j \in I$; V_h is the space

$$v_h : v_h = \sum_{i \in I} V_i \varphi_i^h(x)$$

$$u_h^0 = \sum_{i \in I} U_i^0 \varphi_i^h \qquad U_i^0 = u_0(x_i)$$

and

$$v_h^0 = \sum_{i \in I} V_i^0 \varphi_i^h \qquad V_i^0 = v_0(x_i)$$

are the interpolated functions of u_0, v_0, under the hypotheses $u_0, v_0 \in C^0(\overline{\Omega}) \cap \mathscr{H}$;

$$\mathbb{K}_{1h} = [u_h^0, +\infty] \cap V_h \qquad (\mathbb{K}_{2h} = [u_h^0, v_h^0] \cap V_h)$$

$$\mathbb{K}_{1N} = [U^0, +\infty] \qquad (\mathbb{K}_{2N} = [U^0, V^0])$$

$$u_h = \sum_{i \in I} U_i \varphi_i^h \qquad U = \{U_i\}_{i \in I}$$

are the unknowns of the approximate problem;

$$b = \{b_i\}_{i \in I} \qquad b_i = \int_\Omega f \varphi_i^h \, dx$$

$F(\cdot) = \{F_i(\cdot)\}_{i \in I} : \mathbb{R}^N \to \mathbb{R}^N$ is the approximate operator.

We obtain the following problems in V_h and $\mathbb{R}^N$:

(P_{ch}) $\qquad u_h \in \mathbb{K}_{ch} : \langle Au_h - f, v_h - u_h \rangle \geqslant 0 \qquad \forall v_h \in \mathbb{K}_{ch}$

(P_{ch}) $\qquad U \in \mathbb{K}_{cN} : \langle F(U) - b, V - U \rangle \geqslant 0 \qquad V \in \mathbb{K}_{cN} \quad (c = 1, 2)$

We can verify, at once, that (P_{1N}), (P_{2N}) are, respectively, equivalent to the following complementarity systems:

$$U \geqslant U^0$$
$$W = F(U) - b \geqslant 0 \tag{1}$$
$$W(U - U^0) = 0$$

$$V^0 \geqslant U \geqslant U^0$$
$$W = F(U) - b \tag{2}$$
$$W^+(U - U^0) = W^-(V^0 - U) = 0$$

where $W^+ = \sup(W,0)$, $W^- = \sup(-W,0)$. (1), (2) admit only one solution U (see [19, 10]).

Of course, we meet computational difficulties in solving them and that seems to suggest the construction of more and more efficient algorithms (see [1, 2, 3, 4, 6, 12, 14, 24, 26, 30, 31, 33, 34, 36].

We propose now some interative algorithms for the solution of (1) and (2). They supply q vectors U^n such that $U^0 \leqslant U^1 \leqslant \ldots \leqslant U^q = U$ in the case (1) [26, 31], and two sequences $\{U^n\}$, $\{V^n\}$ converging to the solution U of (2) [30, 31] : $U^n \uparrow U, W^n \downarrow U$.

To carry out the algorithms we use convenient hypotheses very often satisfied (see section 3) in the applications.

2. ALGORITHMS

2.1. Basic Knowledge

With the aim of providing background for the treatment of (1) and (2), we collect some definitions and theorems widely expounded in papers [10, 11, 21, 22, 23, 28, 29].

Definition 1 The mapping $F = \{F_i\}_{i \in I} : H \subset \mathbb{R}^N \to \mathbb{R}^N$ is a P-function if for each $U \neq V$ in H there is an index $i(U, V)$ such that:

$$(F_i(V) - F_i(U)) \cdot (V_i - U_i) > 0$$

We denote by (P_F) the class of non-linear P-functions, and by (P) the class of P matrices belonging to $L(\mathbb{R}^N)$ (see [11]). It's obvious that F strictly monotone implies $F \in (P_F)$.

Theorem 1
Let $F{:}H \to \mathbb{R}^N$ be Gateaux differentiable in the open convex set $H_0 \subset H$ then we have:

$$F'(U) \in (P) \ \text{ in } H_0 \Rightarrow F(U) \in (P_F) \ \text{ in } H_0$$

Definition 2 The mapping $F{:}H \to \mathbb{R}^N$ is *isotone* (*antitone*) *strictly isotone* (*strictly antitone*), if we have, respectively:

$$\begin{matrix} U \leqslant V \Rightarrow F(U) \leqslant F(V) & \Big(F(U) \geqslant F(V)\Big) \\ U < V \Rightarrow F(U) < F(V) & \Big(F(U) > F(V)\Big) \end{matrix} \quad \text{in } H$$

Definition 3 The mapping $F{:}H \to \mathbb{R}^N$ is (*strictly*) *inverse isotone* in H if

$$F(U) \leqslant F(V) \quad (F(U) < F(V)) \Rightarrow U \leqslant V \quad (U < V) \quad \forall\, U, V \in H$$

The matrix $F \in L(\mathbb{R}^N)$ is inverse isotone if it is non-singular and $F^{-1} \geqslant 0$ (see [37]).

Theorem 2
Let $F{:}H \to \mathbb{R}^N$ be G-differentiable in the open convex set $H_0 \subset H$ then

$$F'(U)^{-1} \geqslant 0 \Rightarrow F(U) \text{ is inverse isotone in } H_0$$

Here $e_i = \{\delta_{ij}\}_{j \in I}, i \in I.$

Definition 4 The mapping $F{:}H \to \mathbb{R}^N$ is (*strictly*) *diagonally isotone* if the functions:

$$F_{ii}(t) = F_i(U + tc_i) \qquad \forall\, t \in \mathbb{R}^1,\ U \in \mathbb{R}^N$$

such that

$$U + te_i \in H \;\; i \in I$$

are (*strictly*) isotone.

Definition 5 The mapping $F{:}H \to \mathbb{R}^N$ is (*strictly*) *off-diagonally antitone* if the functions:

$$F_{ij}(t) = F_i(U + te_j) \qquad \forall\, t \in \mathbb{R}^1,\ U \in \mathbb{R}^N$$

such that

$$U + te_j \in H \qquad i \neq j,\ i,j \in I$$

are (*strictly*) antitone.

We denote by (Z_F) the class of off-diagonally antitone non-linear functions, and by (Z) the class of matrices with non-positive off-diagonal elements [10].

Theorem 3
Let $F{:}H \to \mathbb{R}^N$ be G-differentiable in the open convex set $H_0 \subset H$, then we have

$F'_{ii}(U) \geqslant 0\ (> 0)\ \forall\ U \in H_0 \Rightarrow F(U)$ is (*strictly*) diagonally isotone in H_0

$F'_{ij}(U) \leqslant 0\ (< 0)\ i \neq j,\ \forall\ U \in H_0 \Rightarrow F(U) \in (Z_F)$ (in the strict sense) in H_0

Definition 6 The mapping $F{:}H \to \mathbb{R}^N$ is an *M-function* if is inverse isotone and $F \in (Z_F)$. We denote by (M_F) the class of non-linear M-functions, and by (M) the class of Minkowski matrices [37].

Theorem 4
Let $F \in L(\mathbb{R}^N)$, we have (see [10]).

$$F \in (P) \cap (Z_F) \Rightarrow F \in (M)$$

Theorem 5
Let $F \in (P_F) \cap (Z_F)$ on $[U^0, V^0]$ then $F \in (M_F)$ on $[U^0, V^0]$ (see [19]).

Theorem 6
If $F{:}H \to \mathbb{R}^N$ and $F \in (M_F)$ then F is strictly diagonally isotone together with $F^{-1}: F(H) \to \mathbb{R}^N$.

Definition 7 The mapping $F{:}H \to \mathbb{R}^N$ is (*strictly*) *order-convex* on the convex set H, if we have

$$(1-t)\,F(U) + tF(V) - F((1-t)\,U + tV) \geqslant 0 \; (> 0)$$

$$\forall\, t \in \,]0,1[,\ \forall\ U, V \in H \text{ comparable } (U \leqslant V, U \geqslant V)(U \neq V)$$

Theorem 7
Let $F{:}H \to \mathbb{R}^N$ be G-differentiable in the open convex set $H_0 \subset H$, then we have

$$F \text{ (strictly) order-convex} \Longleftrightarrow F(V) - F(U) - F'(U)(V-U) \geqslant 0 \; (> 0)$$

$\forall\ U, V \in H_0$ comparable $(U \neq V)$.

2.2. An algorithm for solving (1)

Let the mapping $F{:}\mathbb{K}_{1N} \to \mathbb{R}^N$ be a continuous, coercive, P-function; then the system (1) has a unique solution U (see [19, 20]). We suppose also $F \in (Z_F)$. In the linear case we merely suppose $F \in (P) \cap (Z)$. We try to compute U.

Step 1 If $F(U^0) - b \geqslant 0$, U^0 is the solution of (1). If $F(U^0) - b \ngeqslant 0$ set

$$J_1 = \{i : i \in I / F_i(U^0) - b_i < 0\}, \qquad J_1' = I \setminus J_1$$

$$V^0 = \left(V_{J_1}, U^0_{J_1'}\right) \qquad \widetilde{U}^0 = \left(0_{J_1}, U^0_{J_1'}\right)$$

Owing to the coerciveness, we have

$$\left(F(V^0) - b, V^0 - \widetilde{U}^0\right)\left[\frac{\left(F_{J_1}(V_{J_1}, U^0_{J_1}), V_{J_1}\right)}{\|V_{J_1}\|} \sim \|b_{J_1}\|\right]\|V_{J_1}\| \geqslant 0$$

$\forall V_{J_1}$ with $\| V_{J_1} \| = r$ large enough.

Because of a well known existence theorem ([18], chapter 1, Lemma 4.3), there exists the solution $U^1_{J_1}$ of the system

$$\{F(U^1_{J_1}, U^0_{J'_1}) - b\}_{J_1} = 0$$

Moreover, $F \in (M_F)$ (see Theorem 5) and so the solution is unique. Putting $U^1 = (U^1_{J_1}, U^0_{J'_1})$ we have $U^0 \leqslant U^1$.

Step n If $F(U^{n-1}) - b \geqslant 0$, U^{n-1} is the solution of (1). If $F(U^{n-1}) - b \not\geqslant 0$ set

$$I_n = \{i : i \in I / F(U^{n-1}) - b < 0\} \qquad J_n = I_n \cup J_{n-1}$$

$$J'_n = I \setminus J_n \qquad V^{n-1} = \left(V_{J_n}, U^{n-1}_{J'_n}\right) \quad \tilde{U}^{n-1} = \left(0_{J_n}, U^{n-1}_{J'_n}\right)$$

Owing to the coerciveness, we have

$$\left(F(V^{n-1}) - b, V^{n-1} - \tilde{U}^{n-1}\right) = \left(F_{J_n}(V_{J_n}, U^{n-1}_{J'_n}) - b_{J_n}, V_{J_n}\right)$$

$$\geqslant \left[\frac{F_{J_n}(V_{J_n}, U^{n-1}_{J'_n}), V_{J_n})}{\| V_{J_n} \|} - \| b_{J_n} \|\right] \| V_{J_n} \| \geqslant 0 \qquad \forall V_{J_n}$$

with $\| V_{J_n} \| = r$ large enough.

There is a unique solution $U^n_{J_n}$ of the system

$$\{F(U^n_{J_n}, U^{n-1}_{J'_n}) - b\}_{J_n} = 0$$

and putting $U^n = (U^n_{J_n}, U^{n-1}_{J'_n})$, we have $U^{n-1} \leqslant U^n$.

Clearly after a finite number $q \leqslant N$ of steps, we reach the solution U of (1) with a monotone non-decreasing procedure $U^0 \leqslant U^1 \leqslant \ldots \leqslant U^{q-1} \leqslant U^q = U$ (see [4, 6, 26, 36])

2.3. An algorithm for solving (2)

We suppose F is a continuous P-function on $\mathbb{K}_{2N}$ so there exists a unique solution of (2) (see [19, 20]).

We also suppose $F \in (Z_F)$. In the linear case we merely suppose $F \in (P) \cap (Z)$.

Preliminary test If $F(U^0) - b \geqslant 0$ or $F(V^0) - b \leqslant 0$, then $U = U^0$ or $U = V^0$

are respectively the solutions of (2).

First iteration Put $I_0 = I$.

Test I If

$$\{F(U^0) - b\}_1 < 0 \qquad \{F(V_1^0, U_{I\setminus 1}^0 - b\}_1 \leqslant 0$$

set

$$U_1^1 = V_1^0 \qquad U^1 = (U_1^1, U_{I\setminus 1}^0) \qquad I_1 = I_0 \setminus 1$$

In fact, $F \in (Z_F)$ implies $\{F(V_1^0, U_{I\setminus 1} - b\}_1 < 0, \forall U_{I\setminus 1}^0 \leqslant U_{I\setminus 1} \leqslant V_{I\setminus 1}^0$.

Test II If

$$\{F(U^0) - b\}_1 < 0 < \{F(V_1^0, U_{I\setminus 1}^0) - b\}_1$$

compute U_1^1 such that

$$\{F(U_1^1, U_{I\setminus 1}^0) - b\}_1 = 0$$

We have $U_1^0 < U_1^1 < V_1^0$ (Theorem 6); set $U^1 = (U_1^1, U_{I\setminus 1}^0), I_1 = I_0$.

Test III If

$$\{F(U^0) - b\}_1 \geqslant 0$$

put $U^1 = U^0, I_1 = I_0$.

nth iteration with $n \leqslant N$ In the preceding iteration we have the ordered set $I_{n-1} \subseteq I_0$ and the configuration U^{n-1} satisfying $U^0 \leqslant U^{n-1} \leqslant V^0$. We consider now the index i_n following i_{n-1} in I_{n-1}.

Test I If

$$\{F(U^{n-1}) - b\}_{i_n} < 0 \qquad \{F(V_{i_n}^0, U_{I\setminus i_n}^{n-1}) - b\}_{i_n} \leqslant 0$$

put $U_{i_n}^n = V_{i_n}^0, U^n = (U_{i_n}^n, U_{I\setminus i_n}^{n-1}), I_n = I_{n-1} \setminus i_n$.

Test II If

$$\{F(U^{n-1}) - b\}_{i_n} < 0 < \{F(V_{i_n}, U_{I\setminus n}^{n-1}) - b\}_{i_n}$$

compute $U_{i_n}^n$ such that

$$\{F(U_{i_n}^n, U_{I\setminus i_n}^{n-1}) - b\}_{i_n} = 0$$

We have $U_{i_n}^{n-1} < U_{i_n}^n < V_{i_n}^0$; set

$$U^n = (U^n_{i_n}, U^{n-1}_{I \setminus i_n}), \quad I_n = I_{n-1}$$

Test III If

$$\{F(U^{n-1}) - b\}_{i_n} \geq 0$$

put $U^n = U^{n-1}$, $I_n = I_{n-1}$.

Continuing in this way we run along a chain of loops all of them consisting of N iterations at most. We can prove that, generally speaking, a sequence $\{U^n\}$ will be obtained monotonically non-decreasing and the following theorem holds.

Theorem 8
The limit U of sequence $\{U^n\}$ exists and coincides with the solution of the system (2).

Proof: We put $W^n = F(U^n) - b$ and consider the vector $\{U^n, W^n\}$. If U^n is the terminal vector of the mth cycle, we obtain the following decomposition:

$$I = T_n \cup J_n \cup G_n \qquad U^n = (U^n_{T_n}, U^n_{J_n}, U^n_{G_n})$$

(a₁) $U^n_{T_n} = V^0_{T_n}$ consists of the components of the initial configurations which assume during the m cycles the values of the upper obstacle owing to the fact that test I is satisfied.

We have

$$W^n_{T_n} = \{F(U^n) - b\}_{T_n} \leq 0$$

The condition $F \in (Z_F)$ implies $W^s_{T_s} \leq 0 \; \forall s \geq n$. The sequence of sets $\{T_n\}$ is non-decreasing;

$$T_{n-1} \subseteq T_n \qquad \lim T_n = \bigcup_n T_n = T$$

exists.
(a₂) $U^n_{J_n} = U^0_{J_n}$ consists of the components of the initial configuration not altered, owing to the fact that test II is satisfied. The sequence $\{J_n\}$ of the sets for which there is contact with the lower obstacle is monotonically non-increasing;

$$J_n \subseteq J_{n-1} \qquad \lim J_n = \bigcap_n J_n = J$$

exists.

(a_3) $U^n_{G_n}$ is the subvector obtained by solution of the equations considered in test II. The sequence $\{G_n\} = \{I \setminus (T_n \cup J_n)\}$ converges to the set $G = I \setminus (T \cup J)$. Owing to the monotonicity of $\{U^n\}$ and the continuity of F,

$$\lim U^n = \widetilde{U} \qquad \lim W^n = \widetilde{W}$$

exist.

Now we want to prove that $\widetilde{U} = U$, $\widetilde{W} = W$ where $\{U, W\}$ is the solution of the system (2).

(b_1) Let us consider an arbitrary index $i \in T$ which belongs to a certain T_{n_i} and consequently to all T_n, with $n > n_i$. If follows that

$$U_i^{n_i} = V_i^0 = U_i^n \quad \text{and then} \quad \widetilde{U}_i = V_i^0$$

$$W_i^n \leqslant W_i^{n_i} \leqslant 0 \quad \text{for } n \geqslant n_i \; \forall i \in T$$

and from this

$$\widetilde{W}_T \leqslant 0$$

(b_2) Let us consider an arbitrary index $i \in J$ which will belong to every $J_{\underline{n}} \cdot U_i^n = U_i^0$ during the whole iterative procedure $\forall\, i \in J$. It results that $\widetilde{U}_j = U_j^0$. From test III the following relations hold

$$W_J^n \geqslant 0 \qquad \forall n$$

and consequently

$$\widetilde{W}_J \geqslant 0$$

(b_3) Now let i belong to G. Because of the equations solved during the cycles, there exists, corresponding to i, a subsequence $\{n_j\} \subseteq \{n\}$ such that

$$W_i^{n_j} = \{F(U^{n_j}) - b\}_i = 0 \;\; \forall\, n_j$$

and then

$$\widetilde{W}_i = \lim_{n \to \infty} W_i^n = \lim_{j \to \infty} W_i^{n_j} = 0 \qquad \forall\, i \in G$$

Therefore we get $\widetilde{W}_G = 0$.

It can be easily concluded that $\{\widetilde{U}, \widetilde{W}\}$ is a solution of (2); thanks to the uniqueness theorem it must coincide with $\{U, W\}$.

We can take V^0 as the initial vector and symmetrically construct a non-increasing algorithm.

2.4. Global Newton monotone algorithms

If F is non-linear, we construct some monotone algorithms for the solution of (2).

Theorem 9
Let $F:H \to \mathbb{R}^n$ be
(j) order-convex on $[U^0, V^0] \subset H$;
(jj) G-differentiable, with continuous derivative F' on $[U^0, V^0] \subset H_0$ (open set) $\subset H$, such that: $F' \epsilon (P) \cap (Z)$.
Then starting from the initial vectors U^0, V^0 we can generate two monotone sequences $\{U^n\}$, $\{V^n\}$ converging to the solution U of (2):

$$U^n \uparrow U \qquad V^n \downarrow V \tag{3}$$

Proof: We verify the first relation in (3). Set

$$G(U) = F'(V^0)(U - U^{n-1}) + F(U^{n-1}) \qquad I_0 = I$$

Step $n = 1$
Test I If

$$\{G(U^0) - b\}_1 < 0 \qquad \{G(V_1^0, U_{I\setminus 1}^0) - b\}_1 \leqslant 0$$

put

$$U_1^1 = V_1^0 \qquad U^1 = (U_1^1, U_{I\setminus 1}^0) \qquad I_1 = I_0 \setminus 1$$

In fact, owing to (j) and Theorem 7, we have

$$\{F(U^1) - b\}_1 \leqslant \{G(U^1) - b\}_1$$

On the other hand, the condition $F \epsilon (Z)$ implies $F \epsilon (Z_F)$ (Theorem 3). $\{F(W) - b\}_1$ is off-diagonally antitone on $[U^0, V^0]$.
Test II If

$$\{G(U^0) - b\}_1 < 0 < \{G(V_1^0, U_{I\setminus 1}^0) - b\}_1$$

compute U_1^1 such that

$$\{G(U^1)-b\}_1 = 0 \qquad \text{where} \qquad U^1 = (U_1^1, U_{I\setminus 1}^0)$$

$F' \in (P)$ implies $U_1^0 < U_1^1 < V_1^0$; set $I_1 = I_0$.

Test III If

$$\{G(U^0)-b\}_1 \geq 0$$

set $U^1 = U^0$, $I_1 = I_0$.

Repeating the iterations in this way with $n = 2, 3, \ldots$, we construct a monotonically non-decreasing sequence $\{U^n\}$ and we can prove that the limit of $\{U^n\}$ exists and coincides of course with the solution U of the system (2).

Symmetrically, starting from the initial vector V^0, we can prove the second relation in (3).

3. UNILATERAL DIRICHLET PROBLEMS RELATED TO SOME OPERATORS

3.1. Unilateral problems for the Laplace operator

Set $V = H_0^1(\Omega)$, $V' = H^{-1}(\Omega)$ and consider the strictly convex functional:

$$J(u) = \frac{1}{2} \int_\Omega |\text{grad } u|^2 \, dx + \frac{1}{2} \int_\Omega |u|^2 \, dx \qquad \forall u \in V$$

J is differentiable on V:

$$J'(u) = Au = -\Delta u + u \qquad (\Delta \text{ Laplace operator})$$

A is bounded, coercive [18]. We apply the finite affine element method relative (for example) to an equilateral triangulation and obtain:

$$F(U) = \{F_i(U)\}_{i \in I} \qquad F_i(U) = \sum_{j \in I} [a_{ij} + h^2 \alpha_{ij}] \, U_j$$

where $a_{ii} = 6$, $a_{ij} = -1$ if the vertices x_i and x_j are contiguous, $a_{ij} = 0$ elsewhere and

$$\alpha_{ij} = \frac{1}{h^2} \int_\Omega \varphi_i^h \varphi_j^h \, dx$$

Thus we have $F \in (P) \cap (Z)$ and we can use the algorithms explained in section 1.2 and 1.3.

3.2. Unilateral problems related to a first non-linear operator

Set

$$V = H_0^1(\Omega) \cap L^p(\Omega) \qquad p > 2$$

$$V' = H^{-1}(\Omega) + L^{p'}(\Omega) \qquad \frac{1}{p} + \frac{1}{p'} = 1$$

and consider the strictly convex functional

$$J(u) = \frac{1}{2} \int_\Omega |\text{grad } u|^2 \, dx + \frac{1}{p} \int_\Omega |u|^p \, dx \qquad \forall u \in V$$

J is twice differentiable on V:

$$J'(u) = Au = -\Delta u + |u|^{p-2} u$$

$$J''(u) = A'(u)\, v = -\Delta v + (p-1)\, |u|^{p-2} v$$

$A(u)$ is bounded, continuous, strictly monotone, strictly convex on $[0, V_0]$. If $n = 2$, owing to the Sobolev imbedding theorem, we have

$$\langle J''(u)\, v, v \rangle \geqslant \|v\|^2_{H_0^1(\Omega)} \qquad \forall u, v \in V$$

We apply the finite element method as in the section 2.1 and obtain:

$$F(U) = \{F_i(U)\}_{i \in I} \qquad F_i(U) = \sum_{j \in I} [a_{ij} + h^2 \alpha_{ij}(U)]\, U_j$$

where, now

$$\alpha_{ij} = \frac{1}{h^2} \int_\Omega |u_h|^{p-2} \varphi_i^h \varphi_j^h \, dx$$

$$F'(U) = \{F'_{ij}(U)\}_{i, j \in I} \qquad F'_{ij}(U) = a_{ij} + (p-1)\, h^2 \alpha_{ij}$$

Thus, we have: (j) $F(U)$ is convex on $[0, V^0]$; (jj) $F'(U) \in (P) \cap (Z)$ and we can use the algorithms explained in section 2.4.

3.3. Unilateral problems for the pseudo-Laplace operator

Set

$$V = W_0^{1,p}(\Omega) \qquad p > 2$$

$$V' = W^{-1,p'}(\Omega)$$

and consider the strictly convex functional

$$J(u) = \frac{1}{p} \|u\|^p$$

We have

$$J'(u) = Au = -\sum_{i=1} \int_\Omega D_i\,(|D_i u|^{p-2} D_i u)\,dx$$

$$J''(u) = A'(u)\,v = -\sum_{i=1}^{n} D_i\,(|D_i u|^{p-2} D_i v)\,dx$$

A (the pseudo-Laplace operator) is bounded, continuous, coercive, strictly monotone in V (see [18, 34]):

$$\|A(v) - A(u)\|_* \leqslant \beta \|v - u\| \,(\|u\|^{p-2} + \|v\|^{p-2}) \qquad \beta > 0$$

$$\langle A(v) - A(u), v - u\rangle \geqslant \alpha \|v - u\|^p \qquad \alpha > 0$$

If, for example, we consider the case $\Omega =]0, 1[\times]0, 1[$ and use the well known finite affine Courant elements, corresponding with the decomposition of Ω in $(m+1)^2$ squares of side h, we obtain the following representation

$$F(U) = \{F_i(U)\}_{i \in I} \qquad F_i(U) = \sum_{j \in I} a_{ij}\, U_j$$

$$a_{ij}(U) = -\frac{1}{h^{p-2}}\, |U_i - U_j|^{p-2} \qquad \text{for } j = i \pm m, i \pm 1 \in I, N = m^2$$

$$a_{ij}(U) = 0 \qquad \text{for the other values of } j \neq i$$

$$a_{ij}(U) = -\sum_{j \neq i}^{1, N} a_{ij}(U)$$

and the following Jacobian matrix $F'(U)$

$$F'_{ij}(U) = (p-1)\, a_{ij}(U) \qquad i, j \in I$$

We have $F \in (Z_F)$ (Theorem 3). The condition A bounded, continuous, coercive, strictly monotone in V, involves F continuous, coercive, strictly monotone in $\mathbb{R}^N$. So the condition $F \in (P_F) \cap (Z_F)$ is fulfilled and we can apply the algorithms of sections 2.2 and 2.3.

REFERENCES

[1] J.P. Aubin.'Approximation des problèmes aux limites non homogènes pour des opérateurs non linéares'. *J. Math. Anal. & Appl.*, **30** (1970), 510-21.

[2] H. Brezis, and M. Sibony. 'Méthodes d'approximation et d'iteration pour les opérateurs monotones'. *Arch. Ration. Mech. & Anal.*, **28** (1968), 59-82.

[3] J. Céa, and R. Glowinski. 'Sur des méthodes d'optimization par relaxion'. *R.A.I. R.O.*, **R. 3** (1973), 5-32.

[4] R. Chandrasekaran. 'A special case of the complementary pivot problem'. *Opsearch,* **7** (1970), 263-8.

[5] V. Comincioli. 'Metodi di rilassamento per la minimizzazione in uno spazio prodotto'. *L.A.N. - C.N.R.* **20** (1971).

[6] R.W. Cottle, and M.S. Goheen. 'A special class of large quadratic programs'. *Tech. Report 76-7,* Dept of Operation Research, Stanford, 1976.

[7] V. De Angelis, A. Fusciardi, and A. Schiaffino. 'Some methods for the numerical solution of variational inequalities with two obstacles'. *Calcolo,* **3-4** (1973), 285-338.

[8] J.F. Durand. 'Résolution numerique dea problèmes aux limites sous-harmoniques'. *Thèse,* Univ. de Montpellier (1968).

[9] J.F. Durand. 'L'algorithme de Gauss-Seidel appliqué à un problème unilatéral non symétrique'. *R.A.I.R.O.*, **R.2** (1972), 23-20.

[10] M. Fiedler, and V. Ptak. 'On matrices with nonpositive off-diagonal elements and positive principal minors'. *Czech.J.*, **12** (1962), 382-400.

[11] M. Fiedler, and V. Ptak. 'Some generalisations of positive definiteness and monotonicity'. *Numer. Math.*, **9** (1966), 163-72.

[12] A. Fusciardi, U. Mosco, F. Scarpini, and A. Schiaffino. 'A dual method for the numerical solution of some variational inequalities'. *J.Math. Anal. & Appl.*, **40** (1972), 471-93.

[13] A. Fusciardi. 'A penalty method for the approximation of projections over cones in Hilbert spaces'. *Calcolo,* **14** (1977), 205-18.

[14] R. Glowinski. 'La méthode de relaxion'. *Quaderni di Matematica,* Ist. Mat. Univ. Roma, 1971.

[15] R. Glowinski, and A. Marocco. 'Sur l'approximation par éléments finis d'ordre un et la résolution par pénalisation dualité d'une classe de problèmes de Dirichlet non linéaires'. *I.R.I.A. Laboria Rapp. de Rech.*, 115 (1975).

[16] R. Glowinski, J.L. Lions, and R. Trémolières. *'Analyse Numérique des inéquations variationelles'.* Dunod, Paris (1976).

[17] C. Levati, F. Scarpini, and G. Volpi. 'Sul trattamento numerico di alcuni problemi variazionali di tipo unilaterale'. *L.A.N. - C.N.R.* 82 (1974).

[18] J.L. Lions. *'Quelques méthodes de résolution des problèmes aux limites non linéaires'.* Dunod, Paris (1969).

[19] J.J. Moré. 'The application of variational inequalities to complementarity problems and existence theorems'. *Tech. Report 71-110,* Cornell. Univ., Ithaca, New York, 1971.

[20] J.J. Moré. 'Coercitivity conditions in nonlinear complementarity problems'. *Tech. Report 72-146,* Cornell Univ., Ithaca, New York, 1972.

[21] J.J. Moré. 'Nonlinear generalisation of matrix diagonal dominance with application to Gauss-Seidel iterations'. *SIAM J. Numer. Anal.*, **9** (1972), 357-78.

[22] J.J. Moré. 'Classes of functions and feasibility conditions in nonlinear complementarity problems'. *Tech. Report 73-174,* Cornell Univ., Ithaca, New York, 1973.

[23] J.J. Moré, and W. Rheinboldt. 'On P and S-functions and related classes of n-dimensional nonlinear mappings'. *Tech. Report 70-120,* Computer Science Center, Univ. of Maryland, 1970.

[24] J.J. Moreau. 'Traitement numérique d'un probléme aux dérivées partielles de type unilateral'. *Univ' Montréal. Sem. d'Anal. Numer.* **15** (1969).

[25] U. Mosco. 'An Introduction to the approximate solution of variational inequalities'. *Constructive Aspects of Functional Analysis,* Corso CIME 1971, Cremonese, Roma (1973).

[26] U. Mosco, and F. Scarpini. 'Complementarity system and approximations of variational inequalities'. *R.A.I.R.O.,* **R.1** (1975), 83-104.

[27] U. Mosco. 'Implicit variational problems and quasi variational inequalities'. Lectures at summer school on '*Nonlinear Operators and the Calculus of Variations*'. *Bruxelles*, Springer-Verlag (1976), 83-156.

[29] W. Rheinboldt. 'On M-functions and their applications to nonlinear Gauss-Seidel iterations and to network flows'. *J. Math. Anal. & Appl.,* **32** (1970), 274-307.

[30] F. Scarpini. 'Some algorithms solving the unilateral Dirichlet problem with two constraints'. *Calcolo,* **12** (1975), 113-49.

[31] F. Scarpini. 'Some nonlinear complementarity systems algorithms and applications to unilateral boundary-value problems'. To appear.

[32] F. Scarpini, and T. Valdinoci. 'Su alcuni sistemi di complementaritá connessi a disequazioni variazionali di tipo ellittico'. *Calcolo,* **9** (1072), 143-82.

[33] M. Sibony. 'Méthodes iteratives pour les équations et inéquations aux dérivées partielles nonlinéares de type monotone'. *Calcolo,* **2** (1970).

[34] M. Sibony. 'Sur l'approximation d'èquations et inéquations aux dérivées partielles non linéaires de type monotone'. *J. Math. Anal. & Appl.,* **34** (1971), 502-64.

[35] G. Strang, and G.I. Fix. '*An Analysis of the Finite Element Method*'. Prentice Hall, Englewood Cliffs, New Jersey (1973).

[36] A. Tamir. 'Minimality and complementarity properties associated with Z-functions and M-functions'. *Math. Program.*, **7** (1974), 17-31.

[37] R.S. Varga. '*Matrix Iterative Analysis*'. Presentice Hall, Englewood Cliffs, New Jersey (1962).

Chapter 22

Error Estimates for the Finite Element Approximation of Some Variational Inequalities

F. Scarpini

1. INTRODUCTION

We shall describe here some techniques by means of which we were able to obtain the error estimates for the finite element approximation of some unilateral problems. We shall consider two stationary and two evolution problems, and in all cases we shall apply the finite affine triangular elements method.

2. A FIRST EXAMPLE OF STATIONARY VARIATIONAL INEQUALITY

We consider now the following problem:

$$(1)\qquad \begin{aligned} & u \in \mathbb{K} \\ & a(u, v-u) \geqslant \langle f, v-u \rangle \qquad v \in \mathbb{K} \end{aligned}$$

For the notation we refer, in general, to [13]; in particular here and in the following (unless otherwise specified), we have: Ω is a bounded open set of $\mathbb{R}^n$ with sufficiently smooth boundary Γ; $H = L^2(\Omega)$, $|\cdot| = \|\cdot\|_H$; $V = H_0^1(\Omega)$, $\|\cdot\| = \|\cdot\|_V$; $V' = H^{-1}(\Omega)$, $\|\cdot\|_* = \|\cdot\|_{V'}$; $W = H^2(\Omega)$, $|||\cdot||| = \|\cdot\|_{H^2(\Omega)}$; $\langle\cdot,\cdot\rangle$ is the duality pairing on $V' \times V$; $(\cdot,\cdot)$ is the scalar product in H; $a(\cdot,\cdot)$ is the bilinear, continuous coercive form

$$a(u, v) = \sum_{i,j}^{1,n} \int_\Omega a_{ij}(x) D_j u D_i v \,dx + \int_\Omega a_0(x) u v \,dx$$

where $\quad a_{ij} \in C^1(\overline{\Omega})$, $a_0 \in L^\infty(\Omega)$, $a_0(x) \geqslant c \geqslant 0$,

$$\sum a_{ij}\,\xi_i\xi_j \geqslant \alpha_0\,|\xi|^2 \quad \alpha_0 > 0 \quad \forall\, \xi \in \mathbb{R}^n$$

$A: V \to V'$ is the operator associated with $a(\cdot,\cdot)$

$$Au = -\sum_{i,j}^{1,\,i} D_i(a_{ij}D_ju) + a_0\,u;$$

$f \in V'$; $\mathbb{K}$ is the closed convex set $\{v : v \in V / v \geqslant \psi$ a.e. (almost everywhere) in $\Omega\}$ defined by a given function $\psi \in H^1(\Omega)$ such that $\psi|_\Gamma \leqslant 0$.

It is well known [13] that the problem (1) has a unique solution u satisfying the dual estimates [16]

$$f \leqslant Au \leqslant f \vee A\psi = \sup\,(f, A\psi)$$

and the following regularity theorem [6]:

$$f \in H \quad \psi \in W \Rightarrow u \in V \cap W$$

We shall always suppose that the regularity hypothesis will be satisfied and, in the case $n = 2$, Ω will be convex, Ω_h a convex polygon inscribed in Ω, whose vertices belong to Γ and whose sides have a length not exceeding $h (0 < h \leqslant 1)$. We shall decompose Ω_h with a 'regular triangulation' [24].

Then we apply the finite affine element method and obtain the following finite-dimensional problem:

$$(2) \qquad \begin{aligned} & u_h \in \mathbb{K}_h \\ & a(u_h, v_h - u_h) \geqslant \langle f, v_h - u_h\rangle \qquad \forall\, v_h \in \mathbb{K}_h \end{aligned}$$

where $I = \{1, \ldots, N\}$ is the set of indices associated with the nodes $x_i \in \Omega$; V_h is the finite-dimensional space

$$\left\{ v_h : v_h = \sum_{i \in I} V_i \varphi_i^h,\ V = \{V_i\} \in \mathbb{R}^N \right\}$$

$\varphi_i^h(x)$ is the continuous function affine in each triangle, such that $\varphi_i^h(x_j) = \delta_{ij}$, $i, j \in I$ (Kronecker delta);

$$\psi_I = \sum_{i \in I} \psi(x_i)\,\varphi_i^h(x)$$

is the interpolated function of ψ; $\mathbb{K}_h = \{v_h : v_h \in V_h \geqslant \psi_I\}$. The inclusion $V_h \subset V$ implies existence and uniqueness of the solution u_h of (2).

We follow here a method of Falk [10] for obtaining an error estimate.

Theorem 1
We have the following estimate for the discretization error

$$\|u - u_h\| = 0(h) \tag{3}$$

between the solution u of (1) and u_h of (2).

Proof: From (1), (2), if we set $v_h = u_I$, we obtain:

$$a(u, u - v) \leqslant \langle f, u - v\rangle \tag{4}$$

$$a(u_h, u_h - u_I) \leqslant \langle f, u_h - u_I\rangle \tag{5}$$

By summing we have

$$a(u - u_h, u - u_h) \leqslant \langle f - Au, u_h - v\rangle + \langle f - Au, u - u_I\rangle + \langle A(u - u_h), u - u_I\rangle \tag{6}$$

Owing to the dual estimates, continuity and coerciveness of A, we obtain

$$\|u - u_h\|^2 \leqslant C(|u_h - v| + |u - u_I| + \|u - u_I\| \cdot \|u - u_h\|)$$

and still more

$$\|u - u_h\|^2 \leqslant C(|u_h - v| + |u - u_I| + \|u - u_I\|^2) \tag{7}$$

We now set

$$v = u_h \vee \psi \tag{8}$$

and observe that it results in

$$\psi(x) \geqslant u_h(x) \geqslant \psi_I(x) \qquad \text{if } v(x) = \psi(x)$$

$$u_h(x) - v(x) = 0 \qquad \text{somewhere else.}$$

From (7), we obtain

$$\|u - u_h\|^2 \leqslant C(|\psi - \psi_i| + |u - u_I| + \|u - u_I\|^2) \tag{9}$$

We introduce in (9) the well known interpolation results [3]:

(10) $$|u - u_I| \leqslant Ch^2 |||u|||$$

(11) $$\|u - u_I\| \leqslant Ch\,|||u|||$$

and obtain (3) . This completes the proof.

Remark 1 Still considering the new closed, convex set

$$\mathbb{K} = \{v : v \in V / \ \psi'' \geqslant v \geqslant \psi' \text{ a.e. in } \Omega\}$$

where ψ'', $\psi' \in H^1(\Omega)$ are given function such that $\psi''|_\Gamma \geqslant 0 \geqslant \psi'|_\Gamma$, we have the existence, uniqueness and regularity theorems about the solution of (1) [23]. Taking into account the following results

(12) $$f \in H \qquad \psi', \psi'' \in W \Rightarrow u \in V \cap W$$

(13) $$\inf(f, A\psi'') = f \wedge A\psi'' \leqslant Au \leqslant f \vee A\psi'$$

and by setting

(14) $$v = u_h \vee \psi' \wedge \psi''$$

we still have (3). Mosco and Strang [15] obtained (3) with one-sided approximation.

3. A SECOND EXAMPLE OF STATIONARY VARIATIONAL INEQUALITY

Let now $V = H^1(\Omega)$; $\mathbb{K} = \{v : v \in V / v \geqslant \psi \text{ on } \Gamma\}$; $a_0(x) \geqslant c > 0$.

We suppose $f \in H$, $\psi \in W$, then [4] the problem (1) has a unique solution $u \in W$. In the following $\partial u / \partial \nu$ will denote the conormal derivative

$$\frac{\partial u}{\partial \nu} = \sum_{i,j}^{1,n} a_{ij}(x) D_i u \cos(n, x_j)$$

and n the exterior normal to Γ.

We can verify [13] that (1) is equivalent to the following internal equation and complementarity system on the boundary:

(15) $$\begin{aligned} &Au = f && \text{in } \Omega \\ &u \geqslant \psi && \frac{\partial u}{\partial \nu} \geqslant 0 \\ &\frac{\partial u}{\partial \nu}(u - \psi) = 0 && \text{on } \Gamma \end{aligned}$$

In the case $n = 2$, Ω convex, $\Gamma \in C^2$, we studied [21] the problem (1). In a short space we cannot give the full details of the proof. We only say, for a general design, that we use the affine, curved, finite element method [3, 24]. We impose the constraints only on the nodes $x_i \in \Gamma$ to yield

$$\mathbb{K}_h = \{v_h : v_h \in V / v(x_i) \geqslant \psi(x_i)\} \tag{16}$$

Using a method similar to Falk's method, by means of the Green formula, the trace theorem and some evaluations on the zone Σ_i between the curve side Γ_i and right side $\overline{x_i x_{i+1}}$, we obtain the following two propositions:

$$-\int_\Gamma \frac{\partial u}{\partial \nu} (u_h - \psi_i) \, d\Gamma \leqslant Ch^{3/2} \tag{17}$$

$$-\int_\Gamma \frac{\partial u}{\partial \nu} [(u - \psi) - (u - \psi)_I] \, d\Gamma \leqslant Ch^{3/2} \tag{18}$$

Introducing (17), (18) in

$$a(u-u_h, u-u_h) \leqslant a(u-u_h, u-u_I) - \int_\Gamma (u_h - \psi_I) \frac{\partial u}{\partial \nu} \, d\Gamma - \int_\Gamma \frac{\partial u}{\partial \nu} [(u-\psi)-(u-\psi)_I] \, d\Gamma \tag{19}$$

we have the following estimate for the discretization error

$$\| u - u_h \| = O(h^{3/4}) \tag{20}$$

Remark 2 Other authors [7] have reached the estimate $\| u - u_h \| = O(h)$ by the use of a different method and a reasonable adjunctive hypothesis.

4. A PARABOLIC EVOLUTION INEQUALITY OF TYPE I

We consider [22] the following problem

$$\begin{aligned} & u \in \mathbb{K} \quad u' \in L^2(0, T; V') \quad u(0, x) = u_0(x) \\ & \int_0^T \langle u', v - u \rangle + a(u, v - u) - \langle f, v - u \rangle \cdot dt \geqslant 0 \qquad \forall v \in \mathbb{K} \end{aligned} \tag{21}$$

where $]0, T[$ is a bounded interval; $Q =]0, T[\times \Omega$; $\Sigma =]0, T[\times \Gamma$;

$$a(u, v) = \sum_{i=1}^n \int_\Omega D_i u D_i v \, dx ;$$

$u_0 \in V; f \in L^2(0, T; V'); \psi_l (l = 1, 2)$ are given functions such that

$$\psi_l \in L^2(0, T; H^1(\Omega)) \qquad \psi_l' \in L^2(0, T; V')$$

$$\psi_1/_\Sigma \leqslant 0 \leqslant \psi_2/_\Sigma \qquad \psi_1 \leqslant \psi_2 \text{ a.e. on } \Omega$$

$$\psi_1(0, x) \leqslant u_0(x) \leqslant \psi_2(0, x)$$

$$\mathbb{K} = \{v : v \in L^2(0, T; V) / \psi_2 \geqslant v \geqslant \psi_1 \text{ a.e. on } Q\}$$

With a translation we can confine ourselves to the case $\psi_2 = \psi$, $\psi_1 = 0$. We have the uniqueness, existence and regularity theorems [22].

Theorem 2
The problem (21) has one solution at most.

Theorem 3
Let

$$\psi \in L^p(0, T; W^{2,p}(\Omega)) \qquad \psi' \in L^p(Q)$$

$$f \in L^p(Q) \qquad u_0 \in V \cap W^{2-2/p,\, p}(Q) \qquad p \geqslant 2$$

then (21) has a unique solution

$$u \in L^p(0, T; W^{2,p}(\Omega)) \qquad u' \in L^p(Q)$$

Theorem 4
Let moreover

$$\psi \in L^\infty(0, T; W) \qquad \psi' \in L^\infty(0, T; V) \qquad \psi'' \in L^2(Q)$$

$$f' \in L^2(Q) \qquad u_0 \in V \cap W$$

then the solution is such that

$$u \in L^\infty(0, T; W) \qquad u' \in L^2(0, T; V) \cap L^\infty(0, T; H)$$

4.1. External approximation of (21)

We construct the finite-dimensional space $S^0_{k,\,h}(Q)$ included in the space of bounded variation functions with values in V : $VB(0, T; V)$. With this purpose we split $\tau = [0, T]$ into m equal subintervals τ_r of size k. We define the characteristic functions θ_r^k on τ_r and the space $S_{k,h}^0$ with base $\{\theta_r^k(t)\varphi_i^h(x)\}$. Then we consider the approximate problem:

$$u_{kh} \in \mathbb{K}_{kh}$$

$$(22) \qquad \langle\langle u'_{kh}, v_{kh} - u_{kh}\rangle\rangle + \int_0^T a(u_{kh}, v_{kh} - u_{kh})\,dt \geqslant \int_0^T \langle f, v_{kh} - u_{kh}\rangle\,dt$$

$$\forall\, v_{kh} \in \mathbb{K}_{kh} \qquad u_{kh}(0,x) = u_{0I}(x) \qquad v(0,x) = u_{0I}(x)$$

where $\mathbb{K}_{kh} = \{v_{kh} : v_{kh} \in S_{kh}^0(Q)/(\psi_2)_I \geqslant v_{kh} \geqslant (\psi_1)_I\}$; $\langle\langle \cdot, \cdot \rangle\rangle$ is the duality pairing on $VB(0,T;V)' \times VB(0,T;V)$.

4.2. Error estimate

With the use of convenient interpolation results we have the following theorem.

Theorem 5
If the solution u of (21) is such that

$$u \in L^\infty(0,T;W) \qquad u' \in L^2(0,T;V)$$

and u_{kh} denotes the solution of (22), then we have

$$(23) \qquad \left(\sum_{r=1}^m |(u - u_{kh})(t_r) - (u - u_{kh})(t_{r-1})|^2 + |u - u_{kh})(T)|^2\right)^{1/2} + \|u - u_{kh}\|_{L^2(0,T;V)} = O(h + k^{1/2})$$

Remark 3 Johnson [12] and Berger and Falk [2] gave, with a different method, a discrete error estimate for the problem (21) with an obstacle ψ independent of the time.

5. A PARABOLIC EVOLUTION INEQUALITY OF TYPE II

We consider the problem

$$u' \in \mathbb{K}$$

$$(24) \qquad \int_0^T \langle u' - \Delta u - f, v - u'\rangle\,dt \geqslant 0 \qquad \forall\, v \in \mathbb{K}, u(0) = u_0$$

where $\psi(t,x)$ is a given smooth function such that $\psi \geqslant 0$ a.e. in Q;

$$\mathbb{K}(t) = \{v(t) : v(t) \in V / \psi(t) \geqslant v(t) \geqslant 0 \text{ a.e. in } \Omega\} \qquad \text{a.e. } t \in]0, T[$$

$$\mathbb{K} = \{v : v \in L^2(0, T; V) / v(t) \in \mathbb{K}(t) \text{ a.e. } t \in]0, T[\,\}$$

u_0, f are given functions in classes that we shall specify; G denotes the Green operator of the problem; $-\Delta u = g$ in Ω; $u|_\Gamma = 0$.

We proved [18] the following existence, uniqueness, regularity theorem.

Theorem 6
Under the hypotheses

$$\psi \in L^\infty(0, T; W) \qquad \psi' \in L^2(0, T; W)$$

$$f, f', f'' \in L^2(Q) \qquad u_0 = Gf(0)$$

the problem (24) has a unique solution u such that

$$u, u' \in L^\infty(0, T; W \cap V) \qquad u'' \in L^2(0, T; V)$$

5.1. Internal approximation of (24)

We construct the finite-dimensional space $S^1_{kh}(Q)$. We define the spline functions $z^k_r(t)$ on the subintervals τ_r. We put $S^1_{kh}(Q)$ the space with base $\{z^k_r \ \varphi^h_i\}$ and consider the following approximated problem

$$u_{kh} \in \mathbb{K}_{kh}$$

$$\int_0^T (u'_{kh}, v'_{kh} - u'_{kh}) + a(u_{kh}, v'_{kh} - u'_{kh}) - (f, v'_{kh} - u'_{kh})\, dt \geqslant 0 \tag{25}$$

$$\forall v_{kh} \in \mathbb{K}_{kh} \qquad u_{kh}(0, x) = u_{0I}(x)$$

where $\mathbb{K}_{kh} = \{v : v \in S^1_{kh}(Q) / \psi_I \geqslant v' \geqslant 0\}$ (the interpolated function ψ_I coincides with ψ_I in section 4).

The problem (25) has a unique solution.

5.2. Error estimate

Theorem 7
If u is the solution of (24) such that

$$u, u' \in L^\infty(0, T; W \cap V) \qquad u'' \in L^2(0, T; V)$$

and u_{kh} denotes the solution of (25), then we have:

(26) $\|u' - u'_{kh}\|_{L^2(Q)} + \|u - u_{kh}\|_{L^\infty(0,T;V)} = O(h + k^{1/2})$

Remark 4 We have also obtained some error estimates for finite element solution of some unilateral Dirichlet problems related to nonlinear operators. See [19] for a stationary problem and [20] for an evolution problem.

REFERENCES

[1] C. Baiocchi (1977). 'Estimations d'erreurs dans L^∞ pour les inéquations a obstacle'. *Math Aspects of Finite Element Method*, Rome, 1975 (*Lecture Notes in Math.* Springer-Verlag).

[2] A.E. Berger, and R. Falk. 'An error estimate for the truncation method for the solution of parabolic obstacle variational inequalities'. To appear.

[3] J.H. Bramble, and M. Zlamal. 'Triangular elements in the finite element method'. *Math. Comput.*, **24** (1970), 809-820.

[4] H. Brezis. 'Problèmes unilatéraux'. *J. Math. Pures et Appl.* **51** (1972), 1-168.

[5] H. Brezis. '*Operateurs maximaux monotones*'. North Holland, New York (1973).

[6] H. Brezis, and G. Stampacchia. 'Sur la régularté de la solution d'inéquations elliptiques'. *Bull. Soc. Math. Fr.*, **96** (1968), 153-180.

[7] F. Brezzi, W. Hager, and P.A. Raviart. 'Error estimate for the finite element solution of variational inequalities'. *Numer. Math.*, **28** (1977), 431-443.

[8] F. Donati, and M. Matzeu. 'Solutions fortes et estimations duales pour des inéquations variationnelles paraboliques non-linéaires'. *C.R. Acad. Sci., Paris,* **285** (1977), 347-350.

[9] G. Duvaut, and J.L. Lions. '*Les inéquations en mecanique et en physique*'. Dunod, Paris (1972).

[10] R. Falk. 'Error estimates for the approximation of a class of variational inequalities'. *Math. Comput.*, **28** (1974), 963-971.

[11] R. Glowinski. 'Introduction to the approximation of elliptic variational inequalities'. *Univ. Paris VI. Lab. Anal. Num.* (1976).

[12] C. Johnson. 'A convergence estimate for an approximation of a parabolic variational inequality'. *S.I.A.M. J. Numer. Anal.* To appear.

[13] J.L. Lions. '*Quelques méthodes de résolution des problemes aux limites non linéaires*'. Dunod, Paris (1969).

[14] J.L. Lions, and E. Magenes. '*Non-homogeneous Boundary Value Problems and Applications*'. Vol. I-III, Springer, Berlin (1972).

[15] U. Mosco, and G. Strang. 'One sided approximation and variational inequalities'. *Bull. Am. Math. Soc.*, **80** (1974), 308-312.

[16] U. Mosco, and G. Troianiello. 'On the smoothness of solutions of unilateral Dirichlet Problems'. *Boll. Unione Mat. Ital.*, **8** (1973), 56-67.

[17] U. Mosco (1975). 'Error estimates for some variational inequalities'. In *Math. Aspects of Finite Element Method,* Rome. (*Lecture Notes in Math,* Springer-Verlag, 1977).

[18] F. Scarpini. 'Sur une inéquation d'evolution parabolique du type II, associeé à un problème d'asserviment thermique'. *Calcolo.* To appear.

[19] F. Scarpini. 'Some nonlinear complementarity systems algorithms and applications to unilateral boundary-value problems'. *Ann. Mat.* To appear.

[20] F. Scarpini. 'Sull'approssimazione numerica di un problema unilaterale parabolico relativo ad alcuni operatori non lineari'. *Unione Mat. Ital.* To appear.

[21] F. Scarpini, and M.A. Vivaldi. 'Error estimates for the approximation of some unilateral problems'. *R.A.I.R.O. Anal. Numer.*, 2 (1977), 197-208.

[22] F. Scarpini, and M.A. Vivaldi. 'Valutation de l'erreur d'approximation pour une inéquation parabolique relative aux convexes dépendants du temps'. *Appl. Math. & Opt. imiz.*, 4 (1978), 121-38.

[23] G. Stampacchia. 'Regularity of solutions of some variational inequalities'. *Proc. Symp. in Nonlinear Anal,* ed. F.E. Browder, *Proc. Am. Math. Soc.*, 18 (1970), 271-81.

[24] G. Strang, and G.I. Fix. '*An Analysis of the Finite Element Method*.' Prentice Hall, New Jersey (1973).

Chapter 23

Bifurcation Analysis of the Free Boundary of a Confined Plasma

J. Sijbrand

1. INTRODUCTION

A great deal of effort is spent nowadays to develop a working prototype of a controlled nuclear fusion reactor. The most common experimental situation is a plasma confined by a torus (a so-called 'tokamak' machine). By applying a magnetic field one induces an intense electric current in the plasma. This current causes a 'pinching' of the plasma so that a high temperature and a high density are obtained. If the temperature is more than 10^8 K and the plasma is sufficiently dense for a long enough period (ions/m^3 x time $> 10^{20}$), a nuclear fusion reaction may be ignited.

Although tokamak machines have been operational since 1965, experimental difficulties, especially instabilities of the plasma column, have inhibited the construction of a working reactor. Therefore, a mathematical investigation of the stability properties of models related to the physical situation seems justified. The first step in this direction is the investigation of stationary problems. One such model [7] as follows.

Let us consider a bounded domain $\Omega \subset \mathbb{R}^n$ with a smooth boundary. We want to find a function u and a subdomain ω defined by $\omega = \{x \in \Omega : u(x) > 0\}$ such that

(1)	in ω	$\mathcal{L} u + \lambda u$	$=$	0
(2)	on $\partial\omega$	u	$=$	0
(3)	in $\Omega \backslash \omega$	$\mathcal{L} u$	$=$	0
(4)	on $\partial\Omega$	u	$=$	constant -1
(5)	across $\partial\omega$	$u, \dfrac{\partial u}{\partial n}$ continuous		

Here u is a flux function, Ω is a smooth cavity (toroidal, for example), ω is the region occupied by the plasma and $\Omega \backslash \omega$ is vacuum. $\mathcal{L}$ is a uniformly elliptic operator of the form

(6) $$\mathcal{L} = \sum_{i,j} \frac{\partial}{\partial x_i} a_{ij}(x) \frac{\partial}{\partial x_j}$$

λ is a positive real constant; $\partial/\partial n$ is the normal derivative on $\partial\omega$.

Let μ_i be the ith eigenvalue of $\mathcal{L}$ on Ω with 0 boundary conditions, the μ_i numbered according to magnitude, multiplicities taken into account. If $\lambda \leqslant \mu_1$, $u = -1$ is the only solution of (1)-(5). A less trivial result was proved for $\mathcal{L} = \Delta$ by Temam [8] ($n=2$) and by Puel [3] (n arbitrary): if $\mu_1 \leqslant \lambda \leqslant \mu_2$, there is exactly one non-constant solution (with a non-empty ω).

For $\lambda > \mu_2$, the situation may become much more interesting. Schaeffer [5] proved for a special geometry in $\mathbb{R}^2$ (Figure 1), that there are $\lambda > \mu_2$

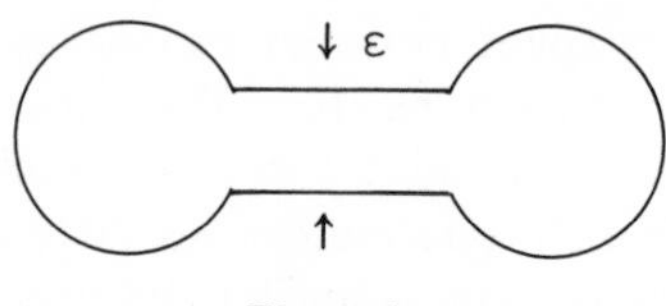

Figure 1

for which more than one non-constant solution exists. A construction of these solutions must be left to the computer. The aim of the research presented in this paper is to give a method to construct analytically, at least for simple geometries, new solutions that are 'branching off' from already known solutions. Moreover, we shall treat a problem for which our method leads to very explicit results.

For a much more detailed analysis, and proofs of the various statements, see Sijbrand [6].

2. METHOD OF SOLUTION

We assume that there is a smooth branch of solutions BB for $\lambda \geqslant \mu_2$ (see Berestycki, and Brézis [1] for the existence of solutions for all λ):

(7) $$BB = \{(\omega_0(\lambda), u_0(\lambda), \lambda)\}$$

where the ω_0 and u_0 satisfy the following smoothness properties:

(i) a neighbourhood of ω_0 can be mapped C^{p_0} to a neighbourhood of the unit ball in $\mathbb{R}^n$; and

(ii) u_0 is twice differentiable on Ω.

Now suppose there is a second branch of solutions $(\tilde{\omega}, \tilde{u}, \lambda)$ which has only the point $(\omega_0(\lambda_0), u_0(\lambda_0), \lambda_0)$ in common with BB. Then λ_0 represents a 'bifurcation point' of the BB.

Let us suppose that the $\tilde{u}$ satisfy (ii) and the $\tilde{\omega}$ satisfy (i), but with C^{p_0} replaced by C^p, $p \leq p_0$ (so we allow loss of smoothness of the plasma domain after the bifurcation). Then we may represent the boundary of $\tilde{\omega}$ by a function $\tilde{\gamma} : \partial\omega_0 \to \mathbb{R}^n$ as follows:

$$\tilde{\gamma} = 1 + \varepsilon \rho \hat{n} \tag{8}$$

where 1 is the identity function on the boundary of $\omega_0(\lambda_0)$, $\hat{n}$ is the unit normal vector on this boundary, ε is a small parameter fixing the 'size' of the perturbation, and ρ is a real C^p-function on $\partial\omega_0$, of order unity, which represents the shape of the deformation of the boundary from $\omega_0(\lambda_0)$ to $\tilde{\omega}$.

We can prove that a C^p coordinate transformation $T(\varepsilon, \rho)$ exists which maps $\omega_0(\lambda_0)$ to $\tilde{\omega}$:

$$T(\varepsilon, \rho) = \sum_{n \geq 0} \varepsilon^n T_n(\rho) \tag{9}$$

Now (1), (2) and the condition $u > 0$ on $\tilde{\omega}$, imply that u is the first eigenfunction on $\tilde{\omega}$. With the transformation (9) we can turn to Kato [2] (Chapter VII, section 6) and we find that functions $\lambda(\varepsilon, \rho)$ and $U(\varepsilon, \rho)(\cdot)$ exist:

$$U = \sum_{n \geq 0} \varepsilon^n U_n(\rho) \tag{10}$$

$$\lambda = \sum_{n \geq 0} \varepsilon^n \lambda_n(\rho) \tag{11}$$

Moreover, $U_n(\rho) \in C^p(\omega_0)$, and u_{in} defined by $U(\varepsilon, \rho, T^{-1}(\varepsilon, \rho)x) = u_{\text{in}}(\varepsilon, \rho)(x)$ satisfies (1), (2) $(x \in \tilde{\omega})$.

By equating powers of ε and using Fredholm's alternative, we can obtain the coefficients U_n and λ_n, still functions of the unknown ρ.

Now we have determined u_{in} inside the plasma domain $\tilde{\omega}$ (in terms of ε and ρ), we turn to all Ω.

We define a new problem: let u_{pre} satisfy

$$\begin{aligned} \mathcal{L} u_{\text{pre}} + \lambda u_{\text{in}} &= 0 \quad \text{in } \tilde{\omega} \\ \mathcal{L} u_{\text{pre}} &= 0 \quad \text{in } \Omega \backslash \tilde{\omega} \\ u_{\text{pre}} &= 0 \quad \text{on } \partial\Omega \end{aligned} \tag{12}$$

Of course, u_{pre} depends (through $\lambda, \tilde{u}, \tilde{\omega}$) on ε and ρ. If $\rho(\varepsilon)$ can be chosen such that for small ε, u_{pre} is a constant $a(\varepsilon)$ on $\partial\tilde{\omega}$, then inside $\tilde{\omega}$ $u_{pre} - a(\varepsilon)$ equals u_{in} which implies that $u_{pre} - a(\varepsilon)$ satisfies (1), (2), (3), (5); hence

$$\tilde{u} = \frac{u_{pre} - a(\varepsilon)}{a(\varepsilon)}$$

satisfies all conditions (1)-(5), and with (11) and (8) we find a branch $(\tilde{u}(\varepsilon), \tilde{\omega}(\varepsilon), \lambda(\varepsilon))$ of solutions. Eventually, we can invert $\lambda(\varepsilon)$ to obtain ε in terms of $\lambda - \lambda_0$; we then have the bifurcation branch in the form $(\tilde{\omega}(\lambda - \lambda_0), \tilde{u}(\lambda - \lambda_0), \lambda)$. So the interesting point is, if we define

$$F(\varepsilon, \rho, x) = -\lambda^{-1} u_{pre}(x)$$

can ρ be chosen such that $F(\varepsilon, \rho, x)$ is a constant function $\alpha(\varepsilon) = -\lambda^{-1} a(\varepsilon)$ for $x \in \partial\tilde{\omega}$? Equivalently, if we define

$$f(\varepsilon, \rho, y) = F(\varepsilon, \rho, T(\varepsilon, \rho)y) \tag{13}$$

for $y \in \partial\omega_0$, then finding a solution for (1)-(5) amounts to finding a $\rho(\varepsilon)$ such that

$$f(\varepsilon, \rho, \cdot) = \alpha(\varepsilon) \tag{14}$$

identically on $\partial\omega_0$.

We now study $f: \rho \to f(\varepsilon, \rho, \cdot)$. The explicit from of f is:

$$f(\varepsilon, \rho)(y) = \int_{\omega_0} G(T(\varepsilon,\rho)(y); T(\varepsilon,\rho)(\eta)) U(\varepsilon,\rho)(\eta) J_T \, d\eta \tag{15}$$

for $y \in \partial\omega_0$, where G is Green's function of $\mathcal{L}$ on Ω and J_T is the Jacobian of $T(\varepsilon, \rho)$. We can prove, if $p \geq n/2$ is not a natural number, and $p > 2$:

Lemma 1 f is a mapping from $C^p(\partial\omega_0)$ to itself.

Lemma 2 f can be developed in an ε series:

$$f = \sum_{m=0}^{p-1} \varepsilon^m f_m(\rho) + 0(\varepsilon^p) \tag{16}$$

where f_m is homogeneous of degree m in ρ.

Lemma 3 For $p \leq p_0 - 1$, f_1 maps $C^p(\partial\omega_0)$ to itself. Moreover, $f_1(\rho) = C \cdot \rho + A(\rho)$, $(y \in \partial\omega_0)$, where $C \in C^{p_0 - 1}(\partial\omega_0)$, $C(y) > 0$ on $\partial\omega_0$ and A is a compact linear operator in $C^p(\partial\omega_0)$. In fact, the function C spoils the regularity, which is easily seen from

(17) $$C(y) = \int_{\omega_0} D_y\, G(y\,;\eta) \cdot \hat{n}\, U_0(\eta) d\eta$$

Now $f =$ constant $\alpha(\varepsilon)$ means

(18) $$f_1 \rho = -(\varepsilon f_2(\rho) + \varepsilon^2 f_3(\rho) + \ldots) + b(\varepsilon)$$

where we have put $\alpha(\varepsilon) = f_0 + \varepsilon\, b(\varepsilon)$. Because we have obtained Lemma 3 for f_1, we can apply the usual bifurcation techniques (see Sattinger [4]).

Theorem 1
If f_1 is invertible, then (18) has a unique solution $\rho(\varepsilon)$, but this solution represents the variation of the basic branch.

Hence, λ_0 will only be a bifurcation point if f_1 is non-invertible, i.e. if a ϕ_0 exists with $f_1 \phi_0 = 0$.

In this case we decompose ρ into a component along ϕ_0 and a residual one so we get

(19) $$\rho = k(\varepsilon)\phi_0 + \psi(\varepsilon)$$

Hence,

(20) $$f_1 \psi = -\varepsilon f_2(k\phi_0 + \psi) - \varepsilon^2 f_3(k\phi_0 + \psi) + \ldots + b$$

which has a solution

(21) $$\psi = f_1^{-1}(-\varepsilon f_2(k\phi_0 + \psi) + \ldots + b)$$

provided

(22) $$P(-\varepsilon f_2(k\phi_0 + \psi) + \ldots + b) = 0$$

where f_1^{-1} is the pseudo-inverse on the ψ-space and P is the projection onto ϕ_0. From (21) we may obtain $\psi = \Psi(\varepsilon, b, k)$; introducing this into (22) we obtain an equation for k, ε, and b which can be solved for b in terms of ε and k. So Ψ is known in terms of ε and k.

If we now substitute $\Psi(\varepsilon, b(\varepsilon, k), k)$ into (19) and the obtained expression for $\rho(\varepsilon, k)$ into (11), we get a relation for λ, ε, and k which is typically of the form

(23) $$\lambda - \lambda_0 = c_1 k\varepsilon + c_2 k^2 \varepsilon^2 + c_3 k^3 \varepsilon^3 + \ldots$$

where c_1, c_2,.. are real constants.

If $c_1 \neq 0$, a proper choice for ε is $\varepsilon = \lambda - \lambda_0$. If $c_1 = 0, c_2 \neq 0$ then we take $\varepsilon = |\lambda - \lambda_0|^{1/2}$, etc. (23) then becomes an equation for k and $\lambda - \lambda_0$ alone, which is easily solved for k in terms of $\lambda - \lambda_0$. By (19), (10) and (12) we now find a bifurcating solution of problem (1)-(5), so we can formulate the next theorem.

Theorem 2

Suppose $p_0 - 1 > p > 3$. Let for increasing λ, a simple eigenvalue of f_1 cross 0. Let this happen for $\lambda = \lambda_0$ and let the corresponding eigenfunction be ϕ_0. If $c_1 \neq 0$ (c_1 defined in (23)), then a (bilaterally) bifurcating branch is given by (10) and

$$\tilde{\gamma} = 1 + \frac{\lambda - \lambda_0}{c_1} \phi_0 + 0(\lambda - \lambda_0)^2$$

If $c_1 = 0, c_2 \neq 0$, then a (unilaterally) bifurcating branch is given by (10) and

$$\tilde{\gamma} = 1 \pm \left(\frac{\lambda - \lambda_0}{c_2} \right)^{1/2} \phi_0 + 0(\lambda - \lambda_0)$$

etc.

The eigenvalues of f_1 and the coefficients $c_1, c_2,..$ are easy to compute if we assume some symmetry properties for Ω and $\mathcal{L}$. For a detailed analysis and computation see Sijbrand [6].

In the next section we give the explicit results we have obtained for a special example, which is an illustration of our technique, although it has no physical importance.

3. EXAMPLE

We study the cylinder

$$\Omega = \{(r,\phi,z) \in \mathbb{R}^3 : 0 \leqslant r < 1; 0 \leqslant \phi < 2\pi\}$$

on which we consider functions u with properties: (i) $u(z + 2\pi) = u(z)$ and (ii) $\partial u/\partial\phi = 0$.

The basic branch is given by

$$\omega_0(\lambda) = \{(r, \phi, z) : 0 \leqslant r < r_0(\lambda); 0 \leqslant \phi < 2\pi\} \qquad r_0 = \frac{\ell_0}{\sqrt{\lambda}}$$

and

$$u_0(\lambda) = J_0\left(\ell_0 \frac{r}{r_0}\right) \text{ inside } \omega_0$$

harmonic outside, where J_0 is the zeroth Bessel function of the first kind, and ℓ_0 is its first zero.

In this example we can express f_1 as follows.

$$f_1(z) = \sum_{n=-\infty}^{+\infty} e^{niz} F_n$$

where the F_n are known real functions of λ. f_1 is non-invertible if an n and a λ_0 exist such that $F_n(\lambda_0) = 0$. We have computed the zeros $\lambda_0^{(n)}$ of the F_n and we have found:

$$\lambda_0^{(1)} \approx 6.7951$$
$$\lambda_0^{(2)} \approx 10.281$$
$$\lambda_0^{(10)} \approx 188$$

The branch bifurcating from $\lambda_0^{(1)}$ is given by the boundary function

$$\gamma(z) = r_0 + 2r_0 d(\lambda - \lambda_0)^{1/2} \cos(z+\alpha) + 0(\lambda - \lambda_0)$$

α is an arbitrary real constant.

The coefficient d is related to c_2 (from (23)) by $d = (2\pi c_2)^{-1/2}$; an explicit analytic expression for c_2 can be obtained and we find after some tedious computations:

$$d \approx 0.0772$$

In Figure 2 below we have drawn the free boundaries of the basic solution and of the bifurcating branch.

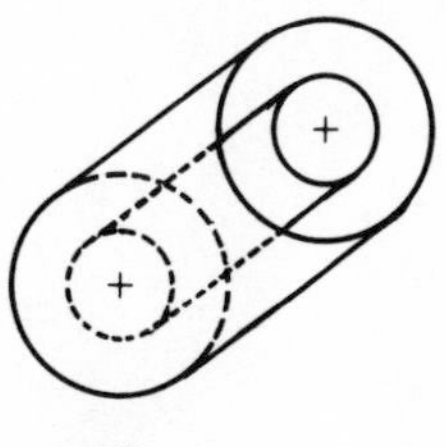

BB

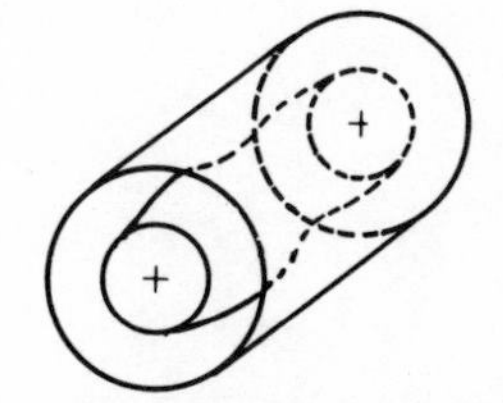

$\lambda > \lambda_0$: *'sausage' solution*

Figure 2

REFERENCES

[1] H. Berestycki, and H. Brezis 'Sur certains problèmes de frontière libre'. *C.R. Acad. Sci., Paris, A,* **283** (1976), 1091.

[2] T. Kato *'Perturbation theory of linear operators'.* Springer, Berlin (1966).

[3] J.P. Puel. 'Sur un problème de valeur propre non linerare et de frontière libre'. *C.R. Acad. Sci., Paris, A,* **284** (1977), 861-3.

[4] D.H. Sattinger. 'Topics in stability and bifurcation theory'. *Springer Lecture Notes,* **309**, (1973).

[5] D. Schaeffer. 'Non uniqueness in the equilibrium shape of a confined plasma'. *Comm. Partial Differ. Equations,* 2 (1977), 587-600.

[6] J. Sijbrand. 'Bifurcation analysis for a class of problems with a free boundary'. *Preprint* Math. Inst., Univ. Utrecht. To appear.

[7] R. Temam. 'A nonlinear eigenvalue problem: the shape at equilibrium of a confined plasma'. *Arch. Ration. Mech. & Anal.*, **60** (1976), 51-73.

[8] R. Temam. 'Remarks on a free boundary value problem arising in plasma physics'. *Comm. Partial Differ. Equations.*, 2 (1977), 563-85.

Chapter 24

Mathematical Problems in Plasticity Theory

R. Temam

1. INTRODUCTION

The determination of the displacement fields in some models of plasticity theory leads to the solution of variational problems which are not coercive or, more precisely, which are coercive on non-reflexive function spaces associated to L^1-spaces: cf. Duvaut and Lions [1] (chapter V, §. 6) and the corresponding references. Considering a simplified problem (the *anti-plane shear*) Strang [7] has proposed a family of model problems containing many crucial aspects of the field displacement problems.

In this lecture we consider the two models of [7] called 'Deformation Theory' and 'Limit Analysis' and our purpose is twofold:

(i) first to make explicit the duality theory for these variational problems following the approach of Rockafellar [6] and Ekeland and Temam [2];

(ii) then, observing the analogy of these problems with the non-parametric minimal surface problem (Plateau problem) we introduce a concept of relaxed solutions for these problems which is similar to the concept of relaxed solutions of the minimal surface problem introduced by De Giorgi-Giusti-Miranda and by Temam [9].

In a progressing work with Strang we are trying to extend these results to the full three-dimensional displacement problem of plasticity (some preliminary results are given in [8]).

2. THE MODEL PROBLEM FOR LIMIT ANALYSIS

After recalling some elements of the convex duality theory in the sense of [2, 6] (section 2.1), we describe the limit analysis problem and its dual (section 2.2) and we then introduce the relaxed problem (section (2.3) for which we give an existence result (section 2.4).

2.1. Elements of duality theory

We consider two pairs of locally convex topological vector spaces which are in duality V, V^* and Y, Y^*. In most of the applications V (or Y) is a reflexive Banach space and V^* (or Y^*) its dual or V (or Y) is a non-reflexive Banach space and V^* (or Y^*) is its dual endowed with the weak topology $\sigma(V^*, V)$ (or $\sigma(Y^*, Y)$).

Let Λ denote a linear continuous operator from V into Y and Λ^*, linear continuous from Y^* into V^*, be its adjoint. We denote by F(resp. G) a convex lower semi-continuous function from V(resp. Y) into $\mathbb{R} \cup \{+\infty\}$, which is not identically equal to $+\infty$. We call primal problem, the optimization problem $(\mathscr{P})$:

(1) $$\inf_{v \in V} \{F(v) + G(\Lambda v)\}$$

The problem $(\mathscr{P}^*)$

(2) $$\sup_{p^* \in V^*} \{-F^*(\Lambda^* p^*) - G^*(-p^*)\}$$

is called the dual problem of (1).

It is well known [2, 6] that

$$-\infty \leqslant \sup \mathscr{P}^* \leqslant \inf \mathscr{P} \leqslant +\infty$$

and if the following conditions are satisfied

(3) $$\inf \mathscr{P} \in \mathbb{R}$$

(4) there exists $u_0 \in V$ with $F(u_0) < +\infty$ and G continuous at Λu_0

then

(5) $$\inf \mathscr{P} = \sup \mathscr{P}^*$$

(6) $\mathscr{P}^*$ posseses at least one solution

If $\mathscr{P}$ and $\mathscr{P}^*$ possess a solution, say $\overline{u}$, $\overline{p}^*$ and if (5) holds then we have the extremality conditions

(7) $$F(\overline{u}) + F^*(\Lambda^* p^*) = \langle \Lambda^* \overline{p}^*, \overline{u} \rangle$$

(8) $$G(\Lambda \overline{u}) + G^*(-\overline{p}^*) = -\langle \overline{p}^*, \Lambda \overline{u} \rangle$$

Many examples in [2] (cf. also [1]) show that this duality framework is particularly adapted to variational problems in the calculus of variation and in mechanics.

2.2. The problem and its dual

Let Ω denote an open Lipschitzian bounded set of $\mathbb{R}^n$ with boundary [(1)] $\partial\Omega = \Gamma$. We denote by $W^{1,1}(\Omega)$ the Sobolev space of functions u in $L^1(\Omega)$ with

$$\frac{\partial u}{\partial x_i} \in L^1(\Omega) \qquad i = 1, \ldots, n;$$

$W_0^{1,1}(\Omega)$ is the subspace of $W^{1,1}(\Omega)$ of functions vanishing on Γ.

Let f be a given continuous function in Ω. A typical limit analysis problem is to find the largest $\xi > 0$ such that

$$\text{(9)} \qquad \inf_{v \in W^{1,1}(\Omega)} \left\{ \int_\Omega |\nabla v| dx - \xi \int_\Omega fv\, dx \right\} > -\infty$$

where $\nabla v = \text{grad}\, v$ and

$$\text{(10)} \qquad |\nabla v(x)| = \left\{ \sum_{i=1}^{n} \left(\frac{\partial v}{\partial x_i} \right)^2 \right\}^{1/2}$$

Therefore we have to study the variational problem

$$\text{(11)} \qquad \inf_{\substack{v \in W_0^{1,1}(\Omega) \\ \int_\Omega fv\, dx \neq 0}} \left\{ \frac{\int_\Omega |\nabla v| dx}{\left| \int_\Omega fv\, dx \right|} \right\}$$

or

$$\text{(12)} \qquad \inf_{\substack{v \in W_0^{1,1}(\Omega) \\ \int_\Omega fv\, dx = 1}} \left\{ \int_\Omega |\nabla v| dx \right\}$$

Our goal is now to write the problem (12) in the form (1) and to find its dual.

(1) $n = 2$ in [7], but the following is valid as well for an arbitrary n.

We set [2]

$$V = W_0^{1,1}(\Omega) \qquad Y = L^1(\Omega)^n \qquad \Lambda = \nabla$$

$$F(v) = \begin{cases} 0 & \text{if } v \in V \text{ and } \int_\Omega fv\,dx = 1 \\ +\infty & \text{otherwise} \end{cases}$$

$$G(p) = \int_\Omega |p(x)|dx \qquad \forall p \in Y$$

V^*(resp. Y^*) is the dual of V(resp. Y) equipped with the weak-star topology. Now (1) is identical to (12). We note that condition (4) is satisfied since G is everywhere continuous from $L^1(\Omega)^n$ into $\mathbb{R}$.

In order to make $\mathscr{P}^*$ explicit, we now compute G^* and F^*. We have, for every $p^* \in Y^* = L^\infty(\Omega)^n$,

$$G^*(p^*) = \sup_{p \in L^1(\Omega)^n} \left\{ \int_\Omega p^* p\,dx - \int_\Omega |p|dx \right\}$$

$$G^*(p^*) = \begin{cases} 0 & \text{if } |p^*(x)| \leqslant 1 \text{ a.e. (almost everywhere)} \\ +\infty & \text{otherwise} \end{cases} \tag{13}$$

For F^* we have the following Lemma.

Lemma 1 For every $v^* \in V^*$

$$F^*(v^*) = \begin{cases} +\infty & \text{if } v^* \text{ is not proportional to } f \\ \lambda & \text{if } v^* = \lambda f \end{cases} \tag{14}$$

and in particular for every $p^* \in y^*$,

$$F^*(\Lambda^* p^*) = \begin{cases} +\infty & \text{if } \operatorname{div} p^* \text{ is not proportional to } f \\ -\lambda & \text{if } \operatorname{div} p^* = \lambda f \end{cases} \tag{15}$$

Proof: By the definition of F^*,

$$F^*(v^*) = \sup_{v \in V} (\langle v^*, v\rangle - F(v)) = \sup_{\substack{v \in V \\ \int_\Omega fv\,dx = 1}} \langle v^*, v\rangle$$

(2) A more appropriate space V will be considered for the existence theory, but at this point the choice for V is not important.

Since we assume that $f \neq 0$, there exists $v_0 \in V$, such that $\langle f, v_0 \rangle = 1$, where we still denote by f the linear form on V:

$$v \to \int_\Omega fv \, dx$$

Whence

$$F^*(v^*) = \langle v^*, v_0 \rangle + \sup_{w \in ker\, f} (\langle v^*, w \rangle)$$

and there are two cases to consider: either there exists $w_0 \in ker\, f$, such that $\langle v^*, w_0 \rangle \neq 0$, in which case the supremum and F^* are equal to $+\infty$, or $\langle v^*, w \rangle = 0$, $\forall w \in ker\, f$ which amounts to saying that v^* is proportional to f, $v^* = \lambda f$ and then we get (14):

$$F^*(v^*) = \langle v^*, v_0 \rangle = \lambda \langle f, v_0 \rangle = \lambda$$

Since V^* is a space of distributions, $\Lambda p^* = -\operatorname{div} p^*$ in the distribution sense, and (15) follows. This completes the proof.

We are now able to make explicit the dual problem (2). We get

(16*a*) $$\sup \{\lambda\}$$

subject to:

(16*b*)
$$\begin{aligned} & p^* \in L^\infty(\Omega)^n \\ & |p^*(x)| \leqslant 1 \quad \text{a.e.} \\ & \operatorname{div} p^* = \lambda f \end{aligned}$$

This means that we are looking for the largest λ such that the set of feasible points p^* is not empty:

(17) $$\{p^* \in L^\infty(\Omega)^n,\ |p^*(x)| \leqslant 1 \text{ a.e., } \operatorname{div} p^* = \lambda f\}$$

Since (3)-(4) are verified we have the next proposition.

Proposition 1

$$\inf \mathscr{P} = \sup \mathscr{P}^* \in \mathbb{R};$$

problem (16) possesses a solution $\bar{p}^*$, and the corresponding $\bar{\lambda}(\bar{\lambda} f = \nabla \bar{p}^*)$ is the supremum in (16) and the infimum in (12).

2.3. The relaxed problem

By analogy with a similar situation for the Plateau problem, we introduce the following problem

$$\inf_{\substack{v \in W^{1,1}(\Omega) \\ \int_\Omega fv\, dx = 1}} \left\{ \int_\Omega |\nabla v|\, dx + \int_\Gamma |v|\, d\Gamma \right\} \tag{18}$$

The motivation which is not totally intuitive becomes apparent in the following results. Note however that we have relaxed in (12) the condition $v=0$ on Γ and we have added in the objective function the term

$$\int_\Gamma |v| d\Gamma$$

which corresponds to a δ-measure on Γ when $v \neq 0$ on Γ. One could replace $W^{1,1}(\Omega)$ by $BV(\Omega)$ but the difference is not essential at this point.

We set problem (18) as a problem (1) by choosing: $V=W^{1,1}(\Omega)$, V^*its dual, F as before

$$F(v) = \begin{cases} 0 & \text{if } \int_\Omega fv\, dx = 1 \\ +\infty & \text{otherwise} \end{cases}$$

Y is a product space, $Y = Y_1 \times Y_2$,

$$Y_1 = L^1(\Omega)^n \qquad Y_2 = L^1(\Gamma)$$

$$Y^* = Y_1^* \times Y_2^* = L^\infty(\Omega)^n \times L^\infty(\Gamma)$$

$$\Lambda u = (\Lambda_1 u, \Lambda_2 u) \qquad \Lambda_1 u = \nabla u$$

$\Lambda_2 u = \gamma_0 u =$ the trace of u on Γ defined in $L^1(\Gamma)$ by Gagliardo's Theorem [3]. Then,

$$G(p) = G_1(p_1) + G_2(p_2) \qquad p = \{p_1, p_2\} \qquad p_i \in Y_i$$

$$G_1(p_1) = \int_\Omega |p_1| dx \qquad G_2(p_2) = \int_\Gamma |p_2| d\Gamma$$

The problem (1) is now identical to (18). We then make explicit the dual (2).

$$G^*(p^*) = G_1^*(p_1^*) + G_2^*(p_2^*)$$

where $p^* = \{p_1^*, p_2^*\}$, $p_1^* \in L^\infty(\Omega)^n$, $p_2^* \in L^\infty(\Gamma)$,

$$G_1^*(p_1^*) = \begin{cases} 0 & \text{if } |p_1^*(x)| \leq 1 \quad dx \text{ a.e.} \\ +\infty & \text{otherwise} \end{cases}$$

$$G_2^*(p_2^*) = \begin{cases} 0 & \text{if } |p_2^*(x)| \leq 1 \quad d\Gamma \text{ a.e.} \\ +\infty & \text{otherwise} \end{cases}$$

For F^*, we prove the following lemma.

Lemma 2 For every $v^* \in V^*$,

$$F^*(v^*) = \begin{cases} +\infty & \text{if } v^* \text{ is not proportional to } f \\ \lambda & \text{if } v^* = \lambda f \end{cases}$$

and for every $p^* = \{p_1^*, p_2^*\} \in Y^*$,

$$F^*(\Lambda^* p^*) = \begin{cases} -\lambda & \text{if } \operatorname{div} p_1^* = \lambda f \text{ and } p_2^* + p_1^* \cdot \upsilon = 0 \text{ on } \Gamma \\ +\infty & \text{otherwise} \end{cases} \tag{19}$$

where υ is the unit outward normal on Γ.

Proof: The first part of the lemma is proved as (14), and it remains to interpret the condition $v^* = \lambda f$, or

$$\langle v^*, v\rangle = \lambda \langle f, v\rangle \qquad \forall v \in V$$

when $v^* = \Lambda^* p^*$. By the definition of Λ^*, the relation is:

$$\begin{aligned} \langle \Lambda^* p^*, v\rangle &= \int_\Omega p_1^* \cdot \nabla v \, dx + \int_\Gamma p_2^*(\gamma_0 v)\, d\Gamma \\ &= \lambda \int_\Omega f v \, dx, \qquad \forall v \in W^{1,1}(\Omega) \end{aligned}$$

Writing this relation for every $v \in W_0^{1,1}(\Omega)$, we obtain $\operatorname{div} p_1^* = -\lambda f$, and then using the generalized Stokes formula [3]:

$$\int_\Omega p_1^* \cdot \nabla v \, dx = \int_\Gamma p_1^* \cdot \upsilon v \, d\Gamma + \int_\Omega (\operatorname{div} p_1^*)\, v \, dx$$

we find

$$\int_\Gamma (p_1^* \cdot \nu + p_2^*)\, v \mathrm{d}\Gamma = 0 \qquad \forall v \in W^{1,1}(\Omega)$$

which implies $p_1^* \cdot \nu + p_2^* = 0$ on Γ.

Conversely if div $p_1^* = -\lambda f$ and $p_1^* \cdot \nu + p_2^* = 0$ on Γ, then $\Lambda^* p^* = \lambda f$ and (19) follows from the first part of the lemma. This completes the proof. We are able to make the dual of (18) (i.e. (2)) explicit:

(20*a*) $$\sup \{\lambda\}$$

subject to:

(20*b*)
$$\begin{array}{ll} p_1^* \in L^\infty(\Omega)^n & p_2^* \in L^\infty(\Gamma) \\ |p_1^*(x)| \leqslant 1 & \mathrm{d}x \text{ a.e.} \\ |p_2^*(x)| \leqslant 1 & \mathrm{d}\Gamma \text{ a.e.} \\ p_1^* \cdot \nu + p_2^* = 0 & \mathrm{d}\Gamma \text{ a.e.} \\ \operatorname{div} p_1^* = \lambda f & \end{array}$$

We observe that p_2^* does not play any role in (20). If we set $q^* = p_1^*$, (20) becomes identical to (16)

(21*a*) $$\sup \{\lambda\}$$

subject to:

(21*b*)
$$\begin{array}{ll} q^* \in L^\infty(\Omega)^n & \\ |q^*(x)| \leqslant 1 & \mathrm{d}x \text{ a.e.} \\ \operatorname{div} q^* = \lambda f & \end{array}$$

We call ($\mathscr{P}$) and ($\mathscr{Q}$) the initial and relaxed problem (i.e. (12) and (18)), ($\mathscr{P}^*$) and ($\mathscr{Q}^*$) their duals (i.e. (16) and (20)-(21)).

Proposition 2

(22) $$\inf \mathscr{P} = \inf \mathscr{Q} = \sup \mathscr{P}^* = \sup \mathscr{Q}^*$$

(23) $\mathscr{P}^*$ and $\mathscr{Q}^*$ are identical, $\mathscr{P}$ and $\mathscr{Q}$ admit the same dual problem

(24) Every minimizing sequence of ($\mathscr{P}$) is a minimizing sequence of $\mathscr{Q}$

Proof: $\inf \mathscr{Q} = \sup \mathscr{Q}^*$ because of (3), (4) and $\sup \mathscr{P}^* = \sup \mathscr{Q}^*$ since these problems are the same as (21); (23), (24) are then obvious.

Remark 1 The fact that ($\mathscr{P}$) and ($\mathscr{Q}$) admit the same dual seems to be important from the mechanical point of view: indeed, the dual problem defines the stress tensor.

Remark 2 We can write the extremality conditions (7), (8) for ($\mathscr{Q}$); (7) is trivial and (8) decouples into two relations

$$G_i^*(\Lambda_i\bar{u}) + G_i^*(-\bar{p}_i^*) = -\langle \bar{p}_i^*, \Lambda_i\bar{u}\rangle \qquad i = 1, 2$$

i.e.

$$|\nabla\bar{u}(x)| + \bar{p}_1^*(x)\cdot\nabla\bar{u}(x) = 0 \qquad dx \text{ a.e.} \tag{25}$$

$$|\gamma_0\bar{u}(x)| + \bar{p}_2^*(x)\cdot\gamma_0\bar{u}(x) = 0 \qquad d\Gamma \text{ a.e.} \tag{26}$$

(we recall that $|\bar{p}_1^*(x)| \leqslant 1$ a.e., $|\bar{p}_2^*(x)| \leqslant 1$ a.e.).

Remark 3 The infimum does not change in ($\mathscr{P}$) and ($\mathscr{Q}$) if we replace $W^{1,1}(\Omega)$ by $BV(\Omega)$, but the duality theory is then more cumbersome.

2.4. An existence result

We are going to establish an existence result (in $BV(\Omega)$) for the relaxed problem ($\mathscr{Q}$). We recall that $BV(\Omega)$ is the space of functions u in $L^1(\Omega)$ whose gradient is a bounded measure.

We define on $L^1(\Omega)$ the following functional:

$$\Phi(u) = \begin{cases} +\infty & \text{if } u \in L^1(\Omega),\ u \notin BV(\Omega) \\ \displaystyle\int_\Omega |\nabla u| + \int_\Gamma |\gamma_0 u|\, d\Gamma & \text{if } u \in BV(\Omega) \end{cases} \tag{27}$$

Here

$$\int_\Omega |\nabla u|$$

is the total variation of ∇u, and $\gamma_0 u$ is the trace of u on Γ which makes sense and belongs to $L^1(\Gamma)$ if $u \in BV(\Omega)$ (cf. the trace theorem of Miranda [5]): the definition of Φ makes sense.

Lemma 3 For every $u \in L^1(\Omega)$

$$\Phi(u) = \sup_\theta \left\{ \int_\Omega (\nabla\theta)\, u\, dx \right\} \tag{28}$$

where the supremum is for all $\theta = (\theta_1, \ldots, \theta_n)$ with $\theta_i \in \mathscr{C}^\infty(\bar{\Omega})$, and

$$|\theta(x)|^2 = \sum_{i=1}^{n} \theta_i(x)^2 \leqslant 1 \qquad (\nabla\theta = \operatorname{div}\theta)$$

Proof: The same as in [10], p. 222.

We infer from (28) that Φ is the upper bound of a family of continuous linear functions on $L^1(\Omega)$ and thus:

(29) Φ is convex and lower semi-continuous on $L^1(\Omega)$

Now the existence result follows:

Proposition 3

The relaxed problem (Problem ($\mathscr{Q}$) or (18)) possesses at least one solution in $BV(\Omega)$.

The cluster points in $L^1(\Omega)$ of any minimizing sequence of ($\mathscr{P}$) or ($\mathscr{Q}$) ((12) or (18)) are solutions in $BV(\Omega)$ of (18).

Proof: If we allow v to vary in $BV(\Omega)$ instead of $W^{1,1}(\Omega)$, the infimum (18) remains unchanged since, for every $v \in BV(\Omega)$, there exists a sequence of smooth functions v_m converging to v in $BV(\Omega)$ with

$$\int_\Omega |\nabla v_m| \, dx \to \int_\Omega |\nabla v|$$

Now let v_m be a minimizing sequence of ($\mathscr{Q}$) (or ($\mathscr{P}$)):

$$v_m \in W^{1,1}(\Omega) \qquad \int_\Omega f v_m \, dx = 1$$

and

$$\int_\Omega |\nabla v_m| \, dx + \int_\Omega |\gamma_0 v_m| \, d\Gamma \to \inf \mathscr{Q} = \inf \mathscr{P}$$

Then,

$$\int_\Omega |\nabla v_m| \, dx \leqslant \text{constant}, \qquad \int_\Gamma (\gamma_0 v_m) \, d\Gamma \leqslant \text{constant}$$

and, by a variation of the Poincaré inequality, the sequence v_m remains bounded in $W^{1,1}(\Omega)$. Since the injection of $W^{1,1}(\Omega)$ into $L^1(\Omega)$ is compact, there exists a subsequence (still denoted v_m) such that

$$v_m \to u \qquad \text{strongly in } L^1(\Omega).$$

It is clear that $u \in BV(\Omega)$ and by (29),

$$\Phi(u) \leqslant \varliminf_{m \to +\infty} \Phi(u_m) = \inf \mathscr{P} = \inf \mathscr{Q} \tag{30}$$

Whence (since $\int_\Omega fu\,\mathrm{d}x = 1$), u is a solution of ($\mathscr{Q}$) in $BV(\Omega)$. This completes the proof.

3. THE MODEL PROBLEM FOR DEFORMATION THEORY

In section 3.1 we make explicit the problem and its dual. The relaxed problem is given in section 3.2 and an existence result for the relaxed problem in section 3.3.

The steps are the same as in section 2 and we only describe the main points.

3.1. The problem and its dual

We consider the real function g defined for every $s \in \mathbb{R}$ by

$$g(s) = \begin{cases} \dfrac{s^2}{2} & \text{if } |s| \leqslant 1 \\ |s| - \dfrac{1}{2} & \text{if } |s| \geqslant 1 \end{cases} \tag{31}$$

The model problem is the following one

$$\inf_{v \in W^{1,1}(\Omega)} \left\{ \int_\Omega g(|\nabla v|)\,\mathrm{d}x - \lambda \int_\Omega fv\,\mathrm{d}x \right\} \tag{32}$$

where f is the same as before and $\lambda > 0$ is given. The values of λ for which the infimum in (32) is finite are given by the limit analysis (problem of section 2) and this relation will be more apparent hereafter (cf. (36)).

We define $V, Y, \ldots,$ with the purpose of writing (32) as a problem (1):

$$V = W_0^{1,1}(\Omega) \qquad Y = L^1(\Omega)^n \qquad \Lambda = \nabla = \text{grad}$$

$$F(v) = -\lambda \int_\Omega fv\,\mathrm{d}x \qquad \forall v \in V$$

$$G(p) = \int_\Omega g(|p(x)|)\,\mathrm{d}x \qquad \forall p \in Y$$

The space V^* is the dual of V equipped with the $\sigma(V^*, V)$ topology, $Y^* = L^\infty(\Omega)^n$. Problem (32) is the same as (1).

Lemma 4 For all $p^* \in Y^*$,

(33)
$$F^*(\Lambda^* p^*) = \begin{cases} 0 & \text{if } \operatorname{div} p^* = \lambda f \\ +\infty & \text{otherwise} \end{cases}$$

(34)
$$G^*(p^*) = \begin{cases} \frac{1}{2} \int_\Omega |p^*(x)|^2 \, dx & \text{if } |p^*(x)| \leqslant 1 \text{ a.e.} \\ +\infty & \text{otherwise} \end{cases}$$

Proof: (33) follows directly from the definition. For (34), we recall (cf. [2]) that

$$G^*(p^*) = \int_\Omega g^*(|p^*(x)|) \, dx$$

where g^* is the conjugate of g:

$$g^*(t) = \sup_{s \in \mathbb{R}} [st - g(s)] \qquad \forall t \in \mathbb{R}$$

It is easy to see that

$$g^*(t) = \begin{cases} \frac{t^2}{2} & \text{if } |t| \leqslant 1 \\ +\infty & \text{otherwise} \end{cases}$$

and (34) follows. This completes the proof.

We are able to make the dual (2) explicit:

(35)
$$\sup_{\substack{p^* \in L^\infty(\Omega)^n \\ |p^*(x)| \leqslant 1 \text{ a.e.} \\ \operatorname{div} p^* = \lambda f}} \left\{ -\frac{1}{2} \int_\Omega |p^*(x)|^2 \, dx \right\}$$

We call ($\mathscr{P}$) and ($\mathscr{P}^*$) the problems (32) and (35). Condition (4) is obviously satisfied and therefore (3) is satisfied ($\inf \mathscr{P} \in \mathbb{R}$) if and only if $\sup \mathscr{P}^* > -\infty$.

Sup $\mathscr{P}^*$ is finite if and only if the set of feasible p^* is not empty and this set is exactly (17). The positive λ for which (17) is not empty are those in the interval $[0, \bar{\lambda}[$, $\bar{\lambda}$ the supremum in (16) and our assumption on λ is therefore

(36)
$$0 \leqslant \lambda < \bar{\lambda}, \quad \bar{\lambda} \text{ the supremum in (16)}$$

Then, we have the following proposition.

Proposition 4
Under the assumption (36),

$$\inf \mathcal{P} = \sup \mathcal{P}^* \qquad (\in \mathbb{R}) \tag{37}$$

and problem ($\mathcal{P}^*$) possesses at least one solution $\bar{p}^*$.

3.2. The relaxed problem

As in section 2.3 we introduce a relaxed problem

$$\inf_{v \in W^{1,1}(\Omega)} \left\{ \int_\Omega g(|\nabla v|)\,dx \int_\Gamma |\gamma_0 v|\,d\Gamma - \int_\Omega \lambda f v\,dx \right\} \tag{38}$$

We set

$$V = W^{1,1}(\Omega) \qquad Y = Y_1 \times Y_2 \qquad Y_1 = L^1(\Omega)^n \qquad Y_2 = L^1(\Gamma)$$

$$\Lambda u = \{\Lambda_1 u, \Lambda_2 u\} \qquad \Lambda_1 u = \operatorname{grad} u$$

$$\Lambda_2 u = \gamma_0 u = \text{the trace of } u \text{ on } \Gamma \qquad \text{(cf. [3])}$$

$$F(v) = -\lambda \int_\Omega f v\,dx$$

$$G(p) = G_1(p_1) + G_2(p_2) \qquad p = \{p_1, p_2\}$$

with

$$G_1(p_1) = \int_\Omega g(|p_1|)\,dx \qquad G_2(p_2) = \int_\Gamma |p_2|\,d\Gamma$$

V^* is the dual, $Y^* = Y_1^* \times Y_2^*$, $Y_1^* = L^\infty(\Omega)^n$, $Y_2^* = L^\infty(\Gamma)$.

Now problem (38) is written as a problem (1) and we look for its dual.

Lemma 5 For every $p^* = \{p_1^*, p_2^*\} \in Y^*$,

$$F^*(\Lambda^* p^*) = \begin{cases} 0 & \text{if } \operatorname{div} p_1^* = \lambda f,\ p_1^* \nu + p_2^* = 0 \text{ on } \Gamma \\ +\infty & \text{otherwise} \end{cases} \tag{39}$$

$$G^*(p^*) = G_1^*(p_1^*) + G_2^*(p_2^*) \tag{40}$$

$$G_1^*(p_1^*) = \begin{cases} \frac{1}{2} \int_\Omega |p_1^*(x)|^2 \, dx & \text{if } |p_1^*(x)| \leqslant 1 \ dx \text{ a.e.} \\ +\infty & \text{otherwise} \end{cases}$$

$$G_2^*(p_2^*) = \begin{cases} 0 & \text{if } |p_2^*(x)| \leqslant 1 \ d\Gamma \text{ a.e.} \\ +\infty & \text{otherwise} \end{cases}$$

Proof: The result is already proved for G_1^* and is easy for G_2^*. For F^*,

$$F^*(\Lambda^* p^*) = \sup_{v \in V} \left\{ \langle p^*, \Lambda v \rangle + \lambda \int_\Omega fv \, dx \right\}$$

$$= \sup_{v \in W^{1,1}(\Omega)} \left\{ \int_\Omega p_1^* \cdot \nabla v \, dx + \int_\Gamma p_2^* (\gamma_0 v) \, d\Gamma + \lambda \int_\Omega fv \, dx \right\}$$

$$\geqslant \sup_{v \in \mathscr{C}_0^\infty(\Omega)} \left\{ \int_\Omega p_1^* \nabla v \, dx + \lambda \int_\Omega fv \, dx \right\}$$

The last supremum (and F^*) is infinite unless div $p^* = \lambda f$. If div $p^* = \lambda f$, then using the generalized Stokes formula we get

$$F^*(\Lambda^* p^*) = \sup_{v \in W^{1,1}(\Omega)} \int_\Gamma (p_1^* \upsilon + p_2^*)(\gamma_0 v) \, d\Gamma$$

and (39) follows. This completes the proof.

Problem (38) is called $(\mathscr{Q})$. Its dual $(\mathscr{Q}^*)$ is:

(41*a*)
$$\sup \left\{ -\frac{1}{2} \int_\Omega |p_1^*(x)|^2 \, dx \right\}$$

subject to:

(41*b*)
$$\begin{aligned} & p_1^* \in L^\infty(\Omega) && p_2^* \in L^\infty(\Gamma) \\ & |p_1^*(x)| \leqslant 1 && dx \text{ a.e.} \\ & |p_2^*(x)| \leqslant 1 && d\Gamma \text{ a.e.} \\ & \text{div } p_1^* = \lambda f && \\ & p_1^* \cdot \upsilon + p_2^* = 0 && \text{on } \Gamma \end{aligned}$$

Again p_2^* does not play any role in (41) and setting $q^* = p_1^*$, this problem reduces to

$$\text{(42a)} \qquad \sup \left\{ -\frac{1}{2} \int_\Omega |q^*(x)|^2 \, dx \right\}$$

subject to:

$$\text{(42b)} \qquad \begin{array}{l} q^* \in L^\infty(\Omega)^n \\ |q^*(x)| \leqslant 1 \qquad \text{a.e.} \\ \operatorname{div} q^* = \lambda f \end{array}$$

and this is exactly (35).

Proposition 5
We assume that λ satisfies (36). Then

$$\text{(43)} \qquad \inf \mathscr{P} = \inf \mathscr{Q} = \sup \mathscr{P}^* = \sup \mathscr{Q}^* \qquad (\in \mathbb{R}).$$

(44) $\mathscr{P}^*$ and $\mathscr{Q}^*$ are identical, $\mathscr{P}$ and $\mathscr{Q}$ admit the same dual problem

(45) Every minimizing sequence of $(\mathscr{P})$ is a minimizing sequence of $(\mathscr{Q})$.

Same proof as Proposition 2.

Remark 4 We can write the extremality relations for $(\mathscr{Q})$. If $\bar{u}$ is a solution of $(\mathscr{Q})$ and $(\bar{p}^*)$ a solution of $(\mathscr{Q}^*)$,

$$\text{(46)} \qquad g(|\nabla \bar{u}(x)|) + g^*(|\bar{p}_1^*(x)|) + \bar{p}_1^*(x) \cdot \nabla \bar{u}(x) = 0 \qquad dx \text{ a.e.}$$

$$\text{(47)} \qquad |\gamma_0 \bar{u}(x)| + \bar{p}_2^*(x) \cdot \gamma_0 \bar{u}(x) = 0 \qquad d\Gamma \text{ a.e.}$$

3.3. An existence result

We consider in $L^1(\Omega)$, the functional

$$\text{(48)} \qquad \Phi(u) = \sup_\theta \left\{ \int_\Omega \left[\frac{1}{2} |\theta(x)|^2 + (\operatorname{div} \theta(x))\, u(x) \right] dx - \lambda \int_\Omega f u \, dx \right\}$$

where the supremum is taken among all the $\theta = (\theta_1, \ldots, \theta_n)$ with $\theta_i \in \mathscr{C}^\infty(\bar{\Omega})$ and

$$|\theta(x)|^2 = \sum_{i=1}^n \theta_i(x)^2 \leqslant 1$$

The functional satisfies:

$$\Phi \text{ is convex lower semi-continuous on } L^1(\Omega) \tag{49}$$

As for Lemma 3, using the techniques of [10] we can show that the following lemma holds.

Lemma 6 For every $u \in L^1(\Omega)$,

$$\Phi(u) = +\infty \qquad \text{if } u \notin BV(\Omega) \tag{50}$$

and, if $u \in BV(\Omega)$,

$$\Phi(u) = \int_\Omega g(|\rho|)\,dx + \int_\Omega |\mu| + \int_\Gamma |\gamma_0 u|\,d\Gamma - \lambda \int_\Omega fu\,dx \tag{51}$$

where

$$|\rho| = \left(\sum_{i=1}^{n} \rho_i^2\right)^{1/2} \qquad |\mu| = |\mu_1| + \ldots + |\mu_n|$$

$$\frac{\partial u}{\partial x_i} = \rho_i\,dx + \mu_i \qquad 1 \leqslant i \leqslant n$$

being the Lebesgue decomposition of the measure $\partial u/\partial x_i$.

We now state the existence result.

Proposition 6

The relaxed problem (problem $(\mathcal{Q})$ or (38)) possesses at least one solution in $BV(\Omega)$.

The cluster points in $L^1(\Omega)$ of any minimizing sequence of $(\mathcal{P})$ or $(\mathcal{Q})$ (i.e. (32) or (38)) are solutions in $BV(\Omega)$ of (38).

Same proof as Proposition 3.

Remark 5 The solutions of the relaxed problem (section 2 or 3) do not necessarily satisfy the boundary condition on Γ, $u = 0$. Lichnewsky [4] has studied the boundary behaviour of the relaxed solutions and given sufficient conditions which guarantee that $u = 0$ on part of Γ.

REFERENCES

[1] G. Duvaut, and J. L. Lions. *'Les inéquations en mécanique et en physique'*. Dunod, Paris (1972).
[2] I. Ekeland, and R. Temam. *'Convex Analysis and Variational Problems'*. North-Holland, Amsterdam (1976).
[3] E. Gagliardo. 'Caratterizzazioni delle tracce sulla frontiera relative ad alcune classi di funzioni di *n* variabili'. *Rend. Sem. Univ. Padova,* **27** (1957), 284-305.
[4] A. Lichnewsky. *C. R. Acad. Sci., Paris* (1978).
[5] M. Miranda. 'Comportamento delle successioni convergenti di frontiere minimali'. *Rend. Sem. Math. Univ. Padova,* (1967), 238-57.
[6] R. T. Rockafellar. 'Duality and stability in extremum problems involving convex functions'. *Pacific J. Math.*, **21** (1967), 167-87.
[7] G. Strang. 'A family of model problems in plasticity'. *Proc. Third Inter. Symp. on Computing Methods in Applied Sciences and Engineering; Lecture Notes in Computer Sciences*, Springer Verlag, Heidelberg. To appear.
[8] R. Temam, and G. Strang. 'Existence de solutions relaxées pour les équations de la plasticité: Etude d'une espace fonctionnel'. *C. R. Acad. Sci., Paris,* (1978).
[9] R. Temam. 'Solutions généralisées de certaines équations du type hypersurfaces minimales'. *Arch. Ration. Mech. & Anal.*, **44** (1971), 121-56.
[10] R. Temam. 'Applications de l'analyse convexe au calcul des variations'. In *Nonlinear operators and the calculus of variations*, eds. Gossez, Lami Dozo, Mawhin, and Waelbroeck, *Lecture Notes in Math.*, no. 543, Springer Verlag, Heidelberg (1976).

NOTE ADDED IN THE PROOFS

For new results concerning the mathematical problems in plasticity theory, the reader is referred to:

[*a*] R. Temam, and G. Strang. 'Functions of Bounded Deformation'. *Arch. Rat. Mech. Anal.* To appear.
[*b*] R. Temam, and G. Strang. 'Duality and Relaxation in the Variational Problems of Plasticity'. *Journal de Mécanique.* To appear.
[*c*] R. Temam. *'Mathematical Problems in Plasticity'.* Proceedings of the 2nd TICOM Conference on Computational Methods in Mechanics, Austin, Texas, March 1979. To appear.

Chapter 25

Convergence of Convex Functions, Variational Inequalities and Convex Optimization Problems

R. J. B. Wets

1. INTRODUCTION

The work of Lions and Stampacchia [8] on variational inequalities motivated the study of the convergence of the solutions of approximating problems. This work was carried out by Mosco [11, 12], Di Giorgi [4] and their students. More recently, Attouch [1, 2] has exploited this approach to establish the convergence of the Yoshida approximates to the resolvant of evolution-type equations. Questions of the same type did arise in optimization, in particular when seeking numerical solutions of infinite-dimensional problems. Here the research was initiated by Wysman [26] in statistical decision theory and by Van Cutsem [22] who applied his results to stochastic optimization problems. This line of investigation is pursued by Zolezzi [27] in his study of an abstract linear-quadratic optimization problem.

The purpose here is to review the salient features of the theory of convergence of convex functions and its implications. The results will be derived for convex functions defined on $\mathbb{R}^n$, the n-dimensional Euclidean space. In this setting, a number of fine distinctions between different types of convergence are swept away; in particular in infinite dimensions it would be imperative to distinguish between convergence to weak and strong solutions.

2. CONVERGENCE OF CONVEX FUNCTIONS

We consider collections $\{f; f_\upsilon, \upsilon \in \mathbf{N}\}$ of *closed convex functions* with domain $\mathbb{R}^n$, the Euclidean space of dimension n, and range $R \cup \{+\infty\} = R^\bullet$; $\underset{\sim}{\mathbf{N}}$ is a countable (ordered) index set, typically the natural numbers. The *effective domain* of a (convex) function f is denoted by

$$D(f) = \{(x \in \mathbb{R}^n : f(x) < +\infty\}$$

and its *epigraph* is

$$E(f) = \{(x, \alpha) \in \mathbb{R}^n \times \mathbb{R} : f(x) \leqslant \alpha\}$$

A convex function is said to be *closed* if it is lower semicontinuous (l.s.c.) and $D(f)$ is non-empty; in particular this implies that $E(f)$ is a closed proper subset of $\mathbb{R}^n \times \mathbb{R}$ that does not contain any 'vertical' line, i.e. parallel to the line $\{0\} \times \mathbb{R}$.

Two different types of convergence for sequences of closed convex functions play a key role in the study of convergence of solutions to variational inequalities and optimization problems: point-wise convergence, called p-*convergence*, and convergence in terms of the epigraphs, called e-*convergence*, to be defined below. We write:

$$f_\upsilon \underset{p}{\rightarrow} f$$

if for all x in $\mathbb{R}^n$

$$\lim_\upsilon f_\upsilon(x) = f(x)$$

where we allow $+\infty$ as limit value, as well as for element of the sequence $\{f_\upsilon(x)\}_{\upsilon \in \mathbf{N}}$. Loosely speaking, point-wise convergence corresponds to convergence of the parameters determining the functions f_υ. Intrinsically, it is the type of convergence which is available in most schemes involving direct approximations of the function f.

A sequence of closed subsets $\{C_\upsilon, \upsilon \in \mathbf{N}\}$ of $\mathbb{R}^n$ is said to converge to a closed set C, if

$$\limsup_\upsilon C_\upsilon = C = \liminf_\upsilon C_\upsilon \tag{1}$$

in which case we write

$$\lim_\upsilon C_\upsilon = C$$

where

$$\limsup C_\upsilon = \{x = \lim x_\mu : x_\mu \in C_\mu, \mu \in \mathbf{M} \subset \mathbf{N}\}$$
$$\liminf C_\upsilon = \{x = \lim x_\upsilon : x_\upsilon \in C_\upsilon, \upsilon \in \mathbf{N}\}$$

By **M** we shall always denote a countable subset of **N**; this will allow us to distinguish between statements that need to hold only for subsequences rather than for sequences. This type of convergence for closed sets is discussed in Kuratowski [7]. A sequence of closed convex functions is said to e-*con-*

verge, written

$$f_\upsilon \underset{e}{\to} f$$

if the epigraphs of the f_υ converge to the epigraph of f that is

$$\text{limsup}\, E(f_\upsilon) = E(f) = \text{liminf}\, E(f_\upsilon) \tag{2}$$

This type of convergence is that required to obtain the convergence of the solutions of variational inequalities (here, their finite-dimensional counterparts) and of the solutions to optimization problems whose optima are known to converge.

3. CONVERGENCE OF SEQUENCES OF CLOSED CONVEX SETS

In this section we review briefly some of the characterizations of convergence for sequences of closed convex sets that will be needed in the following. Let $\delta(C.D)$ denote the *one-sided Hausdorff distance* from C to D; it is defined as follows:

$$\delta(C, D) = \begin{cases} 0 & \text{if } C = D = \emptyset \\ +\infty & \text{if } C \neq D,\ C = \emptyset \ \text{ or } D = \emptyset \\ \sup\limits_{x \in C} d(x, D) & \text{otherwise} \end{cases}$$

where

$$d(x, D) = \inf_{y \in D} d(x, y)$$

is the distance from a point x to the set D with $d(x, y)$ a metric defined on $\mathbb{R}^n$, which we take here to be the Euclidean metric. This is done to avoid introducing a dual metric on $\mathbb{R}^n$; however, our results are not tied to the Euclidean metric. The *Hausdorff distance* is given by

$$h(C, D) = \sup\, [\delta(C, D), \delta(D, C)]$$

For all $\rho \geqslant 0$, the *ρ-distance* $h_\rho(C, D)$ between two sets C and D is by definition the Hausdorff distance between C^ρ and D^ρ, i.e.

$$h_\rho(C, D) = h(C^\rho, D^\rho)$$

where for any set $H \subset \mathbb{R}^n$, H^ρ is the intersection of H with the closed ball of radius ρ and centre at the origin, i.e.

$$H^\rho = H \cap B_\rho(0) = \{x \in H : d(0, x) \leqslant \rho\}$$

The proof of the following theorem, which yields some useful criteria for the convergence of closed sets, can be found in [20].

Theorem 1
Suppose that $\{C; C_\upsilon, \upsilon \in \mathbf{N}\}$ is a collection of closed subsets of $\mathbb{R}^n$. Then the following statements are equivalent:
(i) $\lim_\upsilon C_\upsilon = C$;
(ii) the two following conditions are satisfied:
(ii$_a$) to any compact set $K \subset \mathbb{R}^n$ such that $C \cap K = \emptyset$ there corresponds an index $\upsilon(K)$ such that $C_\upsilon \cap K = \emptyset$ for all $\upsilon \geqslant \upsilon(K)$,
(ii$_b$) to any open set $G \subset \mathbb{R}^n$ such that $C \cap G \neq \emptyset$ there corresponds an index $\upsilon(G)$ such that $C_\upsilon \cap G \neq \emptyset$ for all $\upsilon \geqslant \upsilon(G)$;
(iii) $\lim_{\rho \uparrow \infty} \liminf_\upsilon C_\upsilon^\rho = \lim_{\rho \uparrow \infty} \limsup_\upsilon C_\upsilon^\rho = C$

If the sets are also convex, there is a version of (iii), somewhat stronger, which appears in [19, theorem 4]. Here it is obtained as a corollary to the preceding theorem.

Corollary 1. Suppose that $\{C; C_\upsilon, \upsilon \in \underset{\sim}{\mathbf{N}}\}$ is a collection of closed convex subsets of $\mathbb{R}^n$. Then

$$C = \lim_\upsilon C_\upsilon$$

if and only if
(iii′) there exists $\rho' > 0$ such that

$$\lim_\upsilon h_\rho(C_\upsilon, C) = 0 \qquad \text{for all } \rho \geqslant \rho'$$

Moreover, if the sets $\{C; C_\upsilon, \upsilon \in \underset{\sim}{\mathbf{N}}\}$ are closed convex cones, then

$$C = \lim_\upsilon C_\upsilon$$

if and only if
(iii″) $\lim_\upsilon h_\rho(C_\upsilon, C) = 0$ for any $\rho > 0$

Proof: Clearly, (iii′) implies (iii) in Theorem 1. The converse is obtained as follows. The case C empty is trivial, so let us assume that C is non-empty. First note that the sets in the sequences $\{C^\rho, C_\upsilon^\rho, \upsilon \in \underset{\sim}{\mathbf{N}}\}$ are compact. Hence, to show that

$$\lim_{v} h_\rho(C_v, C) = \lim_{v} h(C_v^\rho, C^\rho) = 0$$

for all ρ larger than some $\rho' > 0$, is equivalent to showing that, for all $\rho > \rho'$,

$$\limsup_{v} C_v^\rho \subset C^\rho \subset \liminf_{v} C_v^\rho$$

Select $\rho' > d(0, C)$. Then, for any $\rho > \rho'$, the first inclusion directly follows the hypothesis

$$\lim_{\rho \uparrow \infty} \limsup_{v} C_v^\rho \subset C$$

since for all ρ,

$$\limsup_{v} C_v^\rho \subset B_\rho(0)$$

For the second inclusion, we need to rely on the convexity of the elements appearing in the sequence. The argument is similar to that appearing in the last part of the proof of theorem 4 of [19] and is not repeated here.

The second assertion follows from the first one, if we simply observe that for C and C_v closed convex cones and $\rho > 0$

$$h_\rho(C_v, C) = \rho h_1(C_v, C)$$

This completes the proof.

The following proposition is needed to derive the main result in the next section. Let D be any subset of $\mathbb{R}^n$, then pos D is the *positive hull* of D, equivalently the convex cone generated by D, i.e.

$$\text{pos } D = \left\{ x \in \mathbb{R}^n : x = \sum_{i=1}^{p} \lambda_i y^i, y^i \in D, \lambda_i \geq 0, p \in \mathbf{N} \right\}$$

In general, this cone is not closed. But if D is a compact set, which does not contain 0 in the boundary of its convex hull, then pos D is closed. If D is convex, then pos $D = \{x : x = \lambda y, y \in D, \lambda \geq 0\}$.

Proposition

Suppose that $\{C; C_v, v \in \underset{\sim}{\mathbf{N}}\}$ is a collection of compact convex sets not containing the origin 0. If

$$C = \lim_{v} C_v$$

then

$$\text{pos } C = \lim_{\upsilon} \text{ pos } C_\upsilon$$

Proof: The case $C = \emptyset$ is trivial. Assume that C is non-empty. The remarks preceding the theorem imply that all sets in the collection {pos C; pos C_υ, $\upsilon \in \tilde{\mathbf{N}}$} are closed convex cones. In view of Corollary 1, it suffices to show that for some $\rho > 0$

$$\lim_{\upsilon} h_\rho(\text{pos } C_\upsilon, \text{pos } C) = 0$$

or equivalently that:

$$\limsup_{\upsilon} (\text{pos } C_\upsilon)^\rho \subset (\text{pos } C)^\rho \subset \liminf_{\upsilon} (\text{pos } C_\upsilon)^\rho$$

This follows directly from the definitions, if we remember that the elements of a converging sequence of connected compact sets are uniformly bounded [19, proposition 3]. This completes the proof.

4. CHARACTERIZING e-CONVERGENCE

As already mentioned above, the restriction to finite dimension does blur some important distinctions that cannot be ignored in infinite dimensions. (Depending on the topology used to compute the limsup and liminf appearing in (2) one refers to e-convergence as M-convergence [13], G-convergence [4, 21], R-convergence [2], Γ-convergence [10], Γ^--convergence, . . .). In finite dimension, however, it is possible to zero in on the prominent properties of e-convergence while bypassing some nontrivial technical developments.

To each closed convex function we can associate a *conjugate* function, denoted by f^*, and defined by

$$f^*(v) = \sup [v \cdot x - f(x) : x \in \mathbb{R}^n]$$

This is another closed convex function and $(f^*)^* = f$. The main feature of e-convergence is that it is preserved under *conjugacy* (polarity, Young-Fenchel transformation, . . .), that is

$$f_\upsilon \xrightarrow{e} f \qquad \text{if and only if} \qquad f_\upsilon^* \xrightarrow{e} f^*$$

This plays a key role in the ensuing development. This theorem—here Corollary 4—was first proved by Wysman [26] and later extended by Mosco [14] and further refined by Joly [5]. The theorem can also be derived from a stronger result of Walkup and Wets [23] for closed convex cones and their polars. (A related result had been derived earlier by Klee [6]). There is no 'short' proof of the general theorem; all known proofs require the derivation of a number of preliminary results. The path followed here has never appeared explicitly in print.

If C is a closed convex cone in $\mathbb{R}^n$, then its *polar*, denoted by pol C, is the closed convex cone defined by

$$\text{pol } C = \{v \in \mathbb{R}^n : v \cdot x \leqslant 0 \text{ for all } x \in C\}$$

Naturally, pol (pol C) = C.

Theorem 2
(See [23].) Suppose that C_1, C_2 are closed convex cones in $\mathbb{R}^n$. Then, for all $\rho > 0$,

$$h_\rho(C_1, C_2) = h_\rho(\text{pol } C_1, \text{pol } C_2)$$

Proof: Since $h_\rho(C, D) = h(C^\rho, D^\rho)$ the theorem will be proved if we show that, for the one-sided Hausdorff distance we have that

$$\delta[(\text{pol } C_2)^\rho, (\text{pol } C_1)^\rho] \geqslant \delta(C_1^\rho, C_2^\rho)$$

This is trivial if $C_1 = C_2$. Suppose that $\delta(C_1^\rho, C_2^\rho) > 0$. Let us start by considering the case when C_1 is a ray and C_2 a half-space not containing C_1. Let c_1 and d_2 be the elements of C_1 and pol C_2 respectively of norm ρ. Then, it is easy to see that

$$\delta(C_1^\rho, C_2^\rho) = d_2 \cdot c_1 = \delta\,[(\text{pol } C_2)^\rho, (\text{pol } C_1)^\rho]$$

In general, i.e. when C_1, C_2 are arbitrary closed convex cones, then, for any ε, $0 < \varepsilon < \delta(C_1^\rho, C_2^\rho)$, there exists $0 \neq c_1 \in C_1^\rho$, such that $\delta(c_1, C_2) = \alpha \geqslant \delta(C_1^\rho, C_2^\rho) - \varepsilon$.

Let D_2 be a closed half-space containing C_2 and such that $\delta(c_1; D_2) = \alpha$. By the above and the inclusions pol $\{\lambda c_1 : \lambda \in \mathbb{R}^n\} \supset$ pol C_1 and pol $C_2 \supset$ pol D_2 we obtain

$$\delta(C_1^\rho, C_2^\rho) - \varepsilon \leqslant \delta(c_1, C_2^\rho) = \delta(c_1, D_2^\rho)$$

$$\leqslant \delta[(\text{pol } D_2)^\rho, (\text{pol } c_1)^\rho] \leqslant \delta\,[(\text{pol } c_2)^\rho, (\text{pol } c_1)^\rho]$$

This holds for all $\varepsilon > 0$, arbitrarily small. This completes the proof.

Corollary 2 Suppose that $\{C; C_\upsilon, \upsilon \in \underset{\sim}{\mathbf{N}}\}$ are closed convex cones. Then $C = \lim\limits_{\upsilon} C_\upsilon$ if and only if pol $C = \lim\limits_{\upsilon}$ pol C_υ.

Corollary 3 (Bicontinuity of conjugacy). Suppose that $\{f; f_\upsilon, \upsilon \in \underset{\sim}{\mathbf{N}}\}$ is a collection of closed convex functions. Then

$$f_v \underset{e}{\rightarrow} f \quad \text{if and only if} \quad f_v^* \underset{e}{\rightarrow} f^*$$

Proof: It clearly suffices to show that

$$E(f) = \lim_v E(f_v)$$

implies that

$$E(f^*) = \lim_v E(f_v^*)$$

or, equivalently, that

$$U = \lim_v U_v$$

implies that

$$V = \lim_v V_v$$

where

$$U = \{-1\} \times E(f) \qquad U_v = \{-1\} \times E(f_v)$$

and

$$V = \{(\eta, -1, \nu) : (\eta, \nu) \in E(f^*)\}, \quad V_v = \{(\eta, -1, \nu) : (\eta, \nu) \in E(f_v^*)\}$$

Note that $f_{(v)}$ and $f_{(v)}^*$ conjugate yields

$$\text{pol cl pos } U_{(v)} = \text{cl pos } V_{(v)}$$

and that if

$$\text{cl pos } V = \lim_v \text{ cl pos } V_v$$

then

$$V = \lim V_v$$

Hence, in view of Corollary 2 it will be sufficient to show that

$$U = \lim_v U_v$$

implies that

$$\text{cl pos } U = \lim_v \text{ cl pos } U_v$$

Note that for all $\rho > 0$, $U^{\rho}_{(v)}$ is a compact convex set bounded away from 0. Corollary 2 and the proposition allow us to conclude that if $U = \lim U_v$ then for some $\rho' > 0$, $\text{pos } U^{\rho} = \lim_v \text{ pos } U^{\rho}_v$ for all $\rho > \rho'$.
There only remains to show that this in turn implies that

$$\lim_{\rho \uparrow \infty} \text{pos } U^{\rho} = \lim_v \lim_{\rho \uparrow \infty} \text{pos}, U^{\rho}_v$$

since clearly

$$\lim_{\rho \uparrow \infty} \text{pos } U^{\rho}_{(v)} = \text{cl} \bigcup_{\rho > \rho'} \text{pos } U^{\rho}_{(v)} = \text{cl pos } U_{(v)}$$

(if needed see [19], proposition 1).
To show that

$$D = \limsup_v \lim_{\rho \uparrow \infty} \text{pos } U^{\rho}_v \subset \text{cl pos } U = C$$

we argue by contradiction. Clearly D and C are closed convex cones. Suppose that $D \not\subset C$. Let y be a point in D but not in C. Then, there exists $\nu \in \text{pol } C$ such that $\nu \cdot y > 0$. Now, observe that $y \in D$. By definition of D, there exists $\{y_\mu : y_\mu \in \text{pos } U^{\rho_\mu}_\mu, \mu \in \underset{\sim}{M}\}$ converging to y. Hence, $\nu \cdot y_\mu > 0$ for μ sufficiently large. On the other hand, from Corollary 2 we know that for all $\rho > \rho'$

$$\text{pol pos } U^{\rho} = \lim_v \text{ pol pos } U^{\rho}_v$$

Since

$$\nu \in \text{pol } C = \lim_{\rho \uparrow \infty} \text{ pol pos } U^{\rho}$$

it follows from a standard diagonalization argument, via the preceding equality, that

$$\nu = \lim_v \{\nu_v : \nu_v \in \text{pol pos } U^{\rho_v}_v, v \in \underset{\sim}{N}\}$$

Considering only $\{\nu_\mu, \mu \in \underset{\sim}{M}\}$, the relation $\nu \cdot y_\mu > 0$ in turn implies that $\nu_\mu \cdot y_\mu > 0$ for μ sufficiently large, with $y_\mu \in \text{pos } U^{\rho_\mu}_\mu$ and $\nu_\mu \in \text{pol pos } U^{\rho_\mu}_\mu$, an evident contradiction.

The proof will be complete if we show that

$$\text{cl pos } U \subset \liminf_v \text{ cl pos } U_v$$

If

$$y \in \text{cl} \bigcup_{\rho > \rho'} \text{pos } U^{\rho}$$

it follows that

$$y = \lim_{\rho} y^{(\rho)}$$

Since each $y^{(\rho)}$ is in

$$\text{pos } U^{\rho} = \lim_{\upsilon} \text{ pos } U^{\rho}_{\upsilon}$$

it follows that

$$y^{(\rho)} = \lim_{\upsilon} y^{(\rho)}_{\upsilon}$$

where

$$y^{(\rho)}_{\upsilon} \in \text{pos } U^{\rho}_{\upsilon} \subset \text{pos } U_{\upsilon}$$

From this double indexed sequence, we rely on the diagonalization procedure to extract a sequence y_{υ} converging to y with $y_{\upsilon} \in \text{pos } U_{\upsilon}$, i.e.

$$y \in \liminf_{\upsilon} \text{ pos } U_{\upsilon} \subset \liminf_{\upsilon} \text{ cl pos } U_{\upsilon}$$

This completes the proof.

The next theorem gives a number of important characterizations of e-convergence. The second one is essentially due to Brezis [3] and Attouch [2], the first one to Mosco [13].

The *subgradient* at x of a closed convex function f is the closed convex set

$$\partial f(x) = \{v \in \mathbb{R}^n : f(y) - f(x) \geq v \cdot (y - x) \text{ for all } y \in \mathbb{R}^n\}$$

By the definition of f^*, this set can also be expressed as

$$\partial f(x) = \{v : v \cdot x = f(x) + f^*(v)\}$$

or equivalently

$$\partial f(x) = \{v : f^*(v) - v \cdot x \leq -f(x)\}$$

This means that $\partial f(x)$ is the α-level set of the function $v \to f^*(v) - v \cdot x$ at $\alpha = -f(x)$. The *graph* $G(f)$ *of the subgradient* of f is given by

$$G(f) = \{(x, v) : v \in \partial f(x)\} = \{(x, v) : x \in \partial f^*(v)\}$$

Theorem 3
Suppose that $\{f;\, f_v,\ v\in \underset{\sim}{N}\}$ are closed convex functions. Then the following statements are equivalent:
(i) $f_v \underset{e}{\to} f$;
(ii) the two following conditions are satisfied
(ii$_a$) if $x = \lim\ [x_\mu,\ \mu \in \underset{\sim}{M}]$ then $f(x) \leq \liminf_\mu f_\mu(x_\mu)$,
(ii$_b$) each $x = \lim\ [x_v,\ v \in \underset{\sim}{N}]$ with $\limsup f_v(x_v) \leq f(x)$;
(iii) $G(f) = \lim_v G(f_v)$
and there exist $(x, v) \in G(f)$ and $\{(x_v, v_v) \in G(f_v), v \in \underset{\sim}{N}\}$ such that

$$(x, v) = \lim_v\ (x_v, v_v) \quad \text{and} \quad f(x) = \lim_v\ f_v(x_v)$$

Proof: It is straightforward to verify that (ii$_a$) is equivalent to $\limsup E(f_v) \subset E(f)$ and that (ii$_b$) is equivalent to $E(f) \subset \liminf E(f_v)$; detailed arguments can be found in [13, p. 536]. This proves the equivalence of (i) and (ii).

We now show that (ii) implies (iii). We first seek to establish that

$$LS = \limsup G(f_v) \subset G(f)$$

Take $\{(x_\mu, v_\mu) \in G(f_\mu),\ \mu \in \underset{\sim}{M}\}$ a sequence converging to (x, v). By definition of $G(f_\mu)$, for all $\mu \in \underset{\sim}{M}$

$$f_\mu(x_\mu) + f_\mu^*(v_\mu) \leq v_\mu \cdot x_\mu$$

Since $f_v \underset{e}{\to} f$ implies that $f_v^* \underset{e}{\to} f^*$, cf. Corollary 4, by (ii$_a$) we have that

$$f(x) + f^*(v) \leq \liminf f_\mu(x_\mu) + \liminf f_\mu^*(v_\mu)$$

$$\leq \liminf\, [f_\mu(x_\mu) + f_\mu^*(v_\mu)] \leq \lim v_\mu \cdot x_\mu = v \cdot x$$

Thus $(x, v) \in G(f)$. Hence $LS \subset G(f)$.

Next we show that (ii$_b$) implies that

$$G(f) \subset \liminf G(f_v) = LI$$

Suppose not. Then, by (ii$_b$) of Theorem 1, there is an open set, which, without loss of generality, can be chosen to be an open ball centred at $(x, v) \in G(f)$ and of radius $\varepsilon > 0$, which meets $G(f)$ but fails to meet the $G(f_v)$ infinitely often, say for $v \in \underset{\sim}{M} \subset \underset{\sim}{N}$. Let $\{x_v,\ v \in \underset{\sim}{N}\}$ and $\{v_v,\ v \in \underset{\sim}{N}\}$ be two sequences converging to x and v respectively and satisfying condition (ii$_b$); again Corollary 4 allows us to conclude that criterion (ii$_b$) must also hold for the conjugate functions. Clearly, for $\mu \in \underset{\sim}{M}$ and sufficiently large

$$f_\mu(x_\mu) + f_\mu^*(v_\mu) > v_\mu \cdot x_\mu + \frac{\varepsilon}{2}$$

Taking limsup on both sides and remembering that these subsequences satisfy (ii_b) we get

$$f(x) + f^*(v) \geqslant \limsup f_\mu(x_\mu) + \limsup f_\mu^*(v_\mu)$$

$$\geqslant \limsup [f_\mu(x_\mu) + f_\mu^*(v_\mu)] \geqslant \lim v_\mu \cdot x_\mu + \frac{\varepsilon}{2} = v \cdot x + \frac{\varepsilon}{2}$$

This contradicts the working hypothesis that $(x, v) \epsilon G(f)$. We have thus shown that

$$G(f) = \lim_v G(f_v)$$

Now take $(x, v) \epsilon G(f)$. By the above there exists $(x_v, v_v) \epsilon G(f_v)$, such that

$$(x, v) = \lim_v (x_v, v_v)$$

From (ii_a) we know that

$$f(x) \leqslant \liminf f_v(x_v)$$

On the other hand, (ii_b) yields the existence of $\{y_v, v \in \mathbb{N}\}$ such that limsup $f_v(y_v) \leqslant f(x)$. The definition of $G(f_v)$ implies that

$$f_v(y_v) \geqslant f_v(x_v) + v_v \cdot (y_v - x_v)$$

Taking limsup on both sides we get that

$$f(x) \geqslant \limsup f_v(y_v) \geqslant \limsup f_v(x_v)$$

Thus

$$f(x) = \lim_v f_v(x_v)$$

Note that a similar argument would show that

$$f^*(v) = \lim_v f_v^*(v_v)$$

There remains to show that (iii) ⇒ (ii). To do this we rely on the following relation due to Rockafellar [15, chapter 4]. Let f be a closed convex function then, for all $x \in \mathbb{R}^n$,

$$f(x) = \sup \left\{ f(x_0) + \sum_{i=1}^{p} (x_i - x_{i-1}) \cdot v_{i-1} : x_i \in D(f), i = 0, \dots p-1, \right.$$
$$\left. x_p = x, v_i \in \partial f(x_i) \text{ and } p \in \mathbf{N} \right\}$$

We start with (ii$_a$). Take $\{y_\mu, \mu \in \mathbf{M}\}$ converging to y. We know that the $G(f_v)$ converge to $G(f)$ and let (x, v) be the point in $G(f)$ having the properties prescribed in (iii). Suppose that $\{(x(i), v(i)), i=1, \dots, m\}$ is a finite collection of points in $G(f)$. By the convergence of the $G(f_\mu)$ to $G(f)$, for $i=1, \dots, m$, there exists $(x(i)_\mu, v(i)_\mu) \in G(f_\mu)$ converging to $(x(i), v(i))$. For these points we have that

$$f_\mu(y_\mu) \geqslant f(x_v) + \sum_{i=1}^{m+1} (x(i)_\mu - x(i-1)_\mu) \cdot v(i-1)_\mu$$

where $x(0)_\mu = x_\mu$; $v(0)_\mu = v$ and $x(m+1)_\mu = y_\mu$. Taking liminf on both sides we get

$$\liminf f_\mu(y_\mu) \geqslant f(x) + \sum_{i=1}^{m+1} (x(i) - x(i-1)) \cdot v(i-1)$$

with $x(0) = x$, $v(0) = v$ and $x(m+1) = y$. The above holds for every finite collection $\{(x(i), v(i))\}$. Taking the sup over all such collections, we get

$$\liminf f_\mu(y_\mu) \geqslant f(y)$$

Now consider $y \in D(f)$ and take $u \in \partial f(x)$. The convergence of the $G(f_v)$ to $G(f)$ implies the existence of $\{y_v, u_v) \in G(f_v), v \in \mathbf{N}\}$ converging to (y, u). Similarly, let $\{(y(i), u(i)), i=1, \dots, m\}$ be a finite collection of elements in $G(f)$ and for $i=1, \dots, m$, $\{(y(i)_v, u(i)_v) \in G(f_v), v \in \mathbf{N}\}$ some corresponding sequences converging to $(y(i), u(i))$. Let $\{x_v, v_v), v \in \mathbf{N}\}$ be a sequence converging to (x, v) with $\lim f_v(x_v) = f(x)$ as provided by the hypothesis of (iii). With $y(m+1)_v = x_v$ and $(y(0)_v, u(0)_v) = (y_v, u_v)$, and since $(y(i-1)_v, u(i-1)_v) \in G(f_v)$ for $i = 1, \dots, m+1$ we have that

$$f_v(y(i)_v) - f_v(y(i-1)_v) \geqslant u(i-1)_v \cdot (y(i)_v - y(i-1)_v)$$

Summing over i, we get

$$f_v(x_v) - f_v(y_v) \geqslant \sum_{i=1}^{m+1} u(i-1)_v \left(y(i)_v - y(i-1)_v \right)$$

Taking limsup on both sides, it follows that

$$f(x) \geqslant \limsup f_v(y_v) + \left(f(y) + \sum_{i=1}^{m+1} (y(i) - y(i-1)) \cdot u(i-1) \right) - f(y)$$

where $y(0) = y$, $u(0) = u$ and $y(m+1) = x$. The inequalities hold for every finite collection $\{(y(i), u(i)), i=1, \dots, m\}$. Hence, by Rockafellar's relation we have that

$$f(y) + f(x) \geqslant \limsup f_\upsilon(y_\upsilon) + f(x)$$

which completes the proof.

5. RELATIONS BETWEEN e-CONVERGENCE AND p-CONVERGENCE

The relations between these two types of convergence have been investigated by Salinetti and Wets [18, 17, 25]. Neither type of convergence implies the other; consider for example the sequence of closed convex functions: for $\upsilon = 1, 2 \ldots$

$$f_\upsilon(x_1, x_2) = \begin{cases} -\upsilon x_1 & \text{if } x_1 \geqslant 0 \text{ and } \upsilon x_1 - x_2 \leqslant 0 \\ +\infty & \text{otherwise} \end{cases}$$

It is easy to verify that

$$f_\upsilon \underset{\mathrm{p}}{\rightarrow} f = \begin{cases} 0 & \text{if } x_1 \geqslant 0,\ x_2 = 0 \\ +\infty & \text{otherwise} \end{cases}$$

and

$$f_\upsilon \underset{\mathrm{e}}{\rightarrow} f' = \begin{cases} -x_1 & \text{if } x_1 \geqslant 0,\ x_2 = 0 \\ +\infty & \text{otherwise} \end{cases}$$

Equivalence between p- and e-convergence demands an equi-continuity condition defined below, that can be viewed as 'local compactness' in the space of closed (convex) functions.

Definition A collection, $\{f;\ f_\upsilon,\ \upsilon \in \mathbf{N}\}$ is said to be *equi-lower semi-continuous*, if the following conditions are satisfied:

(3α) if for all $x \in D(f)$ and all $\varepsilon > 0$, there exists V a neighbourhood of x such that, for all υ in $\mathbf{N}$ (except possibly for a finite number of υ),

$$f_\upsilon(x) \leqslant f_\upsilon(y) + \varepsilon \qquad \text{for all } y \in V;$$

(3β) for all $x \in D(f)$, there exists υ_x such that $x \in D(f_\upsilon)$ for all $\upsilon \geqslant \upsilon_x$;

(3γ) $\{f_\upsilon, \upsilon \in \mathbf{N}\}$ goes uniformly to $+\infty$ on each compact subset of $\mathbb{R}^n \backslash \operatorname{cl} D(f)$, the complement of the closure of $D(f)$.

It is easy to see that the collection $\{f_\upsilon, \upsilon \in \mathbf{N}\}$ in the above example fails to satisfy condition (3α) of equi-lower semicontinuity. Condition (3β) is automatically satisfied by any sequence of functions converging point-wise, but as the following example shows, does not necessarily hold if the convergence is in terms of the epigraphs. Consider the sequence $\{f_\upsilon, \upsilon \in \mathbf{N}\}$ with

$$f_\upsilon(x) = \begin{cases} 0 & \text{if } x \in [\upsilon^{-1}, 1] \\ +\infty & \text{otherwise} \end{cases}$$

then, $f_\upsilon \underset{e}{\to} f$ with

$$f(x) = \begin{cases} 0 & \text{if } x \in [0, 1] \\ +\infty & \text{otherwise} \end{cases}$$

Condition (3β) fails at $x = 0$ and in fact $f_\upsilon \underset{p}{\not\to} f$ since $\lim f_\upsilon(0) = +\infty > 0 = f(0)$.

Similarly, (3γ) is satisfied whenever $f_\upsilon \underset{e}{\to} f$ but does not necessarily hold if the f_υ converge point-wise to f. This is illustrated by the following example due to Patrone. Let

$$f_\upsilon(x_1, x_2) = \begin{cases} 0 & \text{if } \upsilon\, x_2 = x_1 \\ +\infty & \text{otherwise} \end{cases}$$

then, $f_\upsilon \underset{p}{\to} f$ with

$$f(x_1, x_2) = \begin{cases} 0 & \text{if } x_1 = x_2 = 0 \\ +\infty & \text{otherwise} \end{cases}$$

Condition (3γ) fails on the compact arc A,

$$A = \{\ (x_1, x_2)\ :\ x_1^2 + x_2^2 = 1, x_1 \geqslant 0, x_2 \geqslant 0\}$$

Here the e-convergence of the f_υ to f fails, but $f_\upsilon \underset{e}{\to} f'$ with

$$f'(x_1, x_2) = \begin{cases} 0 & \text{if } x_1 = 0 \\ +\infty & \text{otherwise} \end{cases}$$

Note again that $f_\upsilon \underset{p}{\not\to} f'$ precisely because condition (3β) of equi-lower semicontinuity is not satisfied. That the equi-lower semicontinuity of the collection, is in some sense a necessary and sufficient condition for the equivalence between e-convergence and p-convergence is brought home by the theorem below, which was first proved in a reflexive Banach space setting. Lemma 1 fills in a technical point required in the proof of the theorem and its corollary.

Lemma 1 [18, lemma 3] Suppose that $\{f; f_\upsilon\ \upsilon \in \mathbf{N}\}$ are closed convex functions and $f_\upsilon \xrightarrow[\mathrm{p}]{} f$. If property (3$\gamma$) of equi-lower semicontinuity is satisfied, then

$$D(f) \subset \text{prj } LS \subset \text{cl } D(f)$$

when prj $LS = \{x : (x, a) \in \limsup E(f_\upsilon)\}$. If in fact the collection $\{f; f_\upsilon, \upsilon \in \mathbf{N}\}$ is equi-lower semicontinuous, then

$$D(f) = \text{prj } LS$$

Theorem 4

Suppose that $\{f; f_\upsilon, \upsilon \in \mathbf{N}\}$ are closed convex functions. Then any two of the following statements imply the other:

(i) the collection $\{f; f_\upsilon, \upsilon \in \mathbf{N}\}$ is equi-lower semicontinuous;

(ii) $f_\upsilon \xrightarrow[\mathrm{p}]{} f$;

(iii) $f_\upsilon \xrightarrow[\mathrm{e}]{} f$.

In fact, only when equi-lower semicontinuity is satisfied, do p-convergence and e-convergence imply each other.

Proof: (ii)+(iii)$\Rightarrow$(i). Since p-convergence implies (3β) and e-convergence implies (3γ) it suffices to establish (3α). To do this suppose, to the contrary, that there exists x in $D(f)$ for which (3α) fails; that is there exists $\varepsilon > 0$ such that in every neighbourhood V_k, there exists y_k such that $f_k(x) > f_k(y_k) + \varepsilon$. Take $\{V_k, k \in \mathbf{N}\}$ a nested sequence of neighbourhoods of x, such that

$$\bigcap_k V_k = \{x\}$$

From p-convergence it follows that $f(x) \geqslant \liminf f_k(y_k) + \varepsilon$. On the other hand, since the y_k converge to x and the f_υ converge to f, it follows from Theorem 3 (ii) that

$$f(x) \leqslant \liminf f_k(y_k)$$

contradicting the preceding inequality.

(i) + (ii)$\Rightarrow$(iii). It suffices to show that

$$LS = \limsup E(f_\upsilon) \subset E(f)$$

since the inclusions $E(f) \subset \liminf E(f_\upsilon) \subset \limsup E(f_\upsilon)$ follow respectively from the definition of p-convergence and the definitions of liminf and limsup. Take $(x, a) \in LS$ with $(x_\mu, a_\mu) \in E(f_\mu)$ and $(x, a) = \lim_\mu (x_\mu, a_\mu)$.
Equi-lower semicontinuity implies that $x \in D(f)$ – see Lemma 1 – and thus for all $\varepsilon > 0$, there exists V such that, for all y in V,

$$f_\upsilon(x) \leqslant f_\upsilon(y) + \varepsilon \qquad \text{for all } \upsilon \in \mathbf{N}$$

In particular this inequality must hold for all μ sufficiently large. Taking liminf on both sides and remembering that $f_\upsilon \underset{p}{\to} f$ we get

$$f(x) \leqslant \varepsilon + \liminf f_\mu(x_\mu)$$

The above holds for all $\varepsilon > 0$, hence

$$f(x) \leqslant \liminf f_\mu(x_\mu)$$

On the other hand, since $(x_\mu, a_\mu) \in E(f_\mu)$ and $\lim a_\mu = a$, we have that

$$\liminf f_\mu(x_\mu) \leqslant \liminf a_\mu = a$$

The last two relations imply that $(x, a) \in E(f)$ and yield the desired inclusion.
(i) + (iii) $\Rightarrow$ (ii). It suffices to show that

$$\limsup f_\upsilon(x) \leqslant f(x)$$

since the inequalities $f(x) \leqslant \liminf f_\upsilon(x) \leqslant \limsup f_\upsilon(x)$ follow respectively from the definition of e-convergence and the usual relation between the liminf and the limsup of a sequence of (extended) real numbers. We consider only $x \in D(f)$ since otherwise the above inequality is trivially satisfied. e-Convergence implies that – see (ii$_b$) of Theorem 3 – there exists $\{x_\upsilon, \upsilon \in \mathbf{N}\}$ such that $x = \lim x_\upsilon$ and

$$\limsup_{\upsilon} f_\upsilon(x_\upsilon) \leqslant f(x)$$

On the other hand, by equi-lower semicontinuity part (3α), we know that for all $\varepsilon > 0$, there exists υ_ε such that for all $\upsilon \geqslant \upsilon_\varepsilon$

$$f_\upsilon(x) - \varepsilon \leqslant f_\upsilon(x_\upsilon)$$

Thus

$$\limsup f_\upsilon(x) - \varepsilon \leqslant \limsup f_\upsilon(x_\upsilon) \leqslant f(x)$$

This yields the required inequality since it holds for all $\varepsilon > 0$.

The last assertion of the theorem is nothing more than a restatement of the implication (ii) + (iii) $\Rightarrow$ (i). This completes the proof.

In general, it is difficult to verify if a collection of closed convex functions is – or is not – equi-lower semicontinuous. However, there are some criteria that can be usefully exploited in most practical situations. The following corollaries illustrate this point; further results can be found in [18].

Corollary 4 Suppose that $\{f; f_\upsilon, \upsilon \in \mathbf{N}\}$ are closed convex functions, $\operatorname{int} D(f) \neq \emptyset$ and $f_\upsilon \underset{p}{\rightarrow} f$. Then, $f_\upsilon \underset{e}{\rightarrow} f$.

Proof: It suffices to show that

$$LS = \limsup E(f_\upsilon) \subset E(f)$$

since the inclusions $E(f) \subset \liminf E(f_\upsilon) \subset \limsup E(f_\upsilon)$ follow from the p-convergence of the f_υ to f and the definitions of liminf and limsup, respectively.

First consider $x \in \operatorname{int} D(f)$. Then, $x \in \operatorname{int} D(f_\upsilon)$ for υ sufficiently large. It follows that these f_υ are continuous on a neighbourhood V. Hence, we have the uniform convergence of the f_υ to f on V. Suppose that

$$(x, a) = \lim_{\mu \in \mathbf{M}} (x_\mu, a_\mu) \in LS$$

we need to show that $(x, a) \in E(f)$; without loss of generality we may assume that all x_μ lie in V. The continuity of f at x and the uniform convergence of the f_υ to f on V imply that for any $\varepsilon > 0$ and μ sufficiently large.

$$|f(x) - f_\mu(x_\mu)| \leqslant |f(x) - f(x_\mu)| + |f(x_\mu) - f_\mu(x_\mu)| < \varepsilon$$

and thus

$$f(x) \leqslant \limsup f_\mu(x_\mu) \leqslant \lim a_\mu = a$$

i.e. $(x, a) \in E(f)$.

It remains to consider the case $x \notin \operatorname{int} D(f)$. We start by showing that under the hypotheses of this corollary, condition (3γ) of equi-lower semicontinuity is actually satisfied. Suppose not, then there exists

$$\hat{x} = \lim_{\upsilon} x_\upsilon$$

with $\{\hat{x}; x_\upsilon, \upsilon \in \mathbf{N}\} \subset \mathbb{R}^n \backslash \operatorname{cl} D(f)$, such that

$$\lim f_\upsilon(x_\upsilon) = \alpha < +\infty$$

Let V be an open set in $D(f)$. Then, $V \subset D(f_\upsilon)$ for υ sufficiently large, and moreover the f_υ converge uniformly to f on V. Since $f_\upsilon(x_\upsilon) < +\infty$, the points $x_\upsilon \in D(f_\upsilon)$ and thus the sets $W_\upsilon = \text{con}\ \{x_\upsilon, V\}$ are contained in $D(f_\upsilon)$. From this it follows that $W = \text{con}\ \{\hat{x}, V\} \setminus \{\hat{x}\}$ is contained in $D(f)$. To see this take any point w; then, for some $\lambda \in]0, 1]$ and $\nu \in V$

$$w = (1 - \lambda)\hat{x} + \lambda \nu$$

But since V is open and the x_υ converge to $\hat{x}$, for υ sufficiently large $\nu_\upsilon = \lambda^{-1}(1 - \lambda)(\hat{x} - x_\upsilon) + \nu$ is in V, $\nu = \lim \nu_\upsilon$ and

$$w = (1 - \lambda)x_\upsilon + \lambda \nu_\upsilon$$

which shows that w is in $D(f_\upsilon)$; in particular we have that

$$f_\upsilon(w) \leqslant (1 - \lambda)f_\upsilon(x_\upsilon) + \lambda f_\upsilon(\nu_\upsilon)$$

Taking limits on both sides and remembering that the f_υ p-converge to f, uniformly on V, we obtain the inequality:

$$f(w) \leqslant (1 - \lambda)\alpha + \lambda f(\nu)$$

which implies that $f(w)$ is finite. Thus, $W \subset D(f)$ and consequently $\hat{x} \in \text{cl}\ D(f)$ contradicting the working hypothesis that $\hat{x}$, as well as the x_υ, are in the complement as cl $D(f)$. The fact that condition (3γ) is satisfied provides us with the following inclusion:

$$D(f) \subset \{x : (x, a) \in LS\} \subset \text{cl}\, D(f)$$

by means of Lemma 1. The first part of the proof has shown that LS and $E(f)$ 'agree' on int $D(f)$; the above inclusion shows that, if LS and $E(f)$ differ in any way, there must exist (x, a) in $LS \setminus E(f)$ with $x \in \text{bdy}\ D(f)$.

But then $C = \text{con}\ \{(x, a), E(f)\} \subset LS$ and there exists a neighbourhood U of (x, a) disjoint of the closed set $E(f)$ which necessarily intersects int C. Thus, there exists $(x', a') \in \text{int}\ LS$ with $x' \in \text{int}\ D(f)$, $a' < f(x')$, which contradicts the 'agreement' of LS and $E(f)$ on int $D(f)$. This completes the proof.

Corollary 5 Suppose that $\{f; f_\upsilon, \upsilon \in \mathbf{N}\}$ are closed convex functions such that $f_\upsilon \underset{e}{\rightarrow} f$. Then, $f_\upsilon \underset{p}{\rightarrow} f$ on int $D(f)$.

6. A FURTHER CHARACTERIZATION OF e-CONVERGENCE

A collection $\{(g\lambda;\cdot),\lambda\in\mathbb{R}_+\}$of closed convex functions is a *cast* if the following conditions are satisfied:

for all $\lambda>0$, the function $x\to g(\lambda;x)$ is symmetric, inf-compact and (4*a*)

$$g(\lambda;0)=\min_x g(\lambda;x)=0$$

for all $\alpha\in[0,1]$, $\lambda\to\lambda^{-1}[g(\lambda;x)-g(\lambda;\alpha x)]$ is nondecreasing; (4*b*)

$$\lim_{\lambda\uparrow+\infty} g(\lambda;\cdot)=\Psi_{\{0\}} \qquad \text{i.e.} \qquad g(\lambda,\cdot)\underset{\mathrm{p}}{\to}\Psi_{\{0\}} \tag{4c}$$

where inf-compact means that for all $\alpha\in\mathbb{R}$, the level sets $\{x:g(\lambda;x)\leq\alpha\}$ are compact and Ψ_D is the indicator function of the set D, i.e. $\Psi_D(x)=0$, if $x\in D$ and $+\infty$ otherwise. In (4*b*) the usual convention in convex analysis, $(+\infty)+(-\infty)=+\infty$ applies. Elements of a cast are called *casting functions*. Note that (4*a*) implies that for all $\lambda>0$, $g(\lambda;\cdot)\geq 0$ and that, given any $\beta>0$, there exists $r(\beta)>0$ such that $\{x:g(\lambda;x)\leq\beta\}=B_{r(\beta)}(0)$. In fact, if we write $|x|$ for $d(0,x)$, then to each casting function there corresponds a convex function $\gamma:\mathbb{R}_+\times\mathbb{R}_\oplus$ such that

$$\gamma(\lambda;|x|)=g(\lambda;x) \quad \text{with} \quad \gamma(\lambda;0)=\min_{s\in R_\oplus}\gamma(\lambda;s)=0$$

for all $\lambda>0$, $\lim_{s\uparrow+\infty}\gamma(\lambda;s)=+\infty$

(4*b*′) for all $\alpha\in[0,1]$ $\lambda\to\lambda^{-1}[\gamma(\lambda;s)-\gamma(\lambda;\alpha s)]$ is nondecreasing;
and
(4*c*′) for all $s>0$ $\lim_{\lambda\uparrow\infty}\gamma(\lambda;s)=+\infty$

Typical examples of casts are $\{\Psi_{\lambda^{-1}B},\lambda>0\}$, $\{\lambda|\cdot|,\lambda>0\}$and $\{\lambda|\cdot|^2/2,\lambda>0\}$.
There is a more general class of functions that can be used to construct casts; in particular, symmetry is not essential. However, without symmetry the conditions become cumbersome to state.

Casts are used to regularize convex functions. We associate with a closed convex function f a family of closed convex functions $\{f^\lambda;\lambda\in\mathbb{R}_+\}$defined as follows. Let $\{g(\lambda;\cdot)$ be a cast, then

$$E(f^\lambda)=E(f)+E\Big(g(\lambda;\cdot)\Big)$$

or equivalently f^λ is the inf-convolution of f and $g(\lambda;\cdot)$, i.e.

$$f^\lambda(x)=[f\,\Box\, g(\lambda;\cdot)](x)=\min_y\,[f(y)+g(\lambda;x-y)]$$

Since casting functions are inf-compact, f closed convex implies that f^λ is closed convex; moreover, for all x, the infimum of $f + g(\lambda; x - \cdot\,)$ is actually attained provided we accept $+\infty$ as possible value for the minimum. Note that some casts always yields finite-valued functions f^λ, e.g. $\{\lambda\,|\cdot|^2/2, \lambda > 0\}$; but this is not always the case, e.g. $\{\Psi_{\lambda^{-1}B}, \lambda > 0\}$.

Lemma 2 Suppose that $\{g(\lambda; \cdot), \lambda > 0\}$ is a cast. Then the collection $\{\Psi_{\{0\}}, g(\lambda, \cdot), \lambda > 0\}$ is equi-lower semicontinuous; hence, $g(\lambda; \cdot) \vec{e}\ \Psi_{\{0\}}$

Proof: Conditions (3α) and (3β) are trivially satisfied. The condition (3γ) follows directly from ($4b$). The second part of the assertion is then a consequence of the first part and of Theorem 4.

Lemma 3 Suppose that $\{g(\lambda; \cdot), \lambda > 0\}$ is a cast, f is a closed convex function and let $f^\lambda = f \square g(\lambda, \cdot\,)$. Then, for all x, $\lambda \to f^\lambda(x)$ is non-decreasing on $\mathbb{R}_+$.

Proof: Take $0 < \lambda_1 < \lambda_2$. For all $y \in \mathbb{R}^n$

$$f(y) + g(\lambda_1, x - y) \leq f(y) + g(\lambda_2, x - y)$$

This holds in particular for $y = y_{\lambda_2}$ which minimizes the right-hand side of that inequality. We have thus

$$[f \square g(\lambda_1, \cdot\,)](x) \leq f(y_{\lambda_2}) + g(\lambda_1, x - y_{\lambda_2}) \leq [f \square g(\lambda_2, \cdot\,)](x)$$

This completes the proof.

Theorem 5

Suppose that $\{g(\lambda, \cdot\,), \lambda > 0\}$ is a cast and f is a closed convex function. Then,

$$f = \lim_{\lambda \uparrow \infty} f^\lambda$$

where $f^\lambda = f \square g(\lambda; \cdot\,)$

Proof: By y_λ we denote an element that minimizes

$$f(y) + g(\lambda; x - y)$$

and of minimum norm. We start by showing that $\lambda \to |x - y_\lambda|$ is non-increasing when $\lambda \uparrow +\infty$. Take $0 < \lambda_1 < \lambda_2$ with y_{λ_1} and y_{λ_2} minimizing elements as defined above. Then

$$f^{\lambda_1}(x) = f(y_{\lambda_1}) + g(\lambda_1; x - y_{\lambda_1}) \leq f(y_{\lambda_2}) + g(\lambda_1; x - y_{\lambda_2})$$

and

$$f^{\lambda_2}(x) = f(y_{\lambda_2}) + g(\lambda_2; x - y_{\lambda_2}) \leq f(y_{\lambda_1}) + g(\lambda_2; x - y_{\lambda_1})$$

These inequalities yield

$$a=g(\lambda_2\,;x-y_{\lambda_2})-g(\lambda_2\,;x-y_{\lambda_1}) \leqslant f(y_{\lambda_1})\;-f(y_{\lambda_2}) \leqslant g(\lambda_1\,;x-y_{\lambda_2})-g(\lambda_1;x-y_{\lambda_1})$$

If $\lambda \to |x - y_\lambda|$ is non-increasing, then $a \leqslant 0$; suppose not. Then $f(y_{\lambda_1})-f(y_{\lambda_2})>0$ and since $\lambda_2 > \lambda_1$ we have that

$$(\lambda_2)^{-1}[\gamma(\lambda_2;\; |x-y_{\lambda_2}|)-\gamma(\lambda_2\,;\alpha\,|x-y_{\lambda_2}|)]$$

$$< (\lambda_1)^{-1}[\gamma(\lambda_1;\,|x-y_{\lambda_2}|)-\gamma(\lambda_1\,;\alpha\,|x-y_{\lambda_2}|)]$$

where $\alpha = |x - y_{\lambda_1}|\,(|x - y_{\lambda_2}|)^{-1}$; $\alpha \in [0, 1[$ since the working hypothesis is that $|x - y_{\lambda_2}| > |x - y_{\lambda_1}|$. But the above inequality is in contradiction with (4*b*) or equivalently (4*b*$'$). Hence $\lambda \to |x - y_\lambda|$ is non-increasing.

Suppose first that $f(x) = +\infty$. If $x = \lim y_\lambda$, then

$$\lim f^\lambda(x) \geqslant \liminf f(y_\lambda) + \liminf g(\lambda; x - y_\lambda) \geqslant +\infty$$

because f is closed and thus $\liminf f(y_\lambda) = f(x) = +\infty$ and the second term is always non-negative. If $\lim y_\lambda \neq x$, there exists $\delta > 0$ such that $|x - y_\lambda| > \delta > 0$ for all λ. Let $\hat{y}$ be a cluster point of the sequence $\{y_\lambda, \lambda \uparrow +\infty\}$. Note that $|x - \hat{y}| \geqslant \delta > 0$. Restricting the sequence to those values of λ, yielding $\hat{y} = \lim y_\lambda$, we have that

$$f^\lambda(x) = f(y_\lambda) + g(\lambda; x - y_\lambda) \geqslant f(y_\lambda) + g(\lambda; x - \hat{y})$$

Taking liminf on both sides we get:

$$\liminf f^\lambda(x) \geqslant \liminf f(y_\lambda) + \liminf \gamma(\lambda; |x - \hat{y}|) = +\infty$$

since the first term on the right converges to $f(\hat{y})$ and the second to $+\infty$ by (4*c*$'$).

We now assume that $f(x)$ is finite. We show first that $x = \lim y_\lambda$. Suppose not. Since $\lambda \to |x - y_\lambda|$ is non-increasing the sequence y_λ admits a cluster point, say $\hat{y}$. We always have that

$$f^\lambda(x) = f(y_\lambda) + g(\lambda; x - y_\lambda) \leqslant f(x) \tag{5}$$

Thus, for the subsequence $\{y_\lambda, \lambda \in \Lambda\}$ converging to $\hat{y}$ and since $g(\lambda; x - y_\lambda) \geqslant g(\lambda; x - \hat{y})$, we have that

$$f(x) \geqslant \liminf f^\lambda(x) \geqslant \liminf f(y_\lambda) + \liminf g(\lambda; x - \hat{y}) \geqslant +\infty$$

The last inequality follows from (4*c*). This contradicts $f(x)$ finite. Hence, $x = \lim y_\lambda$. Now, from (5) it follows that

$$f(x) \geq \liminf f^\lambda(x) \geq \liminf f(y_\lambda) + \liminf g(\lambda; x - y_\lambda)$$

$$\geq f(x) + \liminf g(\lambda; x - y_\lambda)$$

This implies that $\liminf g(\lambda; x - y_\lambda) = 0$. On the other hand,

$$g(\lambda; x - y_\lambda) \leq f(x) - f(y_\lambda)$$

Taking limsup on both sides shows that $\lim g(\lambda; x - y_\lambda) = 0$, and thus

$$f(x) = \lim f(y_\lambda) + \lim g(\lambda; x - y_\lambda) = \lim f^\lambda(x)$$

This completes the proof.

The next theorem yields a new characterization of e-convergence.

Theorem 6

Suppose that $\{g(\lambda; \cdot), \lambda \in \mathbb{R}_+\}$ is a cast and $\{f; f_\upsilon, \upsilon \in \underset{\sim}{\mathbb{N}}\}$ are closed convex functions. Then, $f_\upsilon \underset{e}{\rightarrow} f$ if and only if

(*a*) $\lim_{\lambda \uparrow +\infty} \liminf f_\upsilon^\lambda \geq f$

and

(*b*) $\lim_{\lambda \uparrow \infty} \limsup f_\upsilon^\lambda \leq f$

Proof: We are going to verify that (*a*) implies and is implied by (ii_a) of Theorem 3, and similarly for (*b*) and (ii_b) of Theorem 3. We only give the proof when $f(x)$ is finite, the case $f(x) = +\infty$ is straightforward.

(ii_a) of Theorem 3 $\Rightarrow$ (*a*). We always have that

$$f_\upsilon(y_{\lambda_\upsilon}) \geq f_\upsilon(y_{\lambda_\upsilon}) + g(\lambda; x - y_{\lambda_\upsilon}) = f_\upsilon^\lambda(x)$$

where y_{λ_υ} minimizes $f_\upsilon + g(\lambda; x - \cdot)$ on $\mathbb{R}^n$. Since $y_{\lambda_\upsilon} \rightarrow x$ as λ goes to $+\infty$, by Theorem 5, for all υ, fix λ_υ such that $|x_\upsilon - x| < \upsilon^{-1}$ with $x_\upsilon = y_{\lambda_\upsilon}$. We have that

$$f_\upsilon(x_\upsilon) \leq f_\upsilon^\lambda(x)$$

for all $\lambda \geq \lambda_\upsilon$ since $\lambda \rightarrow f_\upsilon^\lambda(x)$ is non-decreasing by Lemma 3. We have from (ii_a) of Theorem 3 and the preceding construction

$$f(x) \leq \liminf f_\upsilon(x_\upsilon) \leq \liminf_\upsilon f_\upsilon^{\lambda_\upsilon}(x) \leq \lim_{\lambda \uparrow \infty} \liminf f_\upsilon^\lambda(x)$$

$(a) \Rightarrow (\mathrm{ii}_a)$ of Theorem 3. Take $\{x_\mu, \mu \in \underset{\sim}{M}\}$ a sequence converging to x. By definition of f_υ^λ, for $\lambda > 0$ and all μ

$$f_\mu^\lambda(x) \leqslant f_\mu(x_\mu) + g(\lambda; x - x_\mu)$$

By Theorem 5 if we liminf on μ and the take the limit when λ goes to $+\infty$, we get

$$f(x) \leqslant \lim_{\lambda \uparrow +\infty} \liminf f_\mu^\lambda(x) \leqslant \liminf f_\mu(x_\mu)$$

The first inequality comes from (a).
(ii_b) of Theorem 3 $\Rightarrow$ (b). Take x and $\{x_\upsilon, \upsilon \in \underset{\sim}{N}\}$ the sequence provided by the hypotheses. Then,

$$f_\upsilon^\lambda(x) \leqslant f_\upsilon(x_\upsilon) + g(\lambda; x - x_\upsilon)$$

Taking limsup on υ, using the properties of g as λ goes to $+\infty$ we get from (ii_b) of Theorem 3 that

$$\lim_{\lambda \uparrow \infty} \limsup_{\upsilon} f_\upsilon^\lambda(x) \leqslant \limsup f_\upsilon(x_\upsilon) \leqslant f(x)$$

$(b) \Rightarrow (\mathrm{ii}_b)$ of Theorem 3. We have that

$$f_\upsilon^\lambda(x) = f_\upsilon(y_{\lambda_\upsilon}) + g(\lambda; x - y_{\lambda_\upsilon}) \geqslant f_\upsilon(y_{\lambda_\upsilon})$$

Pick λ_υ such that $|x_\upsilon - x| < \upsilon^{-1}$ where $x_\upsilon = y_{\lambda_\upsilon}$. From Lemma 3 we know that $f_\upsilon^\lambda(x) \geqslant f_\upsilon(x_\upsilon)$ for all $\lambda \geqslant \lambda_\upsilon$. Relying on (b) and Theorem 5 it follows that

$$f(x) \geqslant \lim_{\lambda \uparrow \infty} \limsup f_\upsilon^\lambda(x) \geqslant \limsup_{\upsilon} f_\upsilon(x_\upsilon)$$

Since the x_υ converge to x, this completes the proof.

7. CONVERGENCE OF THE INFIMA

We now apply the results of the preceding sections to convex optimization problems. The purpose here is to illustrate the type of theorems one can derive rather than give a survey of all known results.

Theorem 7
Suppose that $\{f; f_\upsilon, \upsilon \in \underset{\sim}{N}\}$ is a collection of equi-lower semicontinuous inf-compact convex functions. Then, if $f_\upsilon \underset{P}{\rightarrow} f$ it follows that

$$\lim_{\upsilon} (\min f_\upsilon) = \min f$$

Proof: Inf-compactness implies not only that the infima are attained but also that f^* and f_υ^* are continuous at 0, see [9] for example. The rest now follows that from Theorem 4, Corollary 2 and Corollary 5, if we simply observe that continuity at 0 implies that

$$0 \in \text{int dom } f^*_{(\upsilon)} \qquad \text{and that} \qquad f^*_{(\upsilon)}(0) = -\min f_{(\upsilon)}$$

This completes the proof.

Corollary 6 Suppose that $\{u_\upsilon, \upsilon \in \underset{\sim}{\mathbf{N}}\}$ is a collection of points in $\mathbb{R}^n$ converging to u. Suppose that $\{K_\upsilon, \upsilon \in \underset{\sim}{\mathbf{N}}\}$ is a sequence of compact convex sets converging to $K \subset \mathbb{R}^n$. Then,

$$\lim (\min [u_\upsilon \cdot x : x \in K_\upsilon]) = \min [u \cdot x : x \in K]$$

Proof: Consider

$$f_\upsilon(x) = \begin{cases} u_\upsilon \cdot x & \text{if } x \in K_\upsilon \\ +\infty & \text{otherwise} \end{cases}$$

and f defined similarly. One verifies easily that the collection $\{f; f_\upsilon, \upsilon \in \underset{\sim}{\mathbf{N}}\}$ satisfies the conditions of the theorem.

Theorem 8
[18, theorem 6]. (Approximation theorem). Suppose that $\{f; f_\upsilon, \upsilon \in \underset{\sim}{\mathbf{N}}\}$ is a collection of equi-lower semicontinuous closed convex functions such that $f_\upsilon \underset{p}{\rightarrow} f$ and $\inf f = -f^*(0)$ is finite. Then, there exists $\{\nu_\upsilon, \upsilon \in \mathbf{N}\}$, such that

$$\lim [\inf f_\upsilon - (\nu_\upsilon \cdot)] = \inf f$$

Proof: Equi-lower semicontinuity and point-wise convergence imply that $f_\upsilon^* \underset{e}{\rightarrow} f^*$ (cf. Theorem 4 and Corollary 2). By (ii) of Theorem 3 there exists $\{\nu_\upsilon, \upsilon \in \underset{\sim}{\mathbf{N}}\}$ with $\lim \nu_\upsilon = 0$ such that

$$\lim f_\upsilon^*(\nu_\upsilon) = f^*(0) = -\inf f$$

which proves the theorem since

$$-f_\upsilon^*(\nu_\upsilon) = \inf [f_\upsilon(x) - \nu_\upsilon \cdot x]$$

by conjugacy. This completes the proof.

8. CONVERGENCE OF OPTIMAL SOLUTIONS

Even if the collection is equi-lower semicontinuous and the infima of the f_υ converge to the infimum of f there is no guarantee that the set of optimal solutions, argmin f_υ, converge to the set of optimal solutions argmin f.
Consider for example

$$f_\upsilon(x) = \max\,[\upsilon^{-1} x^2, f(x)]$$

and

$$f(x) = (\psi_{[-1,1]} \,\square\, |\cdot|)(x)$$

This collection $\{f; f_\upsilon, \upsilon \in \underset{\sim}{\mathbf{N}}\}$ satisfies the hypotheses of Theorem 7. Thus, the $\min f_\upsilon = 0$ converge to $\min f = 0$ but argmin $f_\upsilon = \{0\}$ and argmin $f = [-1,1]$. This is typically the case when convergence of the optimal solution sets fails.

Theorem 9
Suppose that $\{f; f_\upsilon, \upsilon \in \underset{\sim}{\mathbf{N}}\}$ is a collection of closed convex functions such that $f_\upsilon \underset{e}{\to} f$. Then,

$$\limsup\,[\text{argmin } f_\upsilon] \subset \text{argmin } f$$

Proof: The theorem is trivial if argmin f_υ is empty for all υ (larger than some υ'). Assume that argmin $f_\upsilon \neq \emptyset$ for all $\upsilon \in \underset{\sim}{\mathbf{M}}$ and that there is a sequence $\{x_\mu \in \text{argmin } f_\mu, \mu \in \underset{\sim}{\mathbf{M}}\}$ converging to x. It suffices to show that $x \in$ argmin f. If $x_\mu \in$ argmin f_μ, then $(x_\mu, 0) \in G(f_\mu)$. (iii) of Theorem 3 implies that the $\{G(f_\upsilon), \upsilon \in \underset{\sim}{\mathbf{N}}\}$ converge to $G(f)$. Since

$$(x, 0) = \lim_\upsilon\,(x_\mu, 0)$$

it follows that $(x, 0) \in G(f)$, i.e. $0 \in \partial f(x)$ and hence $x \in$ argmin f. This completes the proof.

Corollary 7 Suppose that $\{f; f_\upsilon, \upsilon \in \underset{\sim}{\mathbf{N}}\}$ is a collection of closed convex functions such that $f_\upsilon \underset{e}{\to} f$. Suppose moreover that the infima of the f_υ and of f are attained and are unique, say x_υ and x respectively. Then

$$x = \lim_\upsilon x_\upsilon$$

We conclude with a result that gives sufficient conditions for the convergence of the Lagrange multipliers associated with the optimal solution of a convex program. We derive this theorem in the framework of the general duality theory for convex program [16].

A convex optimization problem

$$(\underset{\sim}{\mathrm{P}}) \qquad \text{minimize } f(x) \qquad x \in \mathbb{R}^n$$

is embedded in a class of convex optimization problems obtained by perturbing the original problem. We define $F: \mathbb{R}^n \times \mathbb{R}^m \to R^{\bullet}$ with $F(x, 0) = f(x)$ where $\mathbb{R}^m$ is the perturbation space. Here, we only consider the case when the function $(x, u) \to F(x, u)$ is convex.

The perturbation function $\varphi : \mathbb{R}^m \to R^{\bullet}$ is defined by

$$\varphi(u) = \inf_x F(x, u)$$

Let $g : \mathbb{R}^m \to R^{\bullet}$ be the conjugate of φ, i.e.

$$g(\nu) = \sup \{\nu \cdot u - \varphi(u)\}$$

Then, the convex optimization problem

$$(\underset{\sim}{D}) \qquad \text{minimize } g(\nu) \qquad \nu \in \mathbb{R}^m$$

is dual to $\underset{\sim}{\mathrm{P}}$ in the sense that, unless $\underset{\sim}{\mathrm{P}}$ or $\underset{\sim}{\mathrm{D}}$ are unstable, at the optimum $\min f = -\min g$ and the operations that led to the construction of $\underset{\sim}{\mathrm{D}}$ can be 'inverted' to derive $\underset{\sim}{\mathrm{P}}$ from $\underset{\sim}{\mathrm{D}}$. Typically the perturbations are associated with the explicit constraints that appear in the definition of f. The optimal solutions to $\underset{\sim}{\mathrm{D}}$ are then the Lagrange multipliers needed to 'convert' $\underset{\sim}{\mathrm{P}}$ to an unconstrained problem.

The next theorem deals with convergence of the optimal solution to $\underset{\sim}{\mathrm{D}}$ when f is approximated by convex functions f_υ.

Theorem 10

Consider $\{f; f_\upsilon, \upsilon \in \underset{\sim}{\mathrm{N}}\}$ a collection of closed convex functions. Let $\{F; F_\upsilon, \upsilon \in \underset{\sim}{\mathrm{N}}\}$ represent an embedding of the functions $\{f; f_\upsilon, \upsilon \in \underset{\sim}{\mathrm{N}}\}$ into classes of convex problems depending on parameters u in $\mathbb{R}^m$. Suppose that the functions F_υ and F are convex on $\mathbb{R}^n \times \mathbb{R}^m$, that for all u,

$$\{F(u, \cdot\,); F_\upsilon(u, \cdot\,), \upsilon \in \underset{\sim}{\mathrm{N}}\}$$

is a collection of equi-lower semicontinuous inf-compact convex functions, and that for all x, the functions $\{F(\cdot\,, x); F_\upsilon(\cdot\,, x), \upsilon \in \underset{\sim}{\mathrm{N}}\}$ are equi-Lipschitz

Let

$$g(\nu) = \sup_u \inf_x [\nu \cdot u - F(x,u)]$$

and

$$g_\upsilon(\nu) = \sup_u \inf_x [\nu \cdot u - F_\upsilon(x,u)]$$

Suppose that for all υ, g_υ and also g attain their minimum on $\mathbb{R}^m$. Let $\nu_\upsilon \in \operatorname{argmin} g_\upsilon$ and suppose that $\nu = \lim \nu_\upsilon$. Then, $\nu \in \operatorname{argmin} g$.

Proof: Let

$$\varphi_\upsilon(u) = \inf_x F_\upsilon(x,u) \quad \text{and} \quad \varphi(u) = \inf_x F(x,u)$$

Then the hypotheses of the theorem imply, via theorem 7, that $\varphi_\upsilon \underset{\mathrm{p}}{\to} \varphi$. The functions φ and φ_υ, $\upsilon \in \underset{\sim}{\mathrm{N}}$ are convex and equi-Lipschitz as can be readily verified, cf. [24] for example. This implies that the collection $\{\varphi; \varphi_\upsilon, \upsilon \in \underset{\sim}{\mathrm{N}}\}$ is equi-lower semicontinuous, hence $\varphi_\upsilon \underset{\mathrm{e}}{\to} \varphi$ and since the functions g_υ and g are the conjugate of φ_υ and φ it follows that $g_\upsilon \underset{\mathrm{e}}{\to} g$, Corollary 2. We now apply Theorem 9 to obtain the assertion. This completes the proof.

REFERENCES

[1] H. Attouch. 'Convergence de fonctions convexes, des sous-différentiels et semi-groupes associés'. *C. R. Acad. Sci., Paris*, (1977).

[2] H. Attouch. 'Familles d'opérateurs maximaux monotones et mesurabilité'. *Ann. Mat.* To appear.

[3] H. Brezis. 'Opérateurs maximaux monotones et semi-groupes de contractions dans les espaces de Hilbert'. *Lecture Notes* vol. 5, North-Holland, Amsterdam (1972).

[4] E. Di Giorgi. 'Sulla convergenza di alcune successioni d'integrali del tipo dell'area.' *Rend. Matematica*, **8** (1975), 277-94.

[5] J. L. Joly. 'Une famille de topologies et de convergences sur l'ensemble des fonctionelles convexes.' *Thèse*, Grenoble, France (1970).

[6] V. Klee. 'Polyhedral sections of convex bodies'. *Acta Math.*, **103** (1960), 243-67.

[7] C. Kuratowski. *'Topologie'.* Vol I, Panstowowe Wydawnictioo, Naukowe, Warszawa (1958).

[8] J. L. Lions, and G. Stampacchia. 'Variational inequalities' *Commun. Pure & Appl. Math.*, **20** (1967), 493-519.

[9] J. -J. Moreau. 'Sur la fonction polaire d'une fonction semi-continue supérieurement.' *C. R. Acad. Sci., Paris*, **258** (1964), 1128-31.

[10] G. Moscariello. 'Γ-Convergenza negli spazi sequenziali'. *Rend. Accad. Sci. Fis. Mat. Soc. Naz. Sc., Lett. Arti Napoli*, ser. IV, **43** (1976), 333-50.

[11] U. Mosco. 'Approximation of the solutions of some variational inequalities'. *Ann.*

Scu. Norm. Sup. Pisa, **21** (1967), 373-394; **21** (1967), 765.

[12] U. Mosco. 'Perturbazioni di alcune disuguaglianze variazionali non lineari'. In *Atti del Con. Eq. Der. Parz. di Bologna*, Oderisi Gubbio (1967), 110-5.

[13] U. Mosco. 'Convergence of Convex Sets and of Solutions of Variational Inequalities'. *Adv. Math.*, **3** (1969), 510-85.

[14] U. Mosco. 'On the continuity of the Young-Fnechel transform'. *J. Math. Anal. & Appl.*, **35** (1971), 518-35.

[15] R. T. Rockafellar. *'Convex Analysis'.* Princeton University Press, Princeton, N. J. (1969).

[16] R. T. Rockafellar. *'Conjugate Duality and Optimization'.* SIAM monograph Series, Philadelphia (1974).

[17] G. Salinetti. 'Convergence for measurable multifunctions: an application to stochastic optimization'. Tech. Rep. 1978.

[18] G. Salinetti, and R. Wets. 'On the relations between two types of convergence for convex functions'. *J. Math. Anal. & Appl.*, **60** (1977), 211-26.

[19] G. Salinetti, and R. Wets. 'On the convergence of sequences of convex sets in finite dimensions'. *SIAM Rev.*, **21** (1979), 18-33.

[20] G. Salinetti and R. Wets. 'Convergence of sequences of closed sets'. *Tech. Report*, 1978. To appear.

[21] C. Sbordone. 'Su alcune applicazioni di un tipo di convergenza variazionale'. *Ann. Scu. Norm. Sup. Pisa*, serv. IV, **2** (1974), 617-38.

[22] B. Van Cutsem. 'Eléments aléatoires à valeurs convexes compactes'. *Thèse*, Grenoble, France (1971).

[23] D. W. Walkup, and R. Wets. 'Continuity of some convex-cone valued mappings'. *Proc. Am. Math. Soc.*, **18** (1967), 229-35.

[24] R. Wets. 'On inf-compact mathematical programs'. In *5th Conf. on Optimization Techniques; Lecture Notes in Computer Science*, Springer Verlag, Berlin (1974), 426-36.

[25] R. Wets. 'On the convergence of random convex sets'. In *Convex Analysis and its Applications; Lecture Notes in Economics and Mathematical Systems*, 144 Springer Verlag, Berlin (1977) 191-206.

[26] R. A. Wysman. 'Convergence of sequences of convex sets, cones and functions II'. *Trans. Am. Math. Soc.*, **123** (1966), 32-45.

[27] T. Zolezzi. 'Characterizations of some variational perturbations of the abstract linear-quadratic problem'. *Siam J. Control & Optimiz.*, **16** (1978), 106-21.

Subject Index

Alternative, 151
  Fredholm's, 351
  strong, 152
  theorem of, 151, 156, 158-9, 175
  weak, 152, 159
Antitone mapping, 326
Arrow-Debreu-Nash equilibria, 78, 83-4
Augmented Lagrangian, 161, 317-8
Auxiliary function, 163,

Banach space, 5, 315, 358, 390
Bellman's equation, 92
Bifurcation analysis, 349
Bilinear form(s), 1, 15, 80, 137, 215, 290
Bi-matrix game, 218, 224, 227
Binary program, 163
Binary vector, 103
Boundary condition(s), 202, 205-7, 210, 272, 350, 372
  unilateral, 6
Boundary-value problems, 76
Brower fixed-point theorem, 227, 233
Brownian motions, 143

Capacity, 5, 127
Cauchy problem, 196, 294, 298, 301
Clustering approach, 114
Coercive(ness), 6, 197, 203-4, 328, 336, 340, 357
Complementarity (basis, cones, submatrices), 97-8, 100-3
Complementarity problem(s), 76, 101, 151, 159, 175, 177, 213, 215, 217, 220, 251, 252, 256, 258-9, 267-9, 281, 286-8, 291-2, 303, 309-12, 315, 323, 325, 342
  approximate, 261, 263
  generalized, 76
  implicit, 75-6, 79, 84, 271, 282
  linear, 97, 173, 175-6, 213, 217, 220, 223-4, 312
  relations, 197
  strong implicit, 75
  theory, 213, 254
  weak implicit, 77
Computational complexity, 97, 102, 226
Conormal derivative, 342
Continuation, 213, 235
Convergence, 83, 84, 375-7, 380, 388, 394, 398, 400
Convex, 153, 159, 161, 169, 216, 221, 251, 306-7, 314, 324, 328, 334-6, 366, 372, 375
  cone(s), 77, 157, 305, 308
  functions, conjugate, 56, 163, 375-6, 392
  programs, 255-6, 303, 307, 316
  quadratic programming, 101
  valued mappings, 85
Convolution inequalities, 57
Coulomb potential, 53, 55, 62
Covariance (kernel, matrix), 121
Covering characteristic, 110
Cutting plane approach, 181

Dam problem, 28-9, 135-6
Decomposition, 159, 169, 174-5, 177
Deformation theory, 367
Degeneracy, non, 97, 100-1, 130, 133, 204-5, 234, 264
Deterministic strategies, 106
Dirac mass, 67

Dirichlet-Neumann conditions, 136
Dirichlet problem(s), 242, 282, 323, 334, 347
Dual, 359, 361, 365, 367, 369, 371
- estimate (method), 81, 283, 340
- feasibility, 130
- variables, 127, 132, 308
- variational method, 47

Duality, 78, 162-3, 166, 357-8, 365
- bilinear forms, 290
- gap, 182-3
- mapping, 11
- pairing, 77
- strong, 130, 163
- weak, 163

Dynamic programming, 147, 271

Economic equilibria, 227
Economic model, linear, 127, 134
Eigenvalue problems, 195
Elastic membrane, 26
Elasto-plastic problem, 15, 17, 246
Elliptic equation(s), 195, 201, 208
Equation(s)
- Bellman's, 92
- elliptic, 195, 201, 208
- Euler-Lagrange, 55-6, 58, 61
- fixed-point, 76
- heat, 35
- Klein-Gordon, 35
- partial differential, 1, 4, 55, 62, 303, 317
- quasi-linear, 244
- Schrödinger, 35
- Thomas-Fermi, 72
- wave, 35
- Young-Fenchel, 78

Estimator, unbiased, 118
Euler-Lagrange equation, 55-6, 58, 61
Exponential transformation, 155, 160, 162
Extremum problem, vector, 159, 165, 167

Farkas theorem, 156, 159
Fenchel transform, 254, 256
Fixed-boundary, 25, 30, 31
Fixed-point(s), 14, 22, 80, 85, 227, 233, 235
- equation, 76
- methods, 213
- problem, 79

Flow
- around a profile, 30
- in a porous pipe, 139
- non-steady, 138

Free-boundary, 17-9, 25, 141, 148, 195 349
- problem, 8, 25-6, 28, 139-41

Function
- auxiliary, 163
- Bessel, 355
- casting, 394
- conjugate convex, 56
- convex, 253, 256, 375
- graph of a multi-, 252
- maximal monotone multi-, 253-4, 259-60, 268, 316, 318, 320
- monotone multi-, 251, 253, 315
- multimodal, 105
- nonconvex, 105
- quadratic, 85
- saddle, 253
- separation, 152
- spline, 346
- strong separation, 152
- weak separation, 152

Global optimization, 120
Graph of a multifunction, 252

Heat equation, 35
Hilbert space, 11, 27, 294, 304
Hodograph
- mappings, 195, 198, 209
- method, 17
- space, 209
- transforms, 195

Hölder's inequality, 39, 46
Homotopy (methods, approach), 187, 213, 219, 229, 231, 235
Hydraulic problem, 139

Inequality
- convolution, 57
- Hölder's, 39, 46
- Kato's, 71
- Ky-Fan's, 78
- parabolic evolution, 343, 345
- Poincaré, 366
- Sobolev's, 39

Infiltration, 21
Integer programs, 175
Interval problem, 172
Isotone mapping, 326

Jet cavities, 31

Kalman condition, 90
Kato's inequality, 71
Klein-Gordon equation, 35
(Karush-) Kuhn-Tucker conditions, 216, 230, 255, 257, 308
Ky-Fan's inequality, 78

Lagrange multiplier(s), 38, 56, 58, 69, 216, 255, 258, 303, 306-7, 311, 314, 401
Laplace operator, 25, 334-5
Lasalle condition, 92
Legendre transform(s), 195, 198, 201, 205, 209
Lexicografic ordering, 101
Limit analysis, 357, 359, 367
Linear programming, 166, 169, 173, 217, 223
Lipschitzian, equi, 279
Lipschitz norm, 241
Liquid edge, 207

Marcinkiewicz space, 57
Markov inspection sequence, 145-6
Matrices, Minkowski, copositive-plus, 85-98
Maximum principle, 64-5, 224-5, 247
    Pontryagin's, 92
    strong, 71
Measure, 4
    capacitary, 5
    Lebesgue, 116
Membrane(s), 195-6, 203
Minimal surface(s), 13, 207, 357
Minimization (problem), 53
    constrained, 37-38, 42
    convex, 54
Minimum time problem, 89
Minkowski matrices, 85
Min-max principle, 56, 69, 80
Multiplier(s), 319
    Lagrange, 38, 56, 58, 69, 216, 255, 258, 303, 306-7, 311, 314, 401

Newton method, 187, 333
Nonconvex quadratic problems, 175

Obstacle problem, 8, 11, 13, 20, 201, 208, 246, 345
Operator, 276, 352, 359
    abstract, 5
    elliptic, 349
    Laplace, 25, 334-5
    monotonic, 5
    non-linear, 7, 271, 273, 335
    obstacle, 78, 80
    partial differential, 76, 79, 271, 273
    pseudo-monotonic, 6
Overrelaxation method, 286, 293, 296
Oxigen diffusion-absorption, 29

Parabolic evolution inequality, 343, 345
Pareto optimum, 252, 264
Path-following methods, 213, 238
Penalized problem, 148, 243
Penalty, 21, 161, 173
    method(s), 256, 303-4, 316-7
Permeability, variable, 135
Pivot(ing) (method), 101-2, 217-8
Plasma, 195, 202, 349
Plasticity, 224, 357
Plate, thin, 195, 207
Polar, 381
Porous media, infiltration in, 17, 21
Potential (theory, problem), 2, 3, 5, 313
    Coulomb, 53, 55, 62
Pricing, 127
Probabilistic (method, models), 113, 122
Proximal method of multipliers, 319

Quadratic programming, 151, 167, 169, 173, 175, 217

Reflection mappings, 195, 199, 210
Regularity, 8, 11, 15, 17, 19, 20, 28, 43-4, 75, 77-8, 84, 92, 116, 148, 204-5, 210, 233, 243, 272, 275, 306, 340, 342, 344, 346
Runge-Kutta method, 294-6

Schrödinger equation, 35
Semi-group, non-linear, 7
Separation function(s), 152
    strong, 152
    weak, 152
Shadow price, 127
Signorini problem, 289, 291
Simplex method, 101, 218, 223
Simplicial methods, 213, 227
Soap film, 195
Sobolev spaces, 3, 274, 323
    inequality, 39
Smoothness, 195, 198-9, 204, 207, 241, 350-1

Space(s)
- Banach, 5, 315, 358, 390
- Hilbert, 11, 27, 294, 304
- Hodograph, 209
- Marcinkiewicz, 57
- Sobolev, 3, 274, 323

Stability, 83, 97, 100, 130-1, 252, 268
Stationary (problem, point), 135, 217, 347
Steady flow, non-, 138
Stefan problem, 30
Stochastic, 247, 375
- control, 143, 271
- evaluation, 116
- impulse control problems, 77, 80-1
- models, 120

Stopping time, 143-6
Strategies, 108, 143
- deterministic, 106
- optimal, 109, 111, 122
- passive, 108-9, 112
- sequential, 108-112
- stochastic, 105

Subdifferential, 253-4, 315
- twisted, 253

Subgradient, 384
Syntesis problem, 93

Test function, 313
Thomas-Fermi (functional, equation), 53,72
Transfer sets, 89
Transmission conditions, 138
Transversality condition, 140, 189

Unbiased estimator, 118
Unilateral problem, 286, 334-5

Variational (approach, formulation, method, problem), 31, 53, 56, 195, 292, 357, 359
- constrained, 76
- dual, 47

Variational inequalit(ies), 1, 2, 8, 19, 22, 25-6, 84, 135, 137, 159, 164, 195, 213, 215, 246, 251, 253, 258, 286, 296, 301, 303-4, 306-9, 311-2, 323, 339, 375
- derivative of a solution of, 31
- dual estimate for the solution of a, 81
- linear, 173
- monotone, 308
- primal quasi, 79
- quasi, 75-8, 82, 145, 147-8, 271
- (generalized) solution of, 7, 148, 323
- stationary, 339, 342
- strong solutions, 272
- vector, 167
- weak solutions of, 8, 272

Vector extremum problem, 159, 165, 167
- optimal, 165

Viscoelastic unilateral problem, 286

Wave, travelling, 35
Wiener process, 121, 124-5
Worst-case, 102

Young-Fenchel equation, 78